The Emergent Multiverse

The Emergent Multiverse presents a striking new account of the 'many worlds' approach to quantum theory. The point of science, it is generally accepted, is to tell us how the world works and what it is like. But quantum theory seems to fail to do this: taken literally as a theory of the world, it seems to make crazy claims: particles are in two places at once; cats are alive and dead at the same time. So physicists and philosophers have often been led either to give up on the idea that quantum theory describes reality, or to modify or augment the theory.

The Everett interpretation of quantum mechanics takes the apparent craziness seriously, and asks, 'what would it be like if particles really were in two places at once, if cats really were alive and dead at the same time'? The answer, it turns out, is that if the world were like that—if it were as quantum theory claims—it would be a world that, at the macroscopic level, was constantly branching into copies—hence the more sensationalist name for the Everett interpretation, the 'many worlds theory'. But really, the interpretation is not sensationalist at all: it simply takes quantum theory seriously, literally, as a description of the world. Once dismissed as absurd, it is now accepted by many physicists as the best way to make coherent sense of quantum theory.

David Wallace offers a clear and up-to-date survey of work on the Everett interpretation in physics and in philosophy of science, and at the same time provides a self-contained and thoroughly modern account of it—an account which is accessible to readers who have previously studied quantum theory at undergraduate level, and which will shape the future direction of research by leading experts in the field.

David Wallace is Tutorial Fellow in Philosophy of Science at Balliol College, Oxford.

The Emergent Multiverse

Quantum Theory according to the Everett Interpretation

David Wallace

OXFORD
UNIVERSITY PRESS

OXFORD
UNIVERSITY PRESS

Great Clarendon Street, Oxford, OX2 6DP,
United Kingdom

Oxford University Press is a department of the University of Oxford.
It furthers the University's objective of excellence in research, scholarship,
and education by publishing worldwide. Oxford is a registered trade mark of
Oxford University Press in the UK and in certain other countries

Published in the United States of America by Oxford University Press
198 Madison Avenue, New York, NY 10016, United States of America

British Library Cataloguing in Publication Data
Data available
Library of Congress Cataloging in Publication Data
Data available

ISBN 978-0-19-954696-1 (Hbk.)
ISBN 978-0-19-870754-7 (Pbk.)

To Hannah

We have, then, a theory which is objectively causal and continuous, while at the same time subjectively probabilistic and discontinuous. It can lay claim to a certain completeness, since it applies to all systems, of whatever size, and is still capable of explaining the appearence of the macroscopic world. The price, however, is the abandonment of the concept of the uniqueness of the observer, with its somewhat disconcerting philosophical implications.

(Hugh Everett III, draft Ph.D. thesis; cut from submitted version at J. A. Wheeler's request)

What would it have looked like if it had looked like the Earth went round the Sun?

(Attributed to Ludwig Wittgenstein)

Contents

Part III. Quantum Mechanics, Everett Style

Appendices

List of Figures and Boxes

Figures

Boxes

Acknowledgements

I have been working on the Everett interpretation for my whole career, and on this book for most of it, and in the process I've been privileged to have the chance to discuss quantum theory and its philosophy with a great many colleagues and friends. The greatest thanks, by far, are due to Simon Saunders, both for the enormous influence of his own work and, even more importantly, for his tireless engagement with my work as teacher, colleague, and friend. Without his support, this would have been a far weaker piece of work, at best. As likely, it would not have existed at all.

Other than Simon, special thanks are due to Hilary Greaves (whose influence on, in particular, part II of the book was very considerable, over and above her own writings on the topic); to Jeremy Butterfield, Frank Artzenius and Jos Uffink, who patiently read through, and provided detailed and helpful feedback on, most of the manuscript; to David Albert and Wayne Myrvold for extensive and challenging engagement with Everettian ideas, and to Harvey Brown, Olly Pooley and Chris Timpson for a decade and a half of good company and good conversation on matters quantum-mechanical and beyond. But I'm also grateful to the many other colleagues who've discussed the interpretation of quantum mechanics with me over the last decade: Guido Bacciagaluppi, David Baker, Katherine Brading, Jeff Bub, David Deutsch, Cian Dorr, Artur Ekert, Chris Fuchs, Shelly Goldstein, Brian Greene, Robin Hanson, Lucien Hardy, Adrian Kent, Eleanor Knox, James Ladyman, Barry Loewer, Tim Maudlin, John Norton, David Papineau, Huw Price, Richard Healey, Paul Tappenden, Antony Valentini, Lev Vaidman, Alastair Wilson, Wojciech Zurek, and others too many to mention.

For the last six years, I've also been fortunate enough to be teaching Oxford undergraduates and graduates, and much of the argument of this book was honed through teaching encounters with them. Particular thanks are due to Anna Lewis, Johannes Noller, Richard Ollerhead, Paulina Sliwa, Tom Harty, Jasmeen Kanwal, Maddie Geddes-Barton, Chris Fox, and Alex Kaiserman.

For the paperback edition, I am grateful to Gijs Leegwater, Ronnie Hermans, and Fred Muller for the many typos and errors that they patiently identified.

Outside academia (or at least, my part of it), thanks to the many people who've kept me sane—and in particular to Sarah Paul, Kyra Smith, Laura Swift, and above all, Hannah.

Introduction

The basic thesis of this book is that there is no quantum measurement problem.

I do not mean by this that the apparent paradoxes of quantum mechanics arise because we fail to recognize 'that quantum theory does not describe physical reality' (Fuchs and Peres 2000a). Quantum theory describes reality just fine, like any other scientific theory worth taking seriously: describing (and explaining) reality is what the scientific enterprise is about.

Nor do I mean that we have found *solutions* to the problem, ways to modify or supplement quantum mechanics so that the paradoxes are removed. I don't actually think that we have found any very satisfactory solutions in this sense (at least not in the relativistic domain) but that isn't my theme: even if we did have these modifications, they would be ways of changing from quantum physics, which supposedly does have a measurement problem, to some other theory, which supposedly doesn't.

What I mean is that there is actually no conflict between the dynamics and ontology of (unitary) quantum theory and our empirical observations. We thought there was originally, because the theory is subtle, complicated, and highly unintuitive, and because our early attempts to understand it and to relate it to empirical data promoted high-level concepts like 'observation' and 'measurement' to the level of basic posits and confused the issue. But we have now been studying the theory for eighty years, and we are in a position to know better: the work of Mott, Schrödinger, Everett, de Witt, Zeh, Deutsch, Zurek, Gell-Mann, Hartle, Halliwell, Saunders, and a host of others has shown us how the theory connects to the observed world, and how it explains our empirical data.

Everett (1957) was the first to really see clearly, in outline, how this worked, hence the term 'Everett interpretation of quantum mechanics'.

But really the 'Everett interpretation' is just quantum mechanics itself, read literally, straightforwardly—naively, if you will—as a direct description of the physical world, just like any other microphysical theory.

DeWitt (1970) stated explicitly the shocking fact which Everett only hinted at in publications:[1] that if one really takes quantum mechanics seriously, then the various terms in a macroscopic superposition must be taken as describing physically real macroscopic worlds which are for all practical purposes non-interacting. He recognized that quantum mechanics, taken literally, tells us that the macroscopic, approximately classical world we observe around us is only part of a far larger reality. Hence the alternative, and more sensational name, for the Everett interpretation: the Many-Worlds theory.

In this book, I aim to give a detailed, cohesive account of quantum mechanics, when it is interpreted as a literal theory of the world. It is not intended as a historical account: I use 'Everettian' and 'the Everett interpretation' freely, but for present purposes I neither know nor care whether I am describing the historical Everett's own view. (I rather hope that he would be highly sympathetic to what I say, but if so, that's a bonus, not a prerequisite.) Nor do I wish to be read as offering yet one more 'interpretation of quantum mechanics'. The basic strategy I have outlined—of taking quantum theory literally—is, I think, common to the great majority of self-defined 'Everettians', and only confusion ensues if the various authors on the subject are seen as proposing their own incommensurable 'neo-Everett' interpretations, rather than disagreeing about the same thing. I would be neither surprised nor distressed if David Deutsch, or Max Tegmark, or Lev Vaidman were to regard parts of this book as completely wrong; I would be both surprised and distressed if they merely said that they advocated a different Everett interpretation.

In fact, I largely avoid talking about 'interpretations' at all in this book: the term, I think, has had very unfortunate consequences in the philosophy of quantum mechanics, contributing to the idea that there is

---

[1] In fact, so oblique are Everett's mentions of 'many worlds' in his published work that for many years it has been controversial whether he really saw his own interpretation in many-worlds terms: see e.g. Barrett (1999: ch. 3). Examination of Everett's unpublished work, however, has recently made it quite clear that he understood the many-worlds implications of his view, and that he refrained from making them clearer essentially for political reasons. See Byrne (2010) and Bevers (2011) for more on this topic.

one theory with many interpretations and that therefore discussions of the 'correct' interpretation are mere talk. (I will discuss this in more detail in Chapter 1.) In general, I will refer to the subject matter of the book as 'Everettian quantum mechanics', or sometimes '(Everettian) quantum mechanics' or '(Everett-interpreted) quantum mechanics', or just 'unitary quantum mechanics', or even (when confusion seems unlikely to ensue) simply as 'quantum mechanics'.

I also avoid, in large part, two terms which at one point were common in the literature: 'many-worlds' theories and 'many-minds' theories. Both carry connotations I wish to avoid: in the one case, that the many worlds are somehow *fundamental* parts of the theory, rather than an emergent, local, macroscopic phenomenon; in the other, that somehow a detailed theory of the mental is relevant to the understanding of quantum mechanics, or that there is no real multiplicity in nature, just the illusion of multiplicity in our minds. But given this, I should probably note right away that Everettian quantum mechanics really is both a many-worlds and a many-minds theory, in the sense that it entails that there are a great many versions of myself, living in surroundings much like my own and interacting with other versions of *you*rself, elsewhere in physical reality. The other worlds, and their inhabitants, are not abstracta, or fictions, or mere unrealized possibilities: if Everettian quantum mechanics is true, they are as real as I, you, and our mutual surroundings.

Moving on to the details of the book: I have tried to keep its technical prerequisites to a minimum, but that minimum is not trivial: at various points in the text I presuppose knowledge of quantum physics up to about the end of a (UK) undergraduate course (a little background in quantum information theory would help in Chapter 8 and parts of Chapter 10, but is not essential). However, *most* of the book does not require quite this level of background: readers who are familiar with the basic mathematical language of quantum mechanics and who are willing to take some technical results on trust should have no difficulty with most of the book (the main exception is Chapter 3).

Regarding philosophical content: I have tried to avoid presuming any specific philosophical background. There is, however, a strong, but largely tacit, philosophical premise running throughout the books: *naturalism*, of the kind advocated by Quine (1969) and more recently by Ladyman and Ross (2007). Naturalism, in essence, is the doctrine that in studying science and its philosophical implications, we have no tool better than the successful practices of the sciences themselves (and those tools that the sciences themselves use: notably, mathematics and logic). In particular, insofar as our intuitive precepts about how the world works are in conflict with science, so much the worse for those intuitive precepts. There is a view of the philosophy of quantum mechanics according to which it is 'all about being faced with a number of counterintuitive accounts of what's going on and identifying which is the least counterintuitive' (Monton 2004);[2] this view plays no part here.

The book is split into three parts. Parts I and II explain what the Everett interpretation is and develop an account of how Everettian quantum mechanics is explanatory of, and confirmed by, our empirical data. In Part I, I set out the general framework (Chapter 1), give a philosophical account of why quantum mechanics leads to 'many worlds' (Chapter 2), and then explore the technical details of decoherence theory, which accounts quantitatively for the presence and structure of these worlds (Chapter 3).

Part II is exclusively concerned with the issue of probability, frequently described as the most severe problem for the Everett interpretation. I disagree with this assessment: for reasons I explain in Chapter 4, probability is *at least* as explicable in quantum as in classical physics. In fact, though, we can go one better than this: there are good reasons to suppose that we can understand probability much *better* in a quantum universe, and Chapters 5 and 6 explore this in detail, using arguments from decision theory and from the symmetries of the quantum state. This material, though, gets fairly technical and is quite far removed from mainstream issues in physics; it may not be to the taste of all readers.

Part III is less concerned with developing the details of the Everett interpretation, more concerned with exploring some of its consequences. What does Everettian quantum mechanics mean for our everyday concepts of uncertainty, possibility, and identity? For our picture of events playing out against a spacetime background? For the direction of time? And for other issues ranging from cosmological reasoning to time travel?

I also include four Appendices. Appendices A, C, and D just fill in the details of theorems sketched or stated in Chapters 3, 5, and 6. Appendix B,

---

[2] The quoted sentence appears in the electronic version of Monton's paper but not in the print version.

though, is a self-contained account of classical decision theory (including a formal framework and representation theorem which so far as I know is original). This appendix provides relevant background for Chapters 5 and 6; they can be read without it, though, provided a couple of theorems are taken on trust.

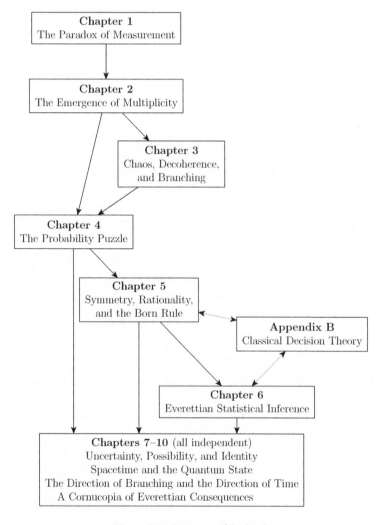

Figure 0.1. Structure of the book

I have tried to sketch the overall interdependence of the various chapters in Fig. 0.1. At one extreme, the book can be read cover to cover (this is, of course, any author's ideal). At the other, Chapters 1, 2, and 4 provide a self-contained and reasonably non-technical account of Everettian quantum mechanics and should equip a reader to study any of Chapters 7–10. Readers mostly interested in the philosophical aspects of the Everett interpretation (and in particular on probability) and who are willing to take some physics on trust might want to skip Chapter 3 and treat Chapter 7 as an integral part of Part II's discussion of probability. Readers less patient with philosophy and more interested in the implications of the Everett interpretation for physics may wish to use Chapters 1–4 as a general introduction to the Everett interpretation, skip Chapters 5–7, and go straight to Chapters 8–10.

Finally, I have included two Interludes. These little dialogues (one after each of Parts I and II) are attempts to engage with, and reply to, some of the (many!) criticisms of Everettian quantum mechanics that I have come across in the last ten years, but which didn't fit naturally into the main structure of the argument. I've resisted the temptation to have the Sceptic make a tearful recantation in the conclusion. Alas, he remains obdurate to the end.

On notation: I basically adopt the Dirac notation, though I occasionally allow myself to write ψ rather than $|\psi\rangle$ in contexts (notably, formal mathematical proofs) where it would be cumbersome to do otherwise. Similarly, I normally adopt the convention of marking Hilbert-space operators with hats ($\widehat{X}$) but I drop this in Chapters 5, 6, and 9, and in Appendices B and C, where it would greatly clutter the text without much improvement in clarity. I write $\neg E$ to denote (according to context) either the orthogonal complement of the Hilbert subspace E, or the set-theoretic complement of the set E, or the negation of the proposition E.

Two notes on style. First, in this book I follow the quantum information convention of using 'quantum mechanics' to refer to quantum theory in general, and not just to the quantum theory of particles; as such, 'quantum theory', 'quantum mechanics', and 'quantum physics' should be read synonymously, and I vary my usage purely for stylistic reasons. (Similarly for 'classical mechanics': classical field theories, for instance, are examples of classical mechanics.) Secondly, where possible I have avoided gender-specific pronouns; where they are unavoidable (as is often the case, in particular, in discussions of rational agents in decision theory) I have chosen the gender of each agent, experimenter, or observer via a quantum-mechanically random process. My apologies to readers unfortunate enough to be in a branch where this has led to an outcome not to their liking.

PART I

The Plurality of Worlds

1

The Paradox of Measurement

The mathematical formalism of the quantum theory is capable of yielding its own interpretation.

Bryce de Witt[1]

1.1 The goals of science

Very little has ever been written, and very little needs to be written, on the interpretation of palaeontology. There is no mystery at all about what the claims of our best theory of dinosaurs are supposed to be: they are supposed to be about *dinosaurs*, they are supposed, quite literally, to be telling us about the giant animals that we believe lived on Earth tens of millions of years ago. Nobody seriously believes that 'dinosaurs' are just a calculational device intended to tell us about fossils; everyone knows that the purpose of palaeontology is to make certain factual claims about certain aspects of the world (in this case, aspects in the past). And furthermore, since we have good reason to think that our best theory of dinosaurs is a pretty *good* theory, we have good reason to think that the claims it makes about the world are more or less *true*.

And almost all of science is like this. Molecular biology tells us about DNA strands. High-energy astrophysics tells us about quasar emission jets. No one has ever observed DNA strands, or quasar jets, or dinosaurs, with their own eyes (for different reasons, in each case), but it is a commonplace amongst working scientists in these various fields that their theories *make claims about* these unobserved phenomena. Nobody studies DNA in order

---

[1] DeWitt (1970).

to better predict the behaviour of the mass spectrometer; nobody studies quasars in order to better predict the X-ray spectrum of the sky. The point of molecular biology, of high-energy astrophysics, of economics, of inorganic chemistry, of science writ large, is to understand more about the way the world is.

In some ways, this moral is clearest in physics, whose claims about the way the world is are normally couched in the language of mathematics. Ever since Newton, physicists have been in the business of constructing mathematical models of aspects of the world, 'models' in the sense that they are—or are intended to be—isomorphic to the features of the world being studied. So classical N-body mechanics, for instance, describes the movement through three-dimensional space of N points—and claims that the movements of real physical particles are correctly described by the movement of those N mathematical points. Field theory assigns a mathematical object (a real number, a complex scalar, a vector, etc.) to each point in four-dimensional spacetime—and claims that a real physical entity, a 'field', inhabits real physical spacetime and has the structure described by the mathematics.

Of course, there is scope for metaphysical disagreement when dealing with entities as alien as 'fields'. (Maxwell, recall, famously worried about the mechanical story underlying the electromagnetic field.) Are fields entities in their own right? Are they merely properties of underlying spacetime points? But while such questions may be significant to metaphysics, they are no longer of much direct concern to science. Physics finds itself content to regard field theory—to regard the theories of mathematical physics in general—as telling us information about the *structure* of our world. Whatever the 'true nature' of the electromagnetic field, it is as real as the dinosaurs, as real as Bolivia, and its strength over *here* is 2.401793 times its strength over *there*.

And of course, if our theories are *wrong*, what they tell us about the world is not to be trusted. Cartesian physics was clearly about the world—it claimed, for instance, that the solar system was a huge rotating vortex of fluid, carrying the planets in its wake—but its claims about the world were flatly untrue. We have excellent reason to think that classical field theory is not true either, so we do not take *its* structural claims about the world as correct—but we also have excellent reason to think that it is approximately true in some regimes, which is to say that in those regimes, what it says about the structure of the particular bit of the world under consideration is roughly *right*. (And, whatever the truth about the metaphysical nature of *field* in classical field theory, the fact that classical field theory is strictly speaking false gives us very little reason to trust that the real world has that metaphysical nature.)

But notwithstanding these points, the moral remains: what scientific theories do is give us information about the universe—about what sort of things there are in it, about how they are structured, about how they come into existence and interact and change and disappear. This is what science is about; this is what every scientific theory does.

Every scientific theory, with perhaps just one exception. But the one exception—quantum mechanics—is the most powerful, most accurate, most fruitful theory that physics has ever devised, and, as our only extant theory of the microscopic world, it underlies practically all of the rest of science. Yet, at least at first sight, it is impossible to understand quantum mechanics as telling us objective facts about the world. At best, it seems to be telling us about the results of experiments we can perform, and to be telling us about them in an unsatisfactorily ad hoc way. At worst, it seems to be telling us plain nonsense, claiming that somehow things can be in two places at once, and that cats can be alive and dead simultaneously.

This, in brief, is the *measurement problem* of quantum mechanics. In the eighty years that we have been wrestling with it, some have argued that it gives us reason to abandon the picture of science as describing an observer-independent world, and to replace it with one where notions of *measurement*, *observation*, and perhaps even *consciousness* play a central role. Others wish to retain the conventional picture of science, but to reject quantum mechanics itself, replacing it with some new theory which lacks quantum mechanics' interpretational problems.

This book, by contrast, takes an extremely conservative approach to quantum mechanics. It supposes—as was first proposed by Hugh Everett, fifty years ago—that neither the mathematical formalism of quantum mechanics nor the standard conception of science is in any need at all of modification. Rather: the unmodified quantum theory can be taken as representing the structure of the world just as surely as any other theory of physics. In other words, quantum mechanics can be taken literally. The only catch is that, when we do take it literally, the world turns out to be rather larger than we had anticipated: indeed, it turns out that our classical 'world' is only a small part of a much larger reality.

The task of this chapter is to expand on these brief remarks: to review the structure of quantum mechanics; to explain just what is meant by 'taking it literally'; to show, in outline, why that leads to a 'many-worlds' theory, and to begin a discussion of the difficulties involved in Everett's proposal. That discussion will continue throughout Parts I and II of the book; in Part III, I explore just what the world is like if that proposal is correct.

1.2 Quantum mechanics: the bare formalism

Whatever the *conceptual* difficulties of quantum mechanics, its mathematical formalism is perfectly well defined. To specify a quantum system, we begin by giving some Hilbert space $\mathcal{H}$, whose rays—that is, its non-zero vectors up to normalization and phase—are intended to represent the possible states of the system. We then provide a dynamics for rays in $\mathcal{H}$: a set of time-indexed unitary transformations which take a state at one time into the states it evolves to at other times (often we specify these unitary transformations by the *Hamiltonian*, the self-adjoint operator which generates them). Finally, we provide some additional structure on the Hilbert space, sufficient to specify the particular system which we are studying.

This last requirement may look unfamiliar, and indeed inelegant (it is certainly not stated with the mathematical cleanness of the other two requirements). Nonetheless it is essential. All Hilbert spaces of the same dimension are isomorphic, so the differences between quantum systems cannot really be understood without additional structure. And absent additional structure, a Hilbert-space ray is just a featureless, unstructured object, whereas the quantum state of a complex system is very richly structured.

In practice, we specify structure for quantum systems in two ways: by providing certain operators on the Hilbert space (or, equivalently, providing certain preferred sets of basis vectors); or by providing a certain decomposition of the Hilbert space into quotient spaces (i.e. by providing a particular decomposition of the system into subsystems). And while this *general* prescription may be somewhat vague, system by system it is perfectly precise.

Non-relativistic particle mechanics offers a simple example. There (ignoring, for simplicity, the existence of identical particles) we specify the system by giving a Hilbert space for each particle (whose tensor product is the Hilbert space of the whole system) and, within each one-particle Hilbert space, specifying a certain set of operators. This can be done in a number of ways: by designating a triple of operators $\widehat{X}_1, \widehat{X}_2, \widehat{X}_3$ as the position operators, for instance, or by picking out a certain representation of states of the system as functions on $\mathbf{R}^3$; the structure provided is essentially the same in both cases.

For a more sophisticated example, consider quantum field theory (QFT). In the so-called 'algebraic' description of QFT,[2] an algebra of operators is associated to each spatial region, so that the operators associated with region $\mathcal{O}$ are intended to represent observables localized in $\mathcal{O}$. At least formally, we can regard this as equivalent to decomposing the Hilbert space into quotient spaces, each one representing the quantum state of a different spatial region (see Wallace 2006b for more details). When a QFT is specified this way, there is no need to say *which* operator at $\mathcal{O}$ corresponds to which observable property of $\mathcal{O}$—once the localization properties of all the operators are known, the quantum system is sufficiently structured to permit physics to be done, and in particular, to understand the emergence of particles. (See Wallace 2001a; 2009 for conceptual discussion of this emergence; for a discussion from the algebraic viewpoint, see Haag 1996: 75–83, 271–92.)

The same phenomenon, where specifying a decomposition of a system into subsystems is sufficient to give that system's structure, can be seen in the very abstract context of quantum computation.[3] There, the system is taken to be composed of *qubits*: elementary quantum systems, with two-dimensional Hilbert spaces. In specifying a quantum computer, it is in general not necessary to pick out any particular basis for the individual qubits: it suffices to give the decomposition of the global Hilbert space into qubit spaces, and to provide the dynamics.

---

[2] By considering an algebraic *description* of QFT, I do not intend any commitment to the 'Algebraic Quantum Field Theory' programme, which aims to replace standard QFT with something formulated on more mathematically rigorous lines. As far as I am concerned, foundationally speaking standard QFT is fine how it is. (See Wallace 2006b; 2011 for more extended discussion of these issues.)

[3] Classic (if slightly dated) references on quantum computation (and information) are Preskill (1999) and Nielsen and Chuang (2000); for a more recent reference see Benenti et al. (2004); for an introduction, see Ekert and Jozsa (1997) or Vedral (2006); Timpson (2008) discusses philosophical and conceptual implications of quantum information theory.

The general point is this: a vector in Hilbert space, in and of itself, is a very unstructured object, a mere line. But the quantum system whose state it represents is in general an enormously complex and richly structured object.

I shall refer to quantum theory specified in these terms—shorn of any mention of probability, observables or measurement—as the *bare quantum formalism*. In practice, nearly all actual calculations performed with quantum theory involve calculating some mathematical property of the bare quantum formalism (normally scattering matrices or expectation values of those operators—usually ones connected to spacetime symmetries—which have been used to specify the quantum theory in question.)

Superficially, the bare quantum formalism bears strong resemblances to the mathematical formalism of *classical* mechanics. There, too, a system can be specified by a space of states (a symplectic manifold rather than a Hilbert space) and a dynamical process on it; there, too, the structural richness of the classical state is hidden when it is represented by a mere point rather than, say, a set of particle positions or a field on space.

In trying to understand how quantum mechanics differs from classical mechanics, then, naively one might look for the structural differences between their respective states and dynamics. And they are certainly there to be found, most obviously in the phenomenon of entanglement. But of course, the really profound difference lies (or appears to lie), not in the structure of the respective formalisms, but in the connection between those formalisms and the way we actually use the theory in practice.

1.3 Quantum mechanics: interpretation and measurement

If the bare *formalism* of quantum mechanics is elegant and clean, the same cannot be said for its standard *interpretation*: if we are to extract empirical content from the mathematics, we seem to have to introduce the notion of measurement as a fundamental concept. Recall that the standard textbook way to do this is to associate measurements with certain self-adjoint operators: if $|\psi\rangle$ is the state of the system, and $\hat{M}$ is a self-adjoint operator whose spectral decomposition is

$$\widehat{M} = \sum_i m_i \widehat{P}_i, \qquad (1.1)$$

then the possible results of the measurement are the m_i, and the probability of a particular result m_j is $\langle \psi | \widehat{P}_j | \psi \rangle$.

Furthermore, the role of the particular m_i (at least from the point of view of the measurement process) is just to label particular outcomes. If we replaced $\widehat{M}$ with $f(\widehat{M})$, where f is any function satisfying $f(m_i) \neq f(m_j)$ for $m_i \neq m_j$, then all we'd have changed is the way measurement outcomes are labelled. So what really matters is only the family of projections $\{\widehat{P}_i\}$: each represents one measurement outcome. For this reason measurements of this type are known as *projection-valued measurements*, or PVMs.

In fact, it is by now widely accepted that the 'standard textbook way' is insufficiently general, and that so-called 'positive operator valued measures' (POVMs) provide a more appropriate formalization of the general notion of measurement.[4] A POVM, like a PVM, is specified by a family of operators $\{\widehat{Q}_i\}$, with each operator corresponding to a possible measurement outcome. And as with PVMs, if a system has state $|\psi\rangle$ then the probability that the measurement yields the outcome corresponding to $\widehat{Q}_j$ is $\langle \psi | \widehat{Q}_j | \psi \rangle$. But unlike PVMs, there is no requirement that the $\widehat{Q}_i$ are projectors. The only requirements on them are the minimal ones needed to make $\langle \psi | \widehat{Q}_j | \psi \rangle$ function like a probability measure: namely:

1. For any state $|\psi\rangle$ and any $\widehat{Q}_j$, we require $\langle \psi | \widehat{Q}_j | \psi \rangle \geq 0$ (hence: 'positive operator'). This ensures that the probability of any outcome is non-negative.
2. We require $\sum_i \widehat{Q}_i = \widehat{1}$, so that probabilities always sum to 1.

POVMs have two main technical advantages over PVMs: they can represent *imprecise* measurements, and they can represent *joint* measurements.

For instance, standard quantum mechanics uses PVMs to represent measurements of position: the relevant projectors are projections of the wavefunction onto regions of configuration space. But this requires us to idealize measurements as having infinitely sharp boundaries, and provides no way to represent approximate joint measurements of position and

---

[4] For technical discussion of POVMs, see any of the quantum information references in n. 3; for a really thorough account, see Busch, Lahti, and Mittelstaedt (1996).

momentum—something we do all the time in experimental practice and in everyday life. The POVM framework allows us to define approximate *phase-space measurements*, which represent approximate joint measurements of position and momentum and which can be coarse-grained in a very wide variety of ways to represent a variety of imprecise measurements. (The details can be found in Box 1.1.)

Box 1.1. A phase-space POVM

For simplicity, let us consider a single one-dimensional particle. In the conventional (PVM) measurement formalism, measurements of position and of momentum are associated (never mind how, for the moment) with operators $\widehat{X}$ and $\widehat{P}$ respectively, obeying the canonical commutation relation: $[\widehat{X}, \widehat{P}] = i\hbar$; in the usual wave-function representation of states as complex functions on the real line,

$$\widehat{X}\psi(x) = x\psi(x) \tag{1.2}$$

and

$$\widehat{P}\psi(x) = -i\hbar\frac{d\psi}{dx}. \tag{1.3}$$

In Dirac's (technically improper) notation, we write the eigenvectors of $\widehat{X}$ and $\widehat{P}$ as $|x\rangle$ and $|p\rangle$ respectively, and the families of projectors $\{|x\rangle\langle x|\}$ and $\{|p\rangle\langle p|\}$ respectively represent measurements of position and momentum. The best that can be done with these projections to model realistic measurements is to coarse-grain the real line into intervals Δ_i, and associate to each region the (now rigorously-definable) projector

$$\Pi_i = \int_{\Delta_i} dx |x\rangle\langle x| \tag{1.4}$$

to represent a measurement which finds the particle to be located in Δ_i.

The easiest way to construct a phase-space measurement POVM is to begin with wave-packet states—paradigmatic examples of states which are approximately localized in both position and momentum. Let $|q, p\rangle$ denote a wavepacket with wavefunction

$$\psi_{q,p}(x) \equiv \langle x|q,p\rangle = \frac{1}{\lambda^{\frac{1}{2}}\pi^{\frac{1}{4}}}e^{-ipx/\hbar}e^{-(x-q)^2/2\lambda^2} \qquad (1.5)$$

Then each $|q,p\rangle\langle q,p|$ is a well-defined projector, and it is a fairly simple matter to check that $\int dq\,dp\,|q,p\rangle\langle q,p| = 2\pi$—that is, that $\{(1/2\pi)|q,p\rangle\langle q,p|\}$ is a POVM.

The constant λ represents an irreducible constraint on joint measurements, stemming from the uncertainty principle; of course, in practice joint measurements of position and momentum will often be much less accurate than this. We can represent this by coarse-graining, as in the PVM case—for instance, we could partition phase space into cells Σ_i and define

$$\widehat{\Pi}_i' = \frac{1}{2\pi}\int_{\Sigma_i} dp\,dq\,|q,p\rangle\langle q,p|. \qquad (1.6)$$

More generally, we could pick any family f_i of positive functions on phase space such that $\sum_i f_i = 1$ everywhere, and define

$$\widehat{\Pi}(f_i) = \int f_i(q,p)dp\,dq\,|q,p\rangle\langle q,p|. \qquad (1.7)$$

But however practically useful POVMs may be in representing measurements, how do we establish *which* PVM or POVM is associated with a given measurement? It is superficially tempting to regard the association as 'primitive', as if measurement devices are delivered from the factory—or gifted to us by God—with a POVM stamped on them. But needless to say, this does a disservice to experimental physicists. Measurement devices are not black boxes which accomplish their mysterious 'measurement' process primitively: they are physical devices, deliberately designed using physical principles.

And now we have a problem: for how can measurement at one and the same time be a primitive component of quantum physics and something analysable within that same quantum physics? We might identify certain POVMs as 'position' or 'momentum' or 'phase space' measurements based on symmetry principles of the dynamics or what-have-you, but as long as these POVMs are *primitively* identified with certain measurements, we have no prospect of using physics to explain *why* a certain physical device does indeed measure position. More practically, we have no way even to make sense of—still less determine the truth of—claims like 'this device is accurate to one part in 10^5'. Yet claims like this are routine in physics.

Consider, for instance, some device that measures the position of a particle in a box. To first approximation we might well regard this device as making a 'perfect measurement of position'—so that if the system being measured has wavefunction ψ, the probability of getting a result in region Δ of the box is

$$\Pr(\Delta) = \int_\Delta |\psi|^2. \tag{1.8}$$

But in practice, the device may not be quite so accurate. Probably the right way to represent the measurement is instead with some POVM (such as the one in Box 1.1), to allow for the imprecisions inherent in the device. But what determines which POVM to choose? In a sense the answer is obvious: the physical makeup of the device determines it. But this move is not available to us as long as it is a primitive *truth* that the device measures position.

And yet in physics we apply quantum physics to measurement devices all the time (ask anyone in quantum optics). In doing so we include the measurement device as part of a larger quantum system: the 'measurement' process then becomes just one more dynamical process, described unitarily in terms of evolution on Hilbert space. Conceptually speaking this should not help at all, since it just pushes the problem back: in order to interpret that larger quantum system we need recourse to a primitive notion of measurement *of that system*. And if we try to model that process of measurement too, we need yet a third primitive notion of measurement, and so on ad infinitum.

As a practical matter, though, the process is not infinite: it terminates when the measurements we are interested in are of *macroscopically large quantities*. If we are measuring something like the readout of a digital display or the position of a needle on a meter, there is in practice no need at all to know the accuracy of the measurement process: it can be assumed to be as accurate as we like. Indeed, we may as well take ourselves to be measuring position *and* momentum of these macroscopic quantities: a phase-space POVM like the one in Box 1.1 will suffice to handle this mathematically, and the ambiguities associated with the variety of POVMs available will be far, far too small to affect our predictions.

This, then, is how physicists can describe the quantum measurement process in practice, and it will be instructive to develop the details further. First, identify some of the system's dynamical variables (i.e. some of its self-adjoint operators) as corresponding to macroscopic degrees of freedom of the system. For a simple system like a pointer, centre-of-mass position and total momentum might suffice; for a fluid or other more complicated system, we might choose the spatial averages of microscopic positions and momenta, averaged over regions large compared to microscopic scales but small compared to human ones.

Having made such a choice, we can decompose the system's Hilbert space $\mathcal{H}$:

$$\mathcal{H} = \mathcal{H}_{macro} \otimes \mathcal{H}_{micro}. \qquad (1.9)$$

To paraphrase John Bell (1987: 166), all measurements of the system are ultimately measurements of the macroscopic degrees of freedom (any measurement whose results are *not* ultimately represented in these degrees of freedom will not actually be usable by human experimentalists) so it suffices to give a theory of measurement for these degrees of freedom—which we can do by specifying a phase-space POVM on $\mathcal{H}_{macro}$, or (somewhat less elegantly) by specifying a position measurement.

To make this clearer, let $\{|\mathbf{q}, \mathbf{p}\rangle\}$ be a basis of coherent states for $\mathcal{H}_{macro}$—each $|\mathbf{q}, \mathbf{p}\rangle$ is a Gaussian wavepacket centred on values $(\mathbf{q}, \mathbf{p})$ of the macroscopic degrees of freedom. (See Box 1.1 for details.) Then a general state of the system can be expressed as

$$|\psi\rangle = \int d\mathbf{q}\, d\mathbf{p}\, \alpha(\mathbf{q}, \mathbf{p}) |\mathbf{q}, \mathbf{p}\rangle \otimes |\psi_{\mathbf{q},\mathbf{p}}\rangle \qquad (1.10)$$

and the rule for extracting observable predictions from the quantum theory is: the probability of the macroscopic degrees of freedom being in region Δ of phase space is

$$\int_{\Delta} d\mathbf{q}\, d\mathbf{p}\, |\alpha(\mathbf{q}, \mathbf{p})|^2 \qquad (1.11)$$

(because the $|\mathbf{q}, \mathbf{p}\rangle$ do not form an orthonormal basis, the function $\alpha(\mathbf{q}, \mathbf{p})$ is not actually uniquely specified by $|\psi\rangle$, but its squared integral over Δ is very nearly uniquely specified, if Δ is large compared to the width of the Gaussian).

Let us call this rule the 'quantum algorithm': applied to the bare quantum formalism, it yields empirical predictions.

(Readers familiar with typical discussions of the measurement problem may be surprised that I have mentioned neither the 'eigenstate–eigenvalue link' nor the 'collapse of the wavefunction'. This is deliberate: neither plays any role in the *formalism* of quantum mechanics, and neither plays any role in the *algorithm* by which we can (pragmatically speaking) extract empirical predictions from quantum mechanics. And neither plays any role in modern discussions of measurement in mainstream quantum physics.[5])

1.4 'Measurement' cannot be represented physically?

This picture of quantum mechanics differs dramatically from the general picture of scientific theories sketched out in section 1.1. There, recall, a theory was taken as giving us objective, observer-independent information about the world. On that view of science there is no fundamental mystery about how the formalism leads to physical predictions, and no *fundamental* role for the notions of 'measurement' and 'observation'. Indeed, measurement is just another physical process, and the structure of the measuring device should in principle be representable in the same way as the structure of any other physical device. There is certainly no need for

---

[5] To expand a little: the eigenstate–eigenvalue link says that a quantum system has a definite value of a property just if its state is an eigenstate of the projector corresponding to that property; the collapse postulate says that upon measurement, a quantum state collapses into one of the eigenstates of the system being measured. The former rule plays no role in calculations, serving only to ground the metaphysical supposition that 'measurements' reveal some pre-existing value, and to ground it only in certain special circumstances; it is sometimes erroneously thought to link the quantum state to our observations in 'unproblematic' cases where there is no macroscopic superposition, but it universally fails to do so at least in the case of position or momentum (no physically realistic quantum system's wavefunction remains bounded in a finite spatial region for more than an instant). The latter rule superficially seems necessary to account for repeated measurements, but in fact those measurements are invariably handled within unitary quantum theory (and are usually quite badly modelled by applying the collapse postulate; cf. Home and Whitaker 1997); it can seem metaphysically required only if the eigenstate–eigenvalue link is already presumed. The obsolescence of both rules is clear from the fact that they cannot even be defined in cases where a POVM, rather than a PVM, is used to represent the measurement process. (I discuss this topic further in Wallace 2008.)

a 'classical algorithm' (still less a 'palaeontological algorithm'!) to connect these theories to the world.

Is there no prospect that something similar is available for quantum mechanics? It seems that there is not, at least if the structure of the system is required to be that of the macroscopic world we observe (as seems unavoidable if quantum mechanics is to fit with experiment and with our experiences). The reasons are deceptively simple: suppose we have some two-state quantum system (the spin of an electron, say) whose possible states are $|+\rangle$ (spin up in some co-ordinate system), $|-\rangle$ (spin down), and superpositions thereof, and let that system's Hilbert space be $\mathcal{H}_s$. And suppose also we have some device whose task it is to measure spin, whose Hilbert space is $\mathcal{H}_m$.

If the quantum state really does represent the structure of the physical world, then, there must be some state $|\text{'ready'}\rangle$ of $\mathcal{H}_m$ that represents the measurement device before any measurement is made.

Furthermore, experience tells us that any spin measurement on such a system has but two possible outcomes: spin up, and spin down. So there must exist states $|\text{'up'}\rangle$ and $'|\text{'down'}\rangle$ of $\mathcal{H}_s \otimes \mathcal{H}_m$ which represent the joint state of measuring device and particle after the measurement, and in which the result given is 'up' and 'down' respectively.

Now, if measurement is really supposed to be a dynamical process, the measurement process itself must be represented by some unitary transformation $\widehat{U}$. And we know the form that $\widehat{U}$ must take, because we know that in the limiting case where the system has state $|+\rangle$ or $|-\rangle$, the measurement deterministically gives 'up' or 'down' respectively. That is, we must have

$$\widehat{U}|+\rangle \otimes |\text{'ready'}\rangle = |\text{'up'}\rangle \qquad (1.12)$$

and

$$\widehat{U}|-\rangle \otimes |\text{'ready'}\rangle = |\text{'down'}\rangle. \qquad (1.13)$$

But unitary dynamics is linear. So we now know what happens if we measure the spin of an arbitrary state of the electron:

$$\widehat{U}\left(\alpha|+\rangle + \beta|-\rangle\right)|\text{'ready'}\rangle = \alpha|\text{'up'}\rangle + \beta|\text{'down'}\rangle. \qquad (1.14)$$

However, we know that the result of the measurement *should be* either 'up' (represented by $|\text{'up'}\rangle$) or 'down' (represented by $|\text{'down'}\rangle$). There is no 'third outcome' for $\alpha|\text{'up'}\rangle + \beta|\text{'down'}\rangle$ to represent. Furthermore, the result of the measurement should be randomly determined, whereas (1.14) is manifestly deterministic.

Can we introduce randomness via our ignorance of the initial conditions of the measuring device? No. Even if there is a multitude of quantum states representing the 'ready' state, *every one of them* must evolve into a superposition of a state representing 'up' and a state representing 'down'. Can we introduce it via consideration of the 'environment'? No, for the simple reason that as much of the environment as you like (up to the entire universe, if necessary) can simply be included in $\mathcal{H}_m$. The conclusion seems inescapable: the mathematical structure of unitary quantum mechanics cannot represent the structure of the physical world. As such, the fundamental, primitive status of 'measurement'—or at least, of macroscopic-scale measurements—*seems* to be an ineliminable part of quantum mechanics.

1.5 The measurement problem and the instrumentalist response

So where does this leave us? It leaves us with a mathematically very elegant theory—the bare quantum formalism—from which predictions about the observable world can be extracted not by understanding how the formalism represents the world, as in the rest of science, but by a rather ad hoc, under-motivated, inherently approximate algorithm: select a phase-space POVM for some subset of the degrees of freedom of the system which we label 'macroscopic', and interpret expectation values with respect to that POVM as being the probability of finding the actual macroscopic world in the appropriate state.

So what? The algorithm may be 'ad hoc, under-motivated, inherently approximate', but it still allows predictions of stunning precision. Isn't the point of science to make predictions? Why care about supposed 'problems' with a theory if it still generates predictions of this degree of accuracy?

But this misses the point of science. As the introduction to this chapter stressed, the purpose of scientific theories is not to predict the results of experiments: it is to describe, explain, and understand the world. And quantum mechanics—as described in the previous sections—fails to do this. No 'description of the world' is to be found in the quantum algorithm of section 1.3. At best, we will find an approximately specified description of the macroscopic degrees of freedom. About the microscopic world, the algorithm is silent.

One robust response to my comments might be: so much the worse for what we thought science was for. Could it not be that our hopes of 'describing, explaining and understanding the world' turn out to be optimistic—even naive? Could quantum mechanics not be telling us to lower our sights, to be content with a more modest picture of science as a mere predictive tool? After all, 'experiments' can be construed quite broadly: the quantum algorithm suffices to predict *all macroscopic phenomena*. Why not be content with that?

In the philosophy of science, views of this sort are termed *operationalist*, or *instrumentalist*, or sometimes *positivist*, philosophies about science.[6] Instrumentalists regard science as a useful predictive tool, whose function is to predict the results of experiments (by contrast, *realists* about science adopt a position more along the lines of my introduction to this chapter). And historically, instrumentalism has certainly tempted many serious students of quantum mechanics, who have taken the difficulties in giving a realist reading of quantum mechanics as evidence that the theory is simply not in the business of giving us a description of the world. Chris Fuchs and Asher Peres make the point very eloquently:

We have learned something new when we can distill from the accumulated data a compact description of all that was seen and an indication of which further experiments will corroborate that description. This is what science is about. If, from such a description, we can *further* distill a model of a free-standing 'reality' independent of our interventions, then so much the better. Classical physics is the ultimate example of such a model. However, there is no logical necessity for a realistic worldview to always be obtainable. If the world is such that we can never identify a reality independent of our experimental activity, then we must be prepared for that, too. ... [Q]uantum theory does *not* describe physical reality. What it does is provide an algorithm for computing *probabilities* for the macroscopic events ('detector clicks') that are the consequences of our experimental interventions. This strict definition of the scope of quantum theory is the only interpretation ever needed, whether by experimenters or theorists. (Fuchs and Peres 2000a)

Todd Brun and Robert Griffiths point out [in Styer et al. 2000] that 'physical theories have always had as much to do with providing a coherent picture of

---

[6] Here I do an injustice to an intricate part of twentieth-century philosophy of science, but the subtle distinctions between these terms will not be relevant for my purposes.

reality as they have with predicting the results of experiment.' Indeed, have always had. This statement was true in the past, but it is untenable in the present (and likely to be untenable in the future). Some people may deplore this situation, but we were not led to reject a freestanding reality in the quantum world out of a predilection for positivism. We were led there because this is the overwhelming message quantum theory is trying to tell us. (Fuchs and Peres 2000b)

But whether or not this 'strict definition' is the only interpretation ever *needed* by physicists, it is certainly not sufficient to explain the *actual practice* of physics. Physics on the Fuchs–Peres proposal would have no need for microscopic talk at all, except as a mere *façon de parler*—yet in quantum physics as much as in the rest of science, physicists talk constantly about the microworld. Particle physicists discuss the mass of the Higgs boson and describe tracks in cloud chambers as made by pions. Solid state physicists discuss the creation and destruction of various species of quasi-particle in various regimes of matter. Quantum information experiments are routinely described as creating microscopic systems in certain states, as sending these states from place to place, as manipulating them in various ways. Molecular biologists and chemists routinely avail themselves of quantum concepts, but this does not stop them talking in a straightforwardly realist way about the molecules that they study. In quantum physics as in the rest of science, what motivates physicists is *the explanation of phenomena*, microscopic or otherwise.

Furthermore, it is hard to see how the split between quantum and classical required by instrumentalism is really coherent. Isn't classical physics supposed to be in some sense a limiting case of quantum physics? Don't we know rather a lot about exactly *how* classical physics emerges from quantum physics? (This last will be discussed more in Chapter 3.) And aren't macroscopic objects like measuring devices supposed to be composed of microscopic objects like atoms, in a way which physicists and materials scientists understand fairly well?

To summarize, the problems with instrumentalism are

1. It fails to do justice to the actual practice of physics.
2. It makes an arbitrary and unmotivated division between the macroscopic and the microscopic worlds.

In fact, such objections are not specific to *quantum-mechanical* forms of instrumentalism. Rather similar problems arose in those versions of instrumentalism advocated in the last century by philosophers of science on purely philosophical grounds, and the consensus view in philosophy is now that instrumentalism is incoherent as a way to make sense of science. Not all serious philosophers of science are *realists* in the sense that they think that our best scientific theories are approximately true, but there is a consensus that the only way to *understand a scientific theory* is to understand it as offering a description of the world.

There is, however, a third reason to reject instrumentalism which is more specific to quantum mechanics. It has been clear for several decades that any formulation of quantum mechanics which gives a special role to measurement will have grave difficulties with quantum cosmology. For measurement presupposes an external measuring process, whereas the subject matter of cosmology is the Universe as a whole. Barring lapses into religion or mysticism, an account of quantum mechanics as being just a method for predicting the results of measurements on a system will be pretty hopeless when that 'system' incorporates everything that there is. Quantum cosmology motivated Everett's original work in 1957; quantum cosmology remains a powerful argument for the Everett interpretation even amongst 'hard-nosed' physicists unmoved by more philosophical considerations.[7]

In any case, much ink has been spilled on the pros and cons of instrumentalism as a philosophy of science, and I have done no more in this section than to gesture at what seem to me to be the hopelessly severe problems that it faces in doing justice to the scientific endeavour. However, for the rest of this book I will be assuming that instrumentalism is an unacceptable way to make sense of quantum mechanics. (Perhaps the unconvinced reader will at least accept Fuchs and Peres' implicit concession that instrumentalism is a counsel of desperation, a position to be adopted only when a realist reading of a scientific theory is 'untenable'. If we can after all find a way to understand quantum mechanics as a theory

---

[7] See e.g. Hawking (1976: 196): 'in order to determine where one is in space-time one has to measure the metric and this act of measurement places one in one of the various different branches of the wave-function in the Wheeler-Everett interpretation of quantum mechanics', and Hartle (2010: 76): 'in quantum cosmology we are interested in classical behavior over cosmological stretches of space and time, and over a wide range of subsystems, independently of whether these subsystems are receiving attention from observers. Certainly our observations of the Moon's orbit, or a bit of the universe's expansion, have little to do with the classical behavior of those systems. Further, we are interested not just in classical behavior as exhibited in a few variables and at a few times of our choosing, but over the bulk of the universe in as refined a description as possible, so that classical behavior becomes a feature of the universe itself and not a choice of observers.' (However, see Fuchs and Peres 2000a for a dissenting view.)

which describes the world (or, indeed, to replace it with a theory which describes the world), then instrumentalism, coherent or not, loses much of its appeal. Maybe one *could* be an instrumentalist about the stegosaurus, but who would want to be?)

1.6 Beyond instrumentalism: vestiges of reality?

If instrumentalism is unacceptable, and straightforward realism is blocked by the problem of measurement, what alternatives are there? Many attempts have been made to steer some sort of 'middle ground' between instrumentalism and realism in quantum mechanics, but the task has proved all but impossible—and no wonder, for the challenge is to find some way in which a physical theory represents the world *other* than by some sort of correspondence between the mathematical description and the physical reality. Proposals made have included:

- Perhaps the quantum state is to be thought of as representing the world only probabilistically, in the same way that, in classical statistical physics, a density function on phase space represents not a single state of the world but a probability measure over possible states of the world. At first sight this is a rather tempting prospect in the case of quantum mechanics: rather than regarding a linear superposition of two distinct measurement outcomes as some weird we-know-not-what, just regard it as stating that the system is in one state or the other, and that the probability of each state is given by the squared modulus of its amplitude in the superposition.

  The problem with this tempting idea is that it is not clear what the probabilities are probabilities *of*. In classical physics, there is an underlying microphysics beneath the statistical-mechanical description, and the latter is valuable only because of the vast complexity of, and because of our ignorance about the precise details of, the former. But what is the underlying microphysics beneath quantum mechanics? Einstein may have hoped that such a microphysics could be found, but the no-go theorems proved by von Neumann (1955),[8]

---

[8] It has become a commonplace in the foundations-of-physics literature (e.g. Maudlin 2006: 1112) to chastise von Neumann for elementary misconceptions in his proof, but this in turn is to misconceive von Neumann's objective. He was concerned not with de Broglie's proposal of additional hidden variables guided by a physically real wavefunction, but with

Bell (1966), Kochen and Specker (1967), Gleason (1957), and others[9] make it clear that such a microphysics would be unacceptably contrived.[10]

Some authors (e.g. Ballentine 1970; 1990) have asserted that there is no need for such a microphysics—that the quantum state can be understood as saying 'the electron has a 30% chance of being in position p' without any commitment to a microphysics in which it makes sense to say that a particular electron *is* in position p. But I for one fail to see how this can avoid both instrumentalism and obscurantism. The option is always open to us to say that we mean just that there is a 30% chance of finding the electron in position p *on measurement*, but that is straightforwardly instrumentalist. If that's not what we're saying, and if we're not saying something about our ignorance of the actual microphysics, then what *are* we saying?

• More radically, perhaps there is some way to regard 'measurement' as a primitive term *without* this bringing some sort of commitment to instrumentalism. On this view (which is a common way of making sense of the 'Copenhagen interpretation' of quantum physics, though it does not seem to have been Niels Bohr's own view[11]), it does not make sense to describe a microscopic system except in a certain experimental context; nonetheless, a microscopic description given in such a context is still to be understood realistically. (For a recent example of such a view, see Bub and Pitowsky (2010), who explicitly describe the idea that measurement cannot be primitive as one of two 'dogmas' of quantum mechanics,[12] but still describe their

Einstein's proposal of a hidden-variable theory which would be to the wavefunction what classical mechanics is to the probability distributions of statistical mechanics. His theorem may not rule that possibility out entirely but it certainly puts strong constraints on it. (I develop this distinction further in Wallace 2008: 63–7.)

[9] For a general analysis of these results and their significance, see Redhead (1987).

[10] For a dissenting view, see Spekkens (2007); also, note that most so-called 'hidden-variable interpretations' in modern philosophy of quantum theory do not provide, and do not aim to provide, such a microphysics; they supplement the wavefunction with additional hidden variables, but do not aim to replace it. (See section 1.7 for further discussion of these theories.)

[11] See Saunders (2005), and references therein, for discussion of Bohr's position.

[12] For the benefit of non-philosophers: this is actually a reference to 'Two dogmas of empiricism', a famous paper of Quine (1953). The terminology does not seem entirely apposite, though: far from the assumption being a 'dogma', it has been highly controversial throughout the development of quantum mechanics.

measurement-dependent approach as 'realist'.) But the difficulties in taking measurement as primitive—sketched in section 1.5—do not seem to be much obviated by this sort of move. It can be made sense of on instrumentalist lines; it can be made sense of on lines which presuppose a fundamental division between macroscopic (classical) and microscopic (quantum) talk and which license the latter only in contexts determined by the former. It is unclear if it can be made sense of otherwise.

- More radically still, rather than the goal of science being to gain information *about* the world, perhaps the world itself *is* just information. Many variants of this idea have been suggested (with varying levels of seriousness) by workers in the new field of quantum information (see e.g. Zeilinger 2005; Clifton et al. 2003, or Fuchs 2002), but I confess that insofar as they mean it literally and not metaphorically, I do not understand what they are talking about. It seems to be in the nature of 'information' that it is information *about something*; how can it make sense to regard the world *itself* as just information? (Compare how little sense it would make to describe the world as made out of *opinion* or *belief* or *rumour*.) The point has been made with force and clarity by Timpson (2008; 2010), who usefully points out that 'quantum information theory' is the quantum-mechanical form of information theory—the theory of information in a quantum universe—and that we misrepresent it if we consider it to be the theory of a new substance, 'quantum information'.

    To be fair, major conceptual shifts do happen in science—perhaps a century ago, talk of energy and matter being the same thing would have seemed equally incoherent. But if this sort of conceptual revolution is what proponents of 'quantum-information' interpretations of quantum mechanics have in mind, then they have barely begun to develop the details, and it is far from clear how their approach, whatever ontological or metaphysical status it gives to *information*, can solve the measurement problem without lapsing back into instrumentalism.[13]

---

[13] Nor is it obvious that an identification of matter and information would provide an *alternative* to the Everett interpretation, rather than a deeper understanding of it. After all, 'information is physical' does not have to mean 'the physical is merely information'; it could equally mean 'information, too, is physical'. See Deutsch (1997) for (speculative) thoughts along these lines.

- Most radically of all, perhaps where our previous assumptions went wrong was that they presupposed the validity of *classical logic*. There is a long tradition of arguing that the true lesson of quantum mechanics is that classical logic, like classical physics, is only approximately valid, and only in certain special regimes (see Putnam 1968 for a classic defence of this position, Dickson 2001 for a contemporary defence, and Bacciagaluppi 2007 for a clear recent review). And if classical logic does not apply to quantum mechanics, all bets are off—the 'no-go theorems' of von Neumann, Gleason et al. which make it well-nigh impossible to interpret the quantum state as a probabilistic entity certainly rely essentially on classical logic.

  Does this even make sense? I'm not sure. Can it legitimately be called *realist*? I'm even less sure. (The philosopher Michael Dummett has argued powerfully that such strategies *do not* amount to any sort of realism about physics; see Dummett 1976 for the details.)

The best that can be said of these strategies, I think, is that they too offer a counsel of desperation: they commit us to a reformulation of the whole nature of science which we would not have contemplated were it not for the paradox of measurement, which doesn't seem to fit the actual practice of physics (in which the quantum state seems taken as straightforwardly physical), and which would seem absurd in any other area of science. As David Deutsch has acerbically remarked (2010), if palaeontologists informed us that dinosaurs were such weird animals that the laws of logic don't apply to them, or that 'palaentology' wasn't really about dinosaurs at all but only about fossils, we'd have a hard time taking them seriously.

1.7 Solving the measurement problem: alternative theories

If that counsel of desperation doesn't appeal, another is on offer: replace quantum mechanics itself with a new theory which we can understand without rethinking the nature of science. Given the tremendous empirical success of quantum mechanics, this would presumably involve (at least as a first step) somehow modifying the existing quantum theory rather than abandoning it altogether, and there are basically two ways to do this: add additional structure to the formalism, sufficient to represent the actual world, so that even if the *quantum state* is in a macroscopic superposition, the additional structure describes a unique classical world; or modify the dynamics of the theory so that macroscopic superpositions do not occur in the first place. As John Bell famously put it, 'Either the wavefunction, as given by the Schrödinger equation, is not everything, or it is not right' (Bell 1987: 201).

Both possibilities have been extensively explored. Theories which supplement the quantum state with additional ontology are generally known as *hidden-variable* theories: the 'hidden variables' (better: the entities whose properties are the hidden variables) obey dynamical equations governed by the quantum state, but do not in turn act back upon the state. In the best-known and best-developed hidden variable theory, the *pilot-wave theory* (or the de Broglie–Bohm theory, or Bohmian mechanics; it was first suggested by de Broglie, in particular at the 1927 Solvay conference (cf. Bacciagaluppi and Valentini 2010) and was then rediscovered by Bohm 1952), the hidden variables are point particles, which at all times have definite positions and which are intended to be the stuff of which macroscopic matter is made. In such theories the role of the quantum state is simply to guide the particles along their trajectories: in the case of a two-slit experiment, for instance, each particle definitely goes through one slit or another, but its trajectory after emerging from the slit is influenced by the portion of the wavefunction that (in a manner of speaking) went through the other slit. In this sense, the term 'hidden variables' is something of a misnomer: far from being hidden, it is these 'variables' (the de Broglie–Bohm particles, in this case) which (are supposed to) constitute the world that we see around us, and it is the quantum state, whose only role is to guide the particles, that is really hidden.

Theories which modify the Schrödinger equation to eliminate macroscopic superpositions are often called *dynamical-collapse theories*, because they attempt to provide a cleanly formulated, microphysical theory which tells us exactly when the Schrödinger equation is violated and how this leads to the suppression of macroscopic superpositions. The best-known theory of this kind, the *GRW* theory (named after its creators, Ghirardi, Rimini, and Weber 1986) postulates that each elementary particle has an independent, and very small, chance per second of spontaneously 'jumping' to a localized state (each particle experiences one of these jumps, on average, every 10^8 years). In isolated microscopic systems, this 'jump' would occur far too rarely to be empirically detectable; in large systems, though, the collapse of one particle into a localized state would cause the collapse of all those particles with which it is entangled, suppressing macroscopic superpositions on very short timescales.[14]

Both hidden-variable theories and dynamical-collapse theories are genuinely new theories of physics, although there is an interesting difference between them: dynamical-collapse theories actually modify the Schrödinger equation and hence are empirically testable (see Leggett 2002 for a review of such attempts), whereas hidden-variable theories leave the entire formalism of quantum mechanics alone and just add the 'hidden variables' as extra structure. (This has led to the criticism that hidden-variable theories are 'parallel-universe theories in a state of chronic denial' (Deutsch 1996: 225), a criticism to which I will return in Chapter 10.)

And from the point of view of practical physics, the most important observation about these theories is that *they only really exist in the nonrelativistic domain*. That is: the hidden-variable and dynamical-collapse theories which have been widely studied (most notably the two described above: the de Broglie–Bohm theory and the GRW theory) are generally agreed to be empirically equivalent (or almost-equivalent, in the case of dynamical-collapse theories) to non-relativistic N-particle quantum mechanics. But non-relativistic quantum mechanics is not the end of the story. To explain the results of particle physics, or even to explain the quantum-mechanical aspects of radiation, we require relativistic quantum field theory: the Standard Model, or at least quantum electrodynamics. But as yet there do not appear to be *any* theories of either kind which reproduce the predictions of these theories.

And that means that, if 'the measurement problem' is the problem that our best current physical theory does not admit of a satisfactory realist interpretation, then neither hidden-variable theories nor dynamical-collapse theories solve, or even come close to solving, the measurement problem. Perhaps we have an interpretation sufficient to non-relativistic quantum mechanics—but then, we had an interpretation sufficient to *classical* mechanics, too. The name of the game is not to reproduce the predictions of quantum physics in certain restricted regimes. It is to

---

[14] It is perhaps worth noting that the original GRW theory itself is empirically inadequate even for non-relativistic quantum mechanics, as it fails to handle indistinguishable particles correctly; however, a modification of the GRW theory due to the original authors and to Philip Pearle is generally accepted to resolve this problem. See Bassi and Ghirardi (2003), and references therein, for more details.

reproduce the predictions of quantum physics, *period*. (And of course, reproducing non-relativistic N-body quantum mechanics is insufficient to reproduce the actual predictions of quantum mechanics even in the non-relativistic regime: electromagnetic radiation, completely essential to any understanding of atomic physics, is not representable in the non-relativistic N-body theory.)

Of course, responsible proponents of modificatory theories are perfectly well aware of this, and have a perfectly reasonable answer: give us time, we're working on it. And indeed much work has been done on the problem, and in recent years some interesting models have been developed.[15] In general they have certain problems in common: a violation of relativistic covariance;[16] a somewhat contrived feel; most importantly, a failure to engage with *renormalized* quantum field theories.

In any case, the fact remains that modificatory strategies for solving the measurement problem involve a commitment to redoing the whole of interacting quantum field theory, and that most of this task remains

[15] The situation differs between hidden-variable theories and dynamical-collapse theories. Quantum-field-theoretic hidden-variable theories have been proposed in recent years in which the hidden variables are, respectively: particles (Dürr et al. 2004; 2005); bosonic fields (Struyve and Westman 2007; Struyve 2007); and fermion number (Colin 2003; Colin and Struyve 2007). All explicitly violate relativistic covariance at the microscopic level. Heuristic arguments have been given in each case to support the suggestion that these theories will empirically reproduce the predictions of quantum field theory. Assessing the claim is a little difficult because of course (as we have seen) these theories are mathematically identical to ordinary unitary quantum mechanics with extra hidden variables added. A reasonable criterion might be: can one prove (to the usual standards of mathematical rigour in theoretical physics) that in the emergent non-relativistic regime of these theories, the hidden variables pick out uniquely one of the quasi-classical terms in the quantum state? (Here I anticipate ideas to be developed in Ch. 3.) So far this has not been achieved.

In the dynamical-collapse case, progress is much less advanced (unsurprisingly, since the theory must be modified and not merely supplemented). There are to my knowledge no relativistic interacting dynamical-collapse theories at present; however, Tumulka (2006) has proposed an extension of the GRW theory to *non-interacting* relativistic quantum particle mechanics. Interestingly, Tumulka's theory is covariant, but achieves this at the expense of making facts about collapse hyperplane-dependent; see Wallace and Timpson (2010) for further discussion of this point. Of course, a non-interacting theory is at most a bare beginning to the project that advocates of dynamical-collapse theories need to complete.

[16] It is normal to point out at this stage that the empirical violation of Bell's inequality (Bell 1981a; for discussions, see Maudlin 2002 or Butterfield 1992, and references therein) means that any empirically adequate theory of physics will be in tension with relativistic covariance. However, Bell's results do not apply to the Everett interpretation (a point to which I return in Ch. 8), so in this context the criticism retains its force.

undone. One of the notable things about discussing the interpretation of quantum mechanics with physicists and with philosophers is that it is the physicists who propose philosophically radical ways of interpreting a theory, and the philosophers who propose changing the physics. One might reasonably doubt that the advocates of either strategy are always fully aware of its true difficulty.

1.8 Everett's insight

Faced with the measurement problem, then, we seem to have only two options:

1. Replace quantum theory—the most predictively powerful, most thoroughly tested, and most widely applicable theory in scientific history—with a new theory which we have not yet constructed.
2. Rethink the nature of the scientific enterprise.

A third[17] could perhaps be added:

3. Hope that the problem goes away when general relativity and quantum mechanics are unified,

and given the unattractiveness of the first two it is easy to find it appealing, but there is not much else to say in its favour, nor much of an argument for it other than: the measurement problem is a mystery, the quantization of gravity is a mystery, therefore they must be connected.[18] (It is also worth noting that both string theory (the current front-runner for a quantum theory of gravity) and loop-space quantum gravity (the current runner-up) are recognizably Hilbert-space quantum theories, and so neither seems to offer much progress with the measurement problem.)

It was Hugh Everett's great insight to recognize that the apparent dilemma is false—that, *contra* the arguments of section 1.4, we can after

---

[17] I suppose technically the third option could be seen as a variant on the first option, but in practice it has a rather different flavour.

[18] A cynic might argue that some (not all) arguments for giving consciousness a role in quantum theory have a similar form: quantum=mystery, consciousness=mystery; therefore, consciousness=quantum.

all interpret the bare quantum formalism in a straightforwardly realist way—without changing either our general conception of science, or modifying quantum mechanics.

How is this possible? Didn't we see in that section that the linearity of quantum mechanics commits us to macroscopic objects being in super-positions, in indefinite states? Didn't we show that the generic outcome of a measurement process is something like

$$\alpha|\text{'up'}\rangle + \beta|\text{'down'}\rangle? \tag{1.15}$$

We did indeed show that states like (1.15) are the generic result of measurement processes in unitary quantum mechanics. But it is actually a non sequitur to go from (1.15) to the claim that macroscopic objects are in indefinite states.

An analogy may help here. In electromagnetism, a certain configuration of the field, say $\mathbf{F}_1(x, t)$ (here $\mathbf{F}$ is the electromagnetic 2-form), might be interpretable as—might be, if you like[19]—a pulse of ultraviolet light zipping between Earth and the Moon. Another configuration, say $\mathbf{F}_2(x, t)$, might represent a different pulse of ultraviolet light zipping between Venus and Mars. What then of the state of affairs represented by

$$\mathbf{F}(x, t) = 0.5\mathbf{F}_1(x, t) + 0.5\mathbf{F}_2(x, t)? \tag{1.16}$$

What weird sort of thing is this? Must it not represent a pulse of ultra-violet light that is *in a superposition* of travelling between Earth and Moon, and of travelling between Mars and Venus? How can a single pulse of ultraviolet light be in two places at once? Doesn't the existence of superpositions of macroscopically distinct light pulses mean that any attempt to give a realist interpretation of classical electromagnetism is doomed?

Of course, this is nonsense. There is a perfectly prosaic description of $\mathbf{F}$: it does not describe a single ultraviolet pulse in a weird superposition, it just describes two pulses, in different places. And *this*, in a nutshell, is what the Everett interpretation claims about macroscopic quantum superpositions:

---

[19] Here I gloss a distinction which some philosophers regard as hugely important (in some contexts, probably rightly): between a physical entity and its mathematical representation. Throughout the book, however, I make rather little attempt to draw this distinction. I leave it to the reader to work out from context which is intended, on the rare occasions where it actually matters.

they are just states of the world in which more than one macroscopically definite thing is happening at once. Macroscopic superpositions do not describe indefiniteness, they describe multiplicity.

The standard terminology of quantum mechanics can be unhelpful here. It is often tempting to say of a given macroscopic system—like a cat, say—that its possible states are all the states in some 'cat Hilbert space', $\mathcal{H}_{cat}$. Some states in $\mathcal{H}_{cat}$ are 'macroscopically definite' (states where the cat is alive or dead, say); most are 'macroscopically indefinite'. From this perspective, it is a very small step to the incoherence of unitary quantum mechanics: quantum mechanics predicts that cats often end up in macroscopically indefinite states; even if it makes sense to imagine a cat in a macroscopically indefinite state, we've certainly never seen one in such a state; so quantum mechanics (taken literally) makes claims about the world that are contradicted by observation.

From an Everettian perspective this is a badly misguided way of thinking about quantum mechanics. This $\mathcal{H}_{cat}$ is presumably (at least in the non-relativistic approximation) some sort of tensor product of the Hilbert spaces of the electrons and atomic nuclei that make up the cat. Some states in this box certainly look like they can represent live cats, or dead cats. Others look like smallish dogs. Others look like the Mona Lisa. There is an awful lot that can be made out of the atomic constituents of a cat, and all such things can be represented by states in $\mathcal{H}_{cat}$, and so calling it a 'cat Hilbert space' is very misleading.

But if so, it is equally misleading to describe a macroscopically indefinite state of $\mathcal{H}_{cat}$ as representing (say) 'a cat in a superposed state of being alive and being dead'. It is far more accurate to say that such a state is a superposition of a live cat and a dead cat.

One might still be tempted to object: very well, but we don't observe the Universe as being in superpositions of containing live cats and containing dead cats, any more than we observe cats as being in superpositions of alive and dead. But it is not at all clear that we *don't* observe the Universe in such superpositions. After all, *cats* are the sort of perfectly ordinary objects that we seem to see around us all the time—a theory that claims that *they* are normally in macroscopically indefinite states seems to make a nonsense of our everyday lives. But the *Universe* is a very big place, as physics has continually reminded us, and we inhabit only a very small part of it, and it will not do to claim that it is just 'obvious' that it is not in a superposition.

This becomes clearer when we consider what actually happens, dynamically, to $\mathcal{H}_{cat}$, to its surroundings, and to those observing it, when it is prepared in a superposition of a live-cat and a dead-cat state. The rest of Part I will be devoted in large part to a detailed discussion of this problem. But in outline, the answer is that the system's surroundings will rapidly become entangled with it, so that we do not just have a superposition of live and dead cat, but a superposition of extended quasi-classical regions—'worlds', if you like—some of which contain live cats and some of which contain dead cats. If the correct way to understand such superpositions is as some sort of multiplicity, then our failure to observe that multiplicity is explained quite simply by the fact that we live in one of the 'worlds' and the other ones don't interact with ours strongly enough for us to detect them.

This, in short, is the Everett interpretation. It consists of two very different parts: a contingent physical postulate, that the state of the Universe is faithfully represented by a unitarily evolving quantum state; and an a priori claim about that quantum state, that if it is interpreted realistically it must be understood as describing a multiplicity of approximately classical, approximately non-interacting regions which look very much like the 'classical world'.

And this is all that the Everett interpretation consists of. There are no additional physical *postulates* introduced to describe the division into 'worlds', there is just unitary quantum mechanics. For this reason, it makes sense to talk about *the* Everett interpretation, whereas it does not to talk about *the* hidden-variables interpretation or *the* quantum-logical interpretation. The 'Everett interpretation of quantum mechanics' is just quantum mechanics itself, 'interpreted' the same way we have always interpreted scientific theories in the past: as modelling the world. Someone might be right or wrong *about* the Everett interpretation—they might be right or wrong about whether it succeeds in explaining the experimental results of quantum mechanics, or in describing our world of macroscopically definite objects, or even in making sense—but there cannot be multiple logically possible Everett interpretations any more than there are multiple logically possible interpretations of molecular biology or classical electrodynamics.[20]

---

[20] Perhaps in *some* sense there are multiple interpretations of classical electromagnetism: perhaps realists could agree that the electromagnetic field is physically real but might disagree about its nature. Some might think that it was a property of spacetime points; others might

This in turn makes the study of the Everett interpretation a rather tightly constrained activity—a rare and welcome sight in philosophy! For it is not possible to solve problems with the Everett interpretation by changing the interpretive rules or changing the physics: if there are problems with solving the measurement problem Everett-style, they can be addressed only by hard study—mathematical and conceptual—of the quantum theory we have.

And it certainly seems that there *are* problems. The rest of this chapter will outline them; the remainder of the book will attempt to show how they are solved.

1.9 The challenges for the Everett interpretation

The main challenge for the Everett interpretation, then, is to show exactly how our classical world, which seems to be so well described by a single quasi-classical universe in which quantum processes occur at random, can be incorporated in the macroscopically indefinite structure of deterministic, unitarily evolving quantum mechanics; and if the last section has (I hope) made the case that it is not *obvious* that this is impossible, still, it barely began to address the positive task of justifying it.

What is required of such a justification? Two problems must be solved; we might call these the *ontological* and *probability* problems. It makes sense to start with the ontological problem: how, exactly, are we justified in regarding quantum superpositions as describing multiplicity? (Analogies with pulses of ultraviolet light only go so far ...)

The more traditional name for this problem is the *preferred basis problem*, a term which comes from asking the following question: if superpositions are supposed to represent distinct worlds, with respect to which basis are these superpositions defined? After all, any quantum state is a superposition with respect to *some* basis. From this perspective, one solves the problem by giving some criterion to pick out the right basis. Maybe it is the basis of definite particle positions (Bell 1981b); maybe it is the

---

regard it as an entity in its own right. I am deeply sceptical as to whether this really expresses a distinction (for reasons I expand upon in section 8.8), but in any case, I take it *this* is not the problem—nor the *sort* of problem—that we have in mind when we talk about the measurement problem.

Schmidt decomposition basis with respect to some decomposition of the system into subsystems (Deutsch 1985a); maybe it is the basis of conscious experiences (Lockwood 1989).

But asking 'which basis' distracts us from the real question: why *any* basis?—that is, what justifies regarding a quantum state as a collection of quasi-classical worlds at all? If the answer is to be compatible with our requirement that the Everett interpretation leaves the quantum formalism alone, the only possibility is that the justification comes in some sense from within the theory—that a realist interpretation of the quantum state compels us to understand it as being (or better, as instantiating) multiple classical worlds. So I resist the 'preferred basis' terminology in this book.

The solution of the ontological problem is the subject matter of Chapters 2 and 3. In brief, though, the answer will be that the 'worlds' are *emergent* objects, higher-order entities more like cats or tables than electrons or spacetime points. As I shall argue, the way in which cats, or tables, or any other such entities exist is as structures within the underlying microphysics. (So a cat is a subsystem of the microphysics structured in cattish ways.) The question of whether the quantum state is actually a collection of quasi-classical worlds is then the question of whether the quantum state has the structure of a collection of quasi-classical worlds—a question which, I shall argue, decoherence theory has now fairly conclusively answered in the affirmative.

The probability problem has, historically at least, generally been seen as the more pressing difficulty. Tim Maudlin, writing in sharp criticism of the Everett interpretation, puts it this way:

> The many-worlds theory is incoherent for reasons which have often been pointed out: since there are no frequencies in the theory there is nothing for the numerical predictions of quantum theory to mean. This fact is often disguised by the choice of fortuitous examples. A typical Schrödinger-cat apparatus is designed to yield a 50 percent probability for each of two results, so the 'splitting' of the universe in two seems to correspond to the probabilities. But the device could equally well be designed to yield a 99 percent probability of one result and 1 percent probability of the other. Again the world splits in two; wherein lies the difference between this case and the last? (Maudlin 2002: 5)

There are in fact two somewhat distinct subproblems lurking here. First, how does talk of probability even *make sense* in the deterministic Everett universe? And secondly, even if it does make sense, what justifies the *particular* probabilities assigned to different outcomes by the quantum formalism? We might call these the *incoherence* and *quantitative* problems with probability.[21]

More recently, a slightly different way to break the problem down has become useful: into the *practical problem* of why we should allow quantum probabilities (that is, the modulus squared of the amplitudes) to guide our actions, and the *epistemic problem* of why those quantum probabilities plug into our evidence-gaining activities in the usual way.[22] The latter has turned out to be more complicated than the former, in particular because of the need to connect the quantum probabilities to the evidence for quantum mechanics itself. For suppose that we really do come to understand why, in a deterministic many-worlds theory, we are nevertheless somehow justified in behaving as though the probability rules are true. This falls short of what we actually need: which is not (just) a theory of how we should act if the Everett interpretation is true, but a theory of why the Everett interpretation is a good explanation of the quantum-mechanical phenomena which we observe experimentally.

I address these problems in some detail in Chapters 4–6, where I argue that the problems stem in large part from failing to recognize the philosophical difficulties that the apparently familiar concept of probability actually suffers from. As we will see, it is no harder to understand probability in Everett-interpreted quantum mechanics than it is in classical physics. Indeed, rather strikingly, it turns out that it may be *easier* to understand probability in the Everettian context. Some of the most intractable problems in the philosophy of probability appear to have answers in Everettian quantum mechanics that are simply not available in single-universe theories.

The ontological and probability 'problems' are not so much objections to the Everett interpretation; rather, they are the necessary components that any account of Everettian quantum mechanics must supply. Three more problems, however, are more straightforwardly objections to Everett: these might be called the *metaphysical problem*, the *incredulity problem*, and the *competition problem*.

---

[21] I introduced the *terms* 'incoherence problem' and 'quantitative problem' in Wallace (2002a; 2003b), though I claim no originality for the *concept*.

[22] This terminology was introduced in Greaves (2007).

The metaphysical problem is the problem of what this 'quantum state', which is supposed to represent the whole of microphysical reality, actually *does* represent. Classical physics, so the objection goes, tells a readily understandable story of the physical world as a collection of entities inhabiting and moving through space. Modifications such as relativity (with its merging of space and time) and string theory (which increases the number of spatial dimensions from three to ten) might make a few modifications to that story, but they do not change it out of recognition. By contrast, Everettian quantum mechanics apparently replaces that spatiotemporal story with a story about the 'quantum state' or 'wavefunction'—an entity which does not seem to live in physical space at all, but in Hilbert space, or on configuration space, or somesuch. How (so the objection goes) are we to understand this as a story about the world?

I am not at all clear that the metaphysical problem deserves to be taken seriously, at least not if the ontological problem can be solved in the way I suggest above and develop in Chapters 2 and 3. Physics is ultimately concerned with making claims about the *structure* of the world, described mathematically, rather than its 'true nature'. (Note that in quantum gravity, it has long been quite routine to suppose that spacetime (of however many dimensions) is not fundamental at all, but is emergent from some fundamentally non-spatio-temporal entity—spin foam, for instance. The criteria for a successful story about this emergence certainly seem to be structural—can we find states of the underlying quantum-gravitational theory structurally isomorphic to quasi-classical spacetimes?—and worries about the 'true nature' of the spin foam do not seem to be crucial, or even well-formed.)

Nonetheless, if the Everett interpretation is correct, then it is certainly of considerable interest to ask what the structure of the quantum world is actually like, and what relation it bears to ordinary spacetime. I will address this question in Chapter 8, where I will argue that the most obvious way of understanding the quantum state—as representing a complex-valued field on a high-dimensional space—is very misleading and not at all perspicuous. Correctly understood, quantum mechanics is as much about regions of physical space and their properties and contents as classical mechanics is—something which may be reassuring to someone who takes the metaphysical problem more seriously than I.

The last two problems are closely related to one another. The incredulity problem is: how on earth can we take something as extraordinary, and as ontologically extravagant, as the Everett interpretation at all seriously? (As Hilary Putnam once memorably remarked (1990: 42), 'alas, we don't find this picture is one we can *believe*. What good is a metaphysical picture one can't believe?') And the competition problem is: why attempt to solve the measurement problem in this (extraordinary, extravagant) way when there are so many other solutions available to us which remain decently confined to a single universe?

In a sense, this chapter should already have answered the competition problem, but the point is important enough to be worth reiterating here. Namely: it is simply false that there are alternative explanatory theories to Everett-interpreted quantum mechanics which can reproduce the predictions of quantum theory.

To be sure, alternative positions of a sort are available. One can lapse into instrumentalism. One can try to find some way of making sense of the theory that is not *exactly* instrumentalism, but that nonetheless does not take the theory at face value. And one can try to find an alternative theory that is as explanatorily and predictively successful as quantum mechanics but whose interpretation is more to one's liking—taking reassurance from the existence of theories of this kind which putatively explain the non-relativistic fragment of quantum theory's empirical successes, and taking the huge gulf between the predictive successes of quantum mechanics and of such extant competitors as a challenge.

However, one can do this with *any* scientific theory that one dislikes. These 'alternative positions' are equally available in the case of special relativity, or of the atomic theory of matter, or (to be a little unfairly provocative, perhaps) of evolution by natural selection, or indeed of the theory that fossils are the remains of dead dinosaurs. The fact remains that, at present, there are no known ways of explaining the quantitative predictions of relativistic quantum theory other than Everett's. The Everett interpretation is the only game in town.

(Notice that this is a lot of what makes studying the Everett interpretation important. Anyone who accepts that scientific theories are supposed to explain and describe the world, but whoever believes that the Everett interpretation fails to solve the measurement problem should recognize that there is a profound problem with pretty much all of twentieth-century physics. In a way, it would actually be quite exciting to believe this—by contrast, the Everett interpretation is actually rather conservative.)

So: according to our best physical theory, the world we see around us is just one of countlessly many such worlds. Isn't this such an absurd claim that, even *if* our best physical theory makes it, we would be justified in rejecting it?

In a sense, there is little more to be said about this view: it amounts (again) only to a form of instrumentalism.[23] But two observations may help doubtful readers at this point.

The first observation is that Ockham's razor, that much-loved principle of scientists and metaphysicians alike, does not actually tell against the existence of the worlds of the Everett interpretation. William of Ockham enjoined us[24] not to multiply entities *beyond necessity*, but if we know of no way to solve the measurement problem other than the Everett interpretation, then the Everett worlds are entirely necessary, and Ockham should have no quarrel with them.

And the second observation is that, if we were to accept the Everett interpretation, it would not be the first time that science led us to vastly increase our estimates of the size of the Universe.

For consider the theory that the Solar System is alone in an otherwise empty void. This theory is extraordinarily successful at accounting for virtually all the phenomena which we have observed, throughout human history. The only thing it cannot handle are some tiny fluctuations in the pattern of radiation which falls upon the Earth, the only humanly observable part of which is a scattering of rather faint points of light in the night sky.

How absurd, then, for physicists to *take seriously* their best theory of the origins of those fluctuations. No matter (surely) that the theory explains the fluctuations with extraordinary accuracy; no matter (surely) that it has made countless novel predictions about them which experiment has confirmed; fundamentally (surely) it is just unbelievable, for it claims the

---

[23] To be fair, it amounts to a rather more philosophically respectable form of instrumentalism, one that says that our scientific theories have to be *understood* in realist terms, but that this understanding does not commit us to belief in the claim of these theories when we use them. This position, developed in particular by Bas van Fraassen (see e.g. van Fraassen 1980) is known as *constructive empiricism*; although I do not think it successful as a philosophy of science, this is not the place to defend that claim. In any case, the constructive empiricist has as much reason as the realist to study the Everett interpretation: it solves the measurement problem for constructive empiricists if and only if it does so for realists.

[24] I speak figuratively: the relation of the historical Ockham to Ockham's razor is a moot point.

existence of a Universe beyond the Solar System that is absurd, stagger-ingly so, compared to what we believed before: not just one Sun, not just a few thousand, but trillions of trillions, incomprehensibly far away, almost of them invisible to the naked eye and 'detectable' (says the theory), at most, by the most advanced instruments.

And to be sure, the ability of this absurd theory to explain (and not merely predict) much about our solar system—the distribution of the elements, the creation of the Sun—is a neat trick. But to take advantage of that trick, we do not need (surely) to believe this theory, for even our best scientific theory *surely* should be disbelieved if it claims that what we hitherto regarded as the whole Universe is relegated to the smallest part of an unthinkably greater reality.

The world presented to us by modern astrophysics may be small com-pared to that presented to us by Everett-interpreted quantum mechanics, but it is still incomprehensibly vaster than anything we can really intuit or properly visualize. And yet we accept it as reality. The moral is clear: our intuitions as to what is 'unreasonable' or 'absurd' were formed to aid our ancestors scratching a living on the savannahs of Africa, and the Universe is not obliged to conform to them.

CHAPTER 1. The Everett 'interpretation' of quantum mechanics is just quantum mechanics itself, taken literally as a description of the Universe. As such, it is the only solution of the quantum measurement problem which is not committed either to a sea change in the goals of the scientific enterprise or to the wholesale rewriting of our current best physics.

CHAPTER 2. Why does 'quantum mechanics, taken literally' entail that we live in a multiverse?

2

The Emergence
of Multiplicity

Functionalism is the idea that handsome is as handsome does, that
matter only matters because of what matter can do. Functionalism
in this broadest sense is so ubiquitous in science that it is tantamount
to a reigning presumption of all of science.

Daniel Dennett[1]

[A]ny given universe is not actually a *thing* as such, but is just a way of
looking at what is technically known as the WSOGMM, or Whole
Sort of General Mish Mash.

Douglas Adams, *Mostly Harmless*[2]

2.1 Worlds as emergent entities

My goal in this chapter is to show how the Everett interpretation solves
the problem of ontology. I will demonstrate how, if the quantum state is a
faithful description of physical reality, then that physical reality consists of
a vast number of distinct 'worlds' (or 'universes', or 'branches'.) These
worlds are dynamically speaking almost independent of one another;
by and large they behave approximately classically; they are constantly
splitting into multiple versions of themselves; our own world is just one
amongst this multitude.

As was briefly mentioned in the previous chapter, this goal has often
been thought impossible. Bare quantum theory does not speak of 'worlds':
it speaks only of a unitarily-evolving quantum state. However suggestive
it may be to write that state as a superposition of (what appear to be)

---

[1] Dennett (2005: 17)   [2] Adams (1992: ch. 3)

classically definite states, we are not justified in speaking of those states as 'worlds' unless they are somehow added into the formalism of quantum mechanics. As Adrian Kent put it in his influential critique of many-worlds interpretations:

one can perhaps intuitively view the corresponding components [of the wave function] as describing a pair of independent worlds. But this intuitive interpretation goes beyond what the axioms justify: the axioms say nothing about the existence of multiple physical worlds corresponding to wave function components. (Kent 1990)

It appears that the Everettian has a dilemma: either the axioms of the theory must be modified to include explicit mention of 'multiple physical worlds', or the existence of these multiple worlds must be some kind of illusion. Neither is particularly satisfactory. The first choice spoils the *raison d'être* of the Everett interpretation: that it is an interpretation of the quantum formalism we have, not a strategy for modifying that formalism. The second commits us to being drastically deceived about the world we live in (so that we are plain wrong, for instance, to claim that macroscopic objects exist in fairly definite locations, or that our surroundings are even approximately classical), and (it turns out) can be made to work only by embracing some very strong and anti-scientific claims about the nature of the human mind.

But the dilemma is false. It is simply untrue that any entity not directly represented in the basic axioms of our theory is an illusion. Rather, science is replete with perfectly respectable entities which are nowhere to be found in the underlying microphysics. Douglas Hofstadter and Daniel Dennett make this point very clearly:

Our world is filled with things that are neither mysterious and ghostly nor simply constructed out of the building blocks of physics. Do you believe in voices? How about haircuts? Are there such things? What are they? What, in the language of the physicist, is a hole—not an exotic black hole, but just a hole in a piece of cheese, for instance? Is it a physical thing? What is a symphony? Where in space and time does 'The Star-Spangled Banner' exist? Is it nothing but some ink trails in the Library of Congress? Destroy that paper and the anthem would still exist. Latin still *exists* but it is no longer a living language. The language of the cavepeople of France no longer exists at all. The game of bridge is less than a hundred years old. What sort of a thing is it? It is not animal, vegetable, or mineral.

These things are not physical objects with mass, or a chemical composition, but they are not purely abstract objects either—objects like the number pi, which is immutable and cannot be located in space and time. These things have birthplaces and histories [The human *conception* of pi has a history, of course, but not pi itself—DW.] They can change, and things can happen to them. They can move about—much the way a species, a disease, or an epidemic can. We must not suppose that science teaches us that every *thing* anyone would want to take seriously is identifiable as a collection of particles moving about in space and time. (Hofstadter and Dennett 1981)

The generic philosophy-of-science term for entities such as these is *emergent*: they are not directly definable in the language of microphysics (try defining a haircut within the Standard Model) but that does not mean that they are somehow independent of that underlying microphysics. A basic claim of this book is that worlds, in the Everett interpretation, are likewise emergent entities—and that this is actually rather a mundane claim, and that it puts Everettian worlds on a par with all manner of unmysterious, scientifically respectable entities. But to defend the claim will require some preparatory work: the task of the next two sections will be to develop the idea of emergence in a bit more detail, and to sketch how it is that emergent entities actually do emerge.

2.2 Emergence in practice

Fig. 2.1 gives a particularly dramatic example of an entity which is neither illusory, nor directly represented in the axioms of microphysics: the Bengal tiger.[3] Tigers are (I take it!) unquestionably real, objective physical objects, but the Standard Model contains quarks, electrons and the like, but no tigers. Instead, tigers should be understood as patterns, or structures, *within* the states of that microphysical theory.

To see how this works in practice, consider how we could go about studying, say, tiger hunting patterns. In principle—and only in principle—the most reliable way to make predictions about these would be in terms of atoms and electrons, applying molecular dynamics directly to the swirl of molecules which make up, say, the Kanha National Park, in India. In practice, however (even ignoring the measurement problem itself) this is clearly insane: no remotely imaginable computer would be able to solve the 10^{35} or so simultaneous dynamical equations which would be needed to predict what the tigers would do.

---

[3] Photograph ©Philip Wallace, 2007. Reproduced with permission.

Figure 2.1. An object not among the basic posits of the Standard Model

Actually, the problem is even worse than this. For in a sense, we *do* have a computer capable of telling us how the positions and momentums of all the molecules in the Kanha National Park change over time. It is called the Kanha National Park. (And it runs in real time!) Even if, *per impossibile*, we managed to build a computer simulation of the Park accurate down to the last electron, it would tell us no more than what the Park itself tells us. It would provide no explanation of any of its complexity. (It would, of course, be a superb vindication of our extant microphysics.)

If we want to understand the complex phenomena of the Park, and not just reproduce them, a more effective strategy can be found by studying the structures observable at the multi-trillion-molecule level of description of this 'swirl of molecules'. At this level, we will observe robust—though not 100% reliable—regularities, which will give us an alternative description of the tiger in a language of cell membranes, organelles, and internal fluids. The principles by which these interact will be derivable from the under-lying microphysics, and will involve various assumptions and approxima-tions; hence very occasionally they will be found to fail. Nonetheless, this slight riskiness in our description is overwhelmingly worthwhile given the enormous gain in usefulness of this new description: the language of cell biology is both explanatorily far more powerful and practically far more useful for describing tiger behaviour than the language of physics.

Nonetheless it is still ludicrously hard work to study tigers in this way. To reach a really practical level of description, we again look for patterns and regularities, this time in the behaviour of the cells that make up individual tigers (and other living creatures which interact with them). In doing so, we will reach yet another language, that of zoology and evolutionary adaptationism, which describes the system in terms of tigers, deer, grass, camouflage and so on. This language is, of course, the norm in studying tiger hunting patterns. Thus another (in practice very modest) increase in the riskiness of our description is happily accepted in exchange for another phenomenal rise in explanatory power and practical utility.

The moral of the story is: there are structural facts about many microphysical systems which, although perfectly real and objective (try telling a deer that a nearby tiger is not objectively real), simply cannot be seen if we persist in describing those systems in purely microphysical language. Zoology is of course grounded in cell biology, and cell biology in molecular physics, but the entities of zoology cannot be discarded in favour of the austere ontology of molecular physics alone. Rather, those entities are structures instantiated within the molecular physics, and the task of almost all science is to study structures of this kind.

Of *which* kind? (After all, 'structure' and 'pattern' are very broad terms: almost any arrangement of atoms might be regarded as some sort of pattern.) The tiger example suggests the following answer, which I have elsewhere (Wallace 2003a) called 'Dennett's criterion' in recognition of the very similar view proposed by Daniel Dennett (1991b):

> **Dennett's criterion.** A macro-object is a pattern, and the existence of a pattern as a real thing depends on the usefulness—in particular, the explanatory power and predictive reliability—of theories which admit that pattern in their ontology.

Dennett's own favourite example is worth describing briefly in order to show the ubiquity of this way of thinking: if I have a computer running a chess program, I can in principle predict its next move from analysing the electrical flow through its circuitry, but I have no chance of doing this in practice, and anyway it will give me virtually no understanding of that move. I can achieve a vastly more effective method of predictions if I know the program and am prepared to take the (very small) risk that it is not being correctly implemented by the computer, but even this method will be practically very difficult to use. One more vast improvement can be gained if I don't concern myself with the details of the program, but simply assume that whatever they are, they cause the computer to play good chess. Thus I move successively from a language of electrons and silicon chips, through one of program steps, to one of intentions, beliefs, plans and so forth—each time trading a small increase in risk of error for an enormous increase in predictive and explanatory power.[4]

Nor is this account restricted to the relation between physics and the rest of science: rather, it is ubiquitous within physics itself. Statistical mechanics provides perhaps the most important example of this: the temperature of bulk matter is an emergent property, salient because of its explanatory role in the behaviour of that matter. (It is a common error in textbooks to suppose that statistical-mechanical methods are used only because in practice we cannot calculate what each atom is doing separately: even if we could do so, we would be missing important, objective properties of the system in question if we abstained from statistical-mechanical talk.) But it is somewhat unusual because (unlike the case of most of the tiger's macroscopic properties) the principles underlying statistical-mechanical claims are (relatively!) straightforwardly derivable from the underlying physics.

For an example from physics which is closer to the cases already discussed, consider the case of quasi-particles in solid-state physics. As is well known, vibrations in a (quantum-mechanical) crystal, although they can in principle be described entirely in terms of the individual crystal atoms and their quantum entanglement with one another, are in practice overwhelmingly simpler to describe in terms of 'phonons'—collective excitations of the crystal which behave like 'real' particles in most respects. And furthermore, this sort of thing is completely ubiquitous in solid-state physics, with different sorts of excitation described in terms of different sorts

---

[4] In the terminology of Dennett (1987b), the first method of prediction involves taking the 'physical stance' towards the computer, the second requires the 'design stance', and the third, the 'intentional stance. (It is, of course, highly contentious to suppose that a chess-playing computer *really* believes, plans, etc. Dennett himself would embrace such claims (see Dennett 1987b for an extensive discussion); I concur. However, for the purposes of this section there is no need to resolve the issue: the computer can be taken only to 'pseudo-plan', 'pseudo-believe', and so on, without reducing the explanatory importance of a description in such terms.)

of 'quasi-particle'—crystal vibrations are described in terms of phonons; waves in the magnetization direction of a ferromagnet are described in terms of magnons, collective waves in a plasma are described in terms of plasmons, etc.[5]

Are quasi-particles real? They can be created and annihilated; they can be scattered off one another; they can be detected (by, for instance, scattering them off 'real' particles like neutrons); sometimes we can even measure their time of flight; they play a crucial part in solid-state explanations. We have no more evidence than this that 'real' particles exist, and so it seems absurd to deny that quasi-particles exist—and yet they consist only of a certain pattern within the constituents of the solid-state system in question.[6]

When *exactly* are quasi-particles present? The question has no precise answer. It is essential in a quasi-particle formulation of a solid-state problem that the quasi-particles decay only slowly relative to other relevant timescales (such as their time of flight) and when this criterion (and similar ones) are met, then quasi-particles are definitely present. When the decay rate is much too high, the quasi-particles decay too rapidly to behave in any 'particulate' way, and the description becomes useless explanatorily; hence, we conclude that no quasi-particles are present. It is clearly a mistake to ask *exactly* when the decay time is short enough ($2.54 \times$ the interaction time?) for quasi-particles not to be present, but the somewhat blurred boundary between states where quasi-particles exist and states when they don't should not undermine the status of quasi-particles as real, any more than the absence of a precise boundary to a mountain undermines the existence of mountains.[7]

---

[5] For examples of this in practice, see e.g. Tsvelik (2003), Nozieres and Pines (1999), or Abrikosov, Gorkov, and Dzyalohinski (1963); a classic discussion from a conceptual point of view, by a major worker in the field, is Anderson (1972).

[6] The point is only sharpened by quantum field theory, in which 'real' particles emerge as structures in essentially the same way as quasi-particles do. (For philosophical discussion of this point, see Wallace 2009; 2011.)

[7] Philosophers will recognise that the correct analysis of the linguistic phenomenon of vagueness remains controversial, and that on some accounts (notably epistemicism, recently championed by Williamson 1994), there is indeed a precise boundary to a mountain. So far as I can see, this does not affect my discussion: those boundaries are epistemically inaccessible, and so presumably a fortiori can play no relevant role in the analysis of emergence. Further discussion, however, lies beyond the scope of this book.

2.3 Instantiation and the relation between theories

Let us get a bit more systematic. Although I have talked about emergence in a rather piecemeal way so far—focusing on the way in which individual objects (like tigers) emerge from lower-level theories—it has been implicit in the discussion that it is not the properties of the tiger-structure *in isolation* which justify regarding it as 'real'. Rather, it is the usefulness of the tiger-level description (i.e. of zoology), a description which includes not only a single tiger, but all the other animals in its vicinity as well as salient features of their shared environment. It will be helpful to make this assumption explicit.

Staying with the case of the tiger, we find ourselves in the following situation: two theories (molecular physics and zoology) are both applicable to the same physical system (the Kanha National Park), and the reason for this is that there is a way of interpreting *high-level* properties of the molecular-physics-level description of the Park as *basic* properties of the zoology-level description of the Park. To borrow some often-used terminology from computer science and philosophy of mind, molecular physics *instantiates* zoology for this particular system.

To be more precise: molecular physics, and zoology, and indeed chemistry and macroeconomics and in fact pretty much any scientific theory we have, are in the business of describing possible states of a system at a certain level of description (positions and momenta of particles/location and properties of animals/chemical composition of liquids in beakers/inflation rates and population sizes), connecting those states together to form possible histories of that system, and inferring various instantaneous and/or dynamical constraints (some deterministic, some probabilistic) on those histories. Classical physics, for instance, says (roughly) that the only allowable histories are those satisfying Newton's laws; zoology says (roughly) that histories in which the lion lies down with the lamb are disallowed; chemistry says (roughly) that the only allowable histories are those in which energy (among other quantities) is conserved; macroeconomics says (roughly) that the only allowable histories are those in which long-run economic growth per capita approximately equals long-run productivity growth.[8] (Some scientific theories, of course, have only probabilistic

---

[8] I don't intend, by these comments, to imply that the rules by which a science allows and disallows various histories can be codified in anything deserving the name 'law'; i.e. I don't

constraints, in which a given history is not simply allowed or disallowed but rather is assigned some probability relative to given initial data.)

Giving a formal definition of instantiation is cumbersome, but one attempt might go something like this:

> Given two theories A and B, and some subset D of the histories of A, we say that A instantiates B over domain D iff there is some (relatively simple) map ρ from the possible histories of A to those of B such that if some history h in D satisfies the constraints of A, then $\rho(h)$ (approximately speaking) satisfies the constraints of B.[9] (It will often be convenient to speak of the history h as instantiating $\rho(h)$, but this should be understood as shorthand for the more detailed definition here.)

The instantiation concept is much easier to illustrate than to define cleanly. Consider the following cases:

1. In the Solar System, molecular quantum physics (ignoring the measurement problem) instantiates classical mechanics: specifically, it instantiates the theory of classical point particles moving under an inverse-square force between them. A bit more precisely: the domain is all quantum states in which the molecules are bound up into roughly spherical, stable agglomerates; quantum theory itself predicts the stability of these agglomerates; the instantiation map associates to each state of n agglomerates a set of n point particles each located where the agglomerate is[10] and with mass equal to the agglomerate's mass.

2. In the computer on which I'm typing this, solid-state electronics instantiates abstract computability theory. The domain is the set of configurations of my computer that conform to the manufacturer's recommendations (those where it is immersed in water, for instance, fall outside the domain); the instantiation map associates a given

---

intend to imply that there are laws of the special sciences. (Whether there are is a continuing controversy in philosophy of science; see e.g. Roberts 2004; Kincaid 2004, and references therein.)

[9] In the case of probabilistic theories, this requirement should be interpreted as the requirement that if h is assigned probability p relative to initial data h_0, then $\rho(h)$ is assigned probability p relative to initial data $\rho(h_0)$.

[10] Read: 'located at the agglomerate's centre of mass' if you want precision, though, as we have already seen, calls for such precision are often misplaced here.

abstract memory state to each physical configuration of the memory card in the computer, and determines an abstract set of computation rules from the physical configuration of my processor and the current physical states of the input channels. (The example works most cleanly if I just let the computer get on with processing; if I continue to provide inputs to it, things get a bit more complicated.)

3. In many zoological cases, zoology instantiates game theory. The domain is the set of behavioural strategies available to the animals in question; the instantiation map is constructed by assuming that the animals (or, for better accuracy, the animals' genes) are selfish, rational actors who want to maximize their reproductive success, and by assigning to them numbers to represent the various parameters which affect that; from this, we can deduce game-theoretically which strategies (the 'evolutionary stable strategies') we will expect to find in nature.[11]

4. More speculatively, in the domain of human beings, neuroscience seems to instantiate some higher-level psychological theory. It is a matter of great controversy exactly what the latter theory is—and whether it is some as-yet-unobtained scientific theory or something closer to the common-sense psychology of human action which all adults tacitly possess—but the power of the description of human action in terms of 'beliefs', 'desires', 'emotions', 'pain', and the like strongly suggests (to put it mildly) that there is some higher-level theory of mind being instantiated here.[12]

This instantiation relation (I claim) is the right way of understanding the relation between different scientific theories—the sense in which one theory may be said to 'reduce' to another. Crucially: this 'reduction', on the instantiation model, is a local affair: it is not that one theory is a limiting case of another *per se*, but that, *in a particular situation*, the 'reducing' theory instantiates the 'reduced' one.

Consider the first example above, for instance. The reason that classical mechanics is applicable to the planets of the Solar System is not because

---

[11] The classic reference is Maynard Smith (1982); Dawkins (1989) gives a nontechnical introduction.

[12] Much of modern philosophy of mind and philosophy of psychology proceeds (tacitly or explicitly) on this basis; for some accounts where the ideas are fairly explicitly present (in very different forms), see Dennett (1981), Fodor (1985), and Stich (1983).

of some *general* result that classical mechanics is a limiting case of quantum mechanics. Rather, the particular system under consideration—the solar system—is such that some of its properties approximately instantiate a classical-mechanical dynamical system. Others do not, of course: it is not that the solar system is approximately classical, it is that it (or a certain subset of its degrees of freedom) instantiates an approximately classical system. We will see more details of this of this in Chapter 3.

Furthermore (although this will not be crucial for our discussion), the instantiation account shows that my earlier talk of 'higher-level' and 'lower-level' theories must be treated with a certain amount of caution. The real story of the relations between scientific theories is not a story about a hierarchy of theories, with particle physics at the bottom and macroeconomics at the top: rather, it is a network of domain-relative instantiations (albeit that virtually all the network ultimately can be expected to rest on the Standard Model of particle physics).

Notice in particular that there is no reason why a theory cannot instantiate *itself* (better: that one instance of a theory cannot instantiate another instance of the same theory), and in fact this happens rather frequently. For instance, the classical model of the Solar System (neglecting moons, comets, dwarf planets,[13] et al.) consists of nine point particles (the Sun and eight planets), all interacting under mutual gravitational interactions. That system instantiates *another* classical-mechanical system: one where only the Sun produces gravitational forces and all the planets orbit the Sun, uncaring of one another. Similarly, one of the approximations of geopolitics (albeit a hugely noisy one) is to treat nations as persons (i.e. to treat them in terms of psychology), with beliefs, desires and so forth; yet since nations are ultimately made up of persons, it is (in part) psychology itself that instantiates that approximation.

How does all this relate to the previous section, where we were concerned with emergence not of theories but of entities? Very often, the instantiation relation will allow entities in the instantiated theory to be identified with particular, localized structures in the instantiating theory. This is certainly the case with tigers and planets, for instance. It is often the case with computers: we can generally say, of a laptop, not only that the laptop as a whole instantiates a given abstract computer but that the particular strip of RAM in its top left corner instantiates that computer's

---

[13] This is the category in which Pluto was reclassified, following the 2006 decision of the International Astronomical Union.

memory. With other cases it is less precise: the British economy, for instance, cannot be localized more precisely than the British Isles (perhaps not even that much).

What of section 2.2's criterion for the reality of structures in terms of their explanatory usefulness? The crucial links between that criterion and the notion of instantiation are the phrases 'reasonably simple' and 'approximately speaking' in my definition of instantiation. Without the former, any number of useless 'instantiations' could be concocted: pick any two deterministic theories whose sets of dynamically allowed histories have the same cardinality as one another; construct an arbitrary map between the two. Without the latter, there would be no 'real' instantiations at all: two planets might collide; my laptop might malfunction; antelope 30147 might make a mistake that causes it to deviate from the game-theoretically optimal strategy. Ultimately, in both cases the criterion we use is explanatory usefulness: 'real' instantiations are those in which there is explanatory value in using the language and methods of the instantiated theory rather than restricting ourselves to the instantiating one. And ultimately Dennett's criterion for which structures we regard as 'real' can be reformulated in terms of the instantiation relation: the 'real' structures present in the states of a given theory T (over and above any entities explicitly postulated by T itself) are those which occur as entities within those theories instantiated by T.

I want to stress that talk of 'explanation' here is not meant to imply that it is just a *pragmatic* matter whether or not, for example, we take seriously tigers and other such macro-objects. Rather (as I hope the many examples given have made plain) it is an objective fact that (say) tiger-talk picks out high-level structural properties of the microphysical system under study. 'Explanatory power' is being used as a criterion for the objectivity of structures, but the structures thus identified are objectively real.[14]

This suggests, to be sure, that we ought to be able to find a more objective and less agent-relative criterion than 'explanatory'. And some interesting work has been, and continues to be, done on this subject. For instance, one strategy which has been popular in both philosophy of physics and mainstream philosophy[15] has been to restrict the nature

---

[14] For further comments on this issue, see Dennett (1991b).

[15] For examples in philosophy of physics, see Allori et al. (2008) and Maudlin (2010); for examples in mainstream philosophy, see Kim (1998) and Schaffer (2009).

of the instantiation relation ρ in some microphysically specified way. For instance, one might require that the entities of a given higher-level theory must be identified with some particular set of microphysical particles (so that a tiger is not just a pattern in the particles of molecular physics, but *is* some particular set of those particles), or with some spacetime region (so that a tiger is not just a pattern localizable to spacetime region R, but *is* spacetime region R).

This sort of strategy, however, must be rejected on the grounds that it does violence to science as we find it. Entities like *voices, economies, beliefs,* even *universities* simply cannot be identified in any remotely plausible way with any conglomeration of microphysical particles or with any region of spacetime. More generally, the actual practice of science has never respected these sorts of a priori constraints on what is or is not an acceptable story about emergence. Science is interested with interesting structural properties of systems, and does not hesitate at all in studying those properties just because they are instantiated 'in the wrong way'. The general term for this is 'functionalism'; as Dennett notes in the quotation which began this chapter, all science is functionalism.

More plausible attempts to improve on the 'explanatory usefulness' criterion try to make more objective the notion of structure, using ideas from information theory and complexity theory. One such proposal has been developed in detail by James Ladyman and Don Ross (Ross 2000; Ladyman and Ross 2007); others might well be possible (the studies of complexity theory pioneered by Stuart Kaufmann, Murray Gell-Mann, and others of the Santa Fe institute can perhaps be seen as another such attempt).

These details, however, are not crucial for our purposes. This is not a book about the philosophy of emergence: it is a book about the measurement problem. And all we shall need for a solution of *that* problem is an understanding of how emergence works in science as we find it, that is clear enough that it lets us identify emergence when we see it. As I noted in the Introduction, a basic premise of this book is naturalism: the thesis that we have no better guide to metaphysics than the successful practice of science. From a naturalistic perspective, we should regard the conceptual puzzles of emergence as worthy of serious study. But it would be a mistake to eschew the use of emergence in solving the quantum measurement problem until those puzzles are fully solved.

2.4 The situation in quantum mechanics

It is time to apply these ideas to quantum theory and to the measurement problem. It will be convenient to begin doing so by considering the time-honoured Schrödinger-cat experiment, in a variant form which might be preferred by an inveterate cat-hater: in this form of the experiment, there is no radioactive decay, and the cat is certain to die.

To be specific: suppose that we put the cat into the box at 11 a.m. with a glass cylinder of cyanide gas, and that, at noon, an automated device measures the spin of some spin-half particle in the z direction and strikes that cylinder with a hammer if it reads the spin as 'up', and suppose that the particle is indeed prepared in the spin-up state, so that the measurement certainly gives that result. What does our best science tell us about what will happen between, say, 11 a.m. and noon? Quantum mechanics tells us very little. We cannot possibly determine the microphysical state of the cat (the locations of all the atoms in its body, say, along with the electron distribution); even if we could, we could not solve the Schrödinger equation for such a complex system; even if we could, all it would give us is a bare prediction of the microphysical state in one hour's time.

Nonetheless, we can say a great deal about the cat:

- From thermodynamics, we can predict that if the cat's box is sealed, the entropy of the box will go up.
- From solid-state physics, we can predict that it will not spontaneously vaporize.
- From animal physiology, we can predict that the cat will not spontaneously die or grow a second tail.
- From cat psychology, we can predict that the cat will not start eating itself, and that it is quite likely to go to sleep.

It is because of the explanatory power of these various higher-level descriptions that we regard them as instantiated in the quantum-mechanical history of the box, and—as such—regard the cat as an objectively real thing, a real structure instantiated by the box's constituents. Furthermore, exactly the same considerations apply after noon. Solid-state physics tells us that the glass cylinder shatters; kinetic theory tells us that the cyanide rapidly diffuses into the room; cat psychology tells us that the cat jumps backwards when the glass shatters; animal physiology tells us that it rapidly dies and starts to decompose. So, right through from 11 a.m. to 1 p.m.

(and, indeed, before that and after that), there is a cat (or at least: a cat's corpse) present in the box.

A little formalism may help here. Assume (absurdly) that the box is completely isolated from the external environment, and let its state at any time t between 11 and 1 be denoted $|\text{box}_d(t)\rangle$. Assuming for simplicity that the 12 noon measurement is non-disturbing, we can factorize this into the state of the spin-half particle and the state of the cat (and cat-killing device):

$$|\text{box}_d(t)\rangle = |\uparrow\rangle \otimes |\text{cat}_d(t)\rangle. \tag{2.1}$$

And so the history $\{|\text{cat}_d(t)\rangle |t\}$ instantiates a 2-hour period of the history of a cat.

And of course, exactly the same thing would be true if we considered the animal-welfare version of the experiment, in which the particle is prepared in a spin-down state and so the cat is definitely *not* killed, and in which the box's quantum state is $|\text{box}_l(t)\rangle$. Again, this state factorizes, to

$$|\text{box}_l(t)\rangle = |\downarrow\rangle \otimes |\text{cat}_l(t)\rangle. \tag{2.2}$$

Again, the history $\{|\text{cat}_l(t)\rangle |t\}$ instantiates a two-hour period of the history of a cat. (Of course, for $t < 12$ noon, $|\text{cat}_l(t)\rangle = |\text{cat}_d(t)\rangle$ and we might as well just write it as $|\text{cat}(t)\rangle$.

To summarize: if we apply the same principles to quantum mechanics as we apply in general through science to identify higher-level ontology, we find that, since both the histories $|\text{cat}_l(t)\rangle$ and $|\text{cat}_d(t)\rangle$ represent a state of affairs where the system in question is *structured* like a cat, they represent a state of affairs where the system in question *is* a cat. We recover, then, what we would expect to recover: that macroscopically definite quantum states represent classical states of affairs in just the way that they are usually taken to.

The crucial question, then, is: what about the *full* Schrödinger-cat experiment, in which the particle to be measured is not in an eigenstate of z-spin? Specifically, let the particle's state be $\alpha |\uparrow\rangle + \beta |\downarrow\rangle$. In the absence of wavefunction collapse (and maintaining the unrealistic assumption that the cat's box is totally isolated from its environment), we know what quantum theory predicts: that the state of the system at time t is

$$\alpha |\text{box}_d(t)\rangle + \beta |\text{box}_l(t)\rangle. \tag{2.3}$$

Before 12 noon, this can be rewritten as

$$(\alpha \mid \uparrow\rangle + \beta \mid \downarrow\rangle) \otimes \mid \text{cat}(t)\rangle. \tag{2.4}$$

This is a macroscopically definite state; just as before, its history represents the history of a cat. After 12 noon, though, the state is

$$\alpha \mid \uparrow\rangle \otimes \mid \text{cat}_d(t)\rangle + \beta \mid \downarrow\rangle \otimes \mid \text{cat}_l(t)\rangle \tag{2.5}$$

and is *not* macroscopically definite. What, if anything, does *this* represent? Here is the crucial observation:

1. $\mid\text{cat}_d(t)\rangle$, after 12 noon, instantiates structure which represents a dead cat, and so—by the general functionalist principle used in science—that state itself represents a system containing a dead cat.
2. Similarly, $\mid\text{cat}_l(t)\rangle$, after 12 noon (and before, in fact), instantiates structure which represents a live cat, and so—by the general functionalist principle used in science—that state itself represents a system containing a live cat.
3. The superposed state (2.5) instantiates all of the structure by virtue of which $\mid\text{cat}_d(t)\rangle$ instantiates a dead cat, *and* all of the structure by virtue of which $\mid\text{cat}_l(t)\rangle$ instantiates a live cat.
4. So applying, again, the same general principles of functionalism, the state (2.5) represents a system containing both a dead cat, and a live cat. Superposition has become multiplicity at the level of structure: (2.5) instantiates two independent lots of macroscopic structure, and so represents two distinct macroscopic systems at once.

The pattern of reasoning here, really, is just the same as with the light pulse discussed in section 1.8. There (recall) the function $\mathbf{F}_1(x, t)$ represented a pulse of ultraviolet light en route from Earth to the moon, while $\mathbf{F}_2(x, t)$ represented a pulse going from Venus to Mars. Pulses of light are emergent ontology: talk of a particular pulse being created in a certain place, following a certain trajectory, being reflected off a certain surface, etc. cannot be carried out directly in the language of field theory, which contains no persistent objects whatever, only field with varying strengths at different spacetime points. Our talk of light pulses relies instead on the observation that under certain conditions, the electromagnetic field in some region is structured *as if* it consists of a discrete pulse with a determinate trajectory. (And under other conditions, such as in a highly dispersive medium, the field will not have that structure and so will not instantiate light pulses.)

And of course—again, as noted in section 1.8—the superposed state

$$\mathbf{F}(x, t) = 0.5\mathbf{F}_1(x, t) + 0.5\mathbf{F}_2(x, t) \tag{2.6}$$

does not represent one light pulse in a weird indeterminate state. It simply, prosaically, represents two light pulses.

It is crucial to note that in neither case—not in electromagnetism, and not in quantum theory—does the presence of multiplicity follow merely from linearity. Indeed, that way lies contradiction: if

$$\mid \uparrow \rangle \otimes \mid \text{cat}_d(t) \rangle + \mid \downarrow \rangle \otimes \mid \text{cat}_l(t) \rangle \tag{2.7}$$

instantiates both a live cat and a dead cat merely because it is a superposition of a live-cat state and a dead-cat state, so presumably does

$$(\mid \uparrow \rangle \otimes \mid \text{cat}_d(t) \rangle + \mid \downarrow \rangle \otimes \mid \text{cat}_l(t) \rangle) + (\mid \uparrow \rangle \otimes \mid \text{cat}_d(t) \rangle - \mid \downarrow \rangle \otimes \mid \text{cat}_l(t) \rangle).$$
$$\tag{2.8}$$

But this (up to normalization) is just another way of writing the dead-cat state. In general, even in a theory with linear equations, like electromagnetism or quantum theory, adding together two states with certain structures might cause those structures to overlap and cancel out, so that the structure of the resultant state cannot just be read off from the structures of the components. Indeed, in both electromagnetism and quantum theory, the technical term for this 'cancelling out' is the same: interference.

In general, however—and in particular in the case of the cat—there is no interference between the live-cat and dead-cat states, and so both lots of structure continue to be present. The reason we can be confident of this is because of decoherence, which in general prevents the macroscopic degrees of freedom of quantum systems from interfering, and so guarantees that structures instantiated by the macroscopic degrees of freedom of quantum systems are not erased when those systems are in superpositions of macroscopically definite states. Understanding the details of this will be our task in Chapter 3.

For now, though, let us take the technical details on trust and return to the general observation: once the relation between higher-level ontology (like cats) and the lower-level theories which instantiate them is correctly understood, we can see that (2.5) does not represent a single cat in an

THE EMERGENCE OF MULTIPLICITY 63

indefinite superposition of alive and dead. It simply, prosaically, represents two cats.[16]

And of course, in reality, no cat-containing box can be isolated from its surroundings. The room in which the box sits will get entangled with the box—and then there will be two rooms, and soon after that, two planets, and soon after that, two solar systems. And so unitary quantum mechanics, interpreted realistically, is a many-worlds theory—not because the 'worlds' are present in some microphysically fundamental sense but because the quantum state instantiates many different macroscopic systems.

In this context, it is instructive to recall John Bell, who in an influential critique of the Everett interpretation wrote:

> at the microscopic level there is no such asymmetry in time as would be indicated by the existence of branching and the nonexistence of de-branching. Thus the structure of the wavefunction is not fundamentally tree-like. (Bell 1981b: 135)

Quite so: it is not *fundamentally* tree-like. (No more is a table *fundamentally* solid, or a tiger *fundamentally* tiger-like, or (come to that) a tree *fundamentally* tree-like.) But it is tree-like all the same. The branches are not fundamental, but they are not any less real for this restriction.

CHAPTER 2. If we apply to quantum mechanics the same principles we apply right across science, we find that a multiplicity of quasi-classical worlds are emergent from the underlying quantum physics. These worlds are structures instantiated within the quantum state, but they are no less real for all that.

CHAPTER 3. How, in detail, does the dynamics of quantum mechanics give rise to this multiplicity of worlds, and why is that multiplicity correctly described as 'branching'?

---

[16] As we will see in section 3.11, this simplified model is misleading in one important way: in realistic examples, there are far more—indefinitely more—than two cats present in the Schrödinger-cat experiment.

3

Chaos, Decoherence, and Branching

Classicality simply does not follow 'as $\hbar \to 0$' in most *physically* inter-esting cases... The Planck constant is $\hbar = 1.05459 \times 10^{-27}$ erg s and—*licentia mathematica* to vary it notwithstanding—it is a *constant*.

Wojciech Zurek and Juan Pablo Paz[1]

3.1 Introduction: emergent classicality

In Chapter 2 we saw how, in outline, the quasi-classical 'worlds' of the Everett interpretation emerge from the underlying quantum mechanics. They do so because

1. Certain quantum-mechanical histories of certain systems instanti-ate—simulate, if you like—a quasi-classical history.
2. Superpositions of those histories then instantiate multiple quasi-clas-sical histories—always assuming that interference between histories can be neglected.

The purpose of this chapter is to go from this rather hand-waving description of emergence of worlds to something much more quantitative and precise. The basic theme is this: in what circumstances do the macro-scopic degrees of freedom of dynamical systems succeed, approximately, in instantiating a branching dynamical process? (Throughout, 'macroscopic' will be a somewhat vague term; for the reasons spelled out at length in Chapter 2, this vagueness is unproblematic.) I begin (sections 3.2 and 3.3) by considering the dynamics of isolated systems with relatively small

---

[1] Zurek and Paz (1995b).

numbers of degrees of freedom, and casting some doubt on the idea that those dynamics are unproblematically classical even when the systems are macroscopic in scale. In sections 3.4–3.6 I explain why, for both conceptual and technical reasons, we have to consider the interactions between systems and their environments (whether an external environment, or the internal 'environment' of a system's microscopic degrees of freedom). In the remainder of the chapter (sections 3.7–3.11), I explore just why these interactions give the quantum-mechanical universe a branching structure of approximately classical worlds.

3.2 Emergent quasi-classicality in simple isolated systems

We begin by considering the textbook example of emergent quasi-classicality in quantum physics: a single, *isolated* system whose characteristic action is large compared with $\hbar$.

Consider, therefore, a massive point particle of mass m, moving in some potential $V(\mathbf{x})$. The Hamiltonian of this particle is then

$$\widehat{H} = \frac{\widehat{P}^2}{2m} + V(\widehat{X}). \tag{3.1}$$

Under what circumstances does this system behave approximately classically? That is (in the language of Chapter 2): under what circumstances does it instantiate a classical dynamical system? There is a fairly standard answer: it does so when the state of the system is a wavepacket, reasonably localized in position and momentum, and when the centre of that wavepacket follows an approximately classical trajectory. Furthermore, we know from Ehrenfest's theorem[2] that

$$\langle \dot{\widehat{X}} \rangle = \frac{1}{m} \langle \widehat{P} \rangle \tag{3.2}$$

and

$$\langle \dot{\widehat{P}} \rangle = -\langle \nabla V(\widehat{X}) \rangle. \tag{3.3}$$

---

[2] For an account of Ehrenfest's theorem, see Joos et al. (2003: 87–8) or any textbook discussion, such as Cohen-Tannoudji, Diu, and Laloë (1977: 240–45), Sakurai (1994: 84–7), or Townsend (1992: 153–6).

As long as the wave packet is narrow (and V does not vary too quickly), we will have $\langle \nabla V(\widehat{X}) \rangle \simeq \nabla V(\langle \widehat{X} \rangle)$, and so the expectation values of $\widehat{P}$ and $\widehat{X}$ evolve in the same way as their classical counterparts. And so the former condition—that the wavepacket remains localized—suffices to ensure the latter.

So far, so banal; but let us dwell on it a little longer. What justifies our regarding a localized wavepacket following an approximately classical trajectory as an approximately classical state? Sometimes it can seem that some sort of tacit 'hidden variable' theory is present: that the state is approximately classical because the probabilities it predicts for particle location are highly peaked around a certain classical trajectory. But this will not do, of course (at least, not unless we are actually trying to develop that hidden-variable theory). Rather, the real reason that we can regard the quantum state as approximately classical is that it is dynamically isomorphic, very nearly, to a system of one classical point particle.

It may help to consider in more detail how that isomorphism works. We could understand it in the position representation: the trajectory of the centre of a localized wavepacket defines a line in configuration space, and that line is (very nearly) a solution to the classical dynamical equations for a mass-m point particle. It is somewhat more perspicuous when viewed using one of the phase-space POVMs discussed in Chapter 1: a wavepacket thus viewed defines a small region (of volume $\sim \hbar^3$) in phase space, and because its average phase-space position evolves classically (by Ehrenfest's theorem) and its spread around that phase-space position remains small, the trajectory followed by that small region is itself a solution to the classical dynamical equations in Hamiltonian form. (I call this 'more perspicuous' because it makes transparent the fact that an instantaneous quantum state suffices to pick out the corresponding classical trajectory; in the position representation, the needed momentum information is unhelpfully encoded in the phase structure of the wavepacket.)

In either case, both rules:

$|\langle \mathbf{x} | \psi(t) \rangle|^2 \simeq 0$ unless $\mathbf{x} \simeq \mathbf{q}(t)$

$\leftrightarrow$ Wavepacket is centred at $\mathbf{q(t)}$

$\leftrightarrow |\psi(t)\rangle$ instantiates classical particle with trajectory $\mathbf{q}(t)$ (3.4)

and

$$\langle\psi|\widehat{\Pi}(\mathbf{q}',\mathbf{p}')|\psi\rangle \simeq 0 \text{ unless } (\mathbf{q}',\mathbf{p}') \simeq (\mathbf{q},\mathbf{p})$$

$\leftrightarrow$ Wavepacket is centred at $(\mathbf{q},\mathbf{p})$

$\leftrightarrow$ $|\psi\rangle$ instantiates classical particle at phase-space location $(\mathbf{q},\mathbf{p})$ (3.5)

ultimately pick out the same structure[3] in the quantum system. Notice also that we see again the emptiness of questions like 'which is the correct phase-space POVM?' For within broad limits, any such POVM will succeed in picking out the structure we are interested in (and, outside those broad limits, we simply are not using a POVM which makes manifest that structure; it's still *there*).

To see another important property of this emergent dynamics, let us consider a particular (overcomplete) basis $|\mathbf{q},\mathbf{p}\rangle$ of wavepacket states centred at phase-space point $|\mathbf{q},\mathbf{p}\rangle$, one of which is the actual wavepacket of the system. To a very good approximation, then, if the phase-space point $(\mathbf{q},\mathbf{p})$ evolves over time to $(\mathbf{q}(t),\mathbf{p}(t))$ then the corresponding quantum state evolves to $|\mathbf{q}(t),\mathbf{p}(t)\rangle$ over the same period. (Perhaps the wavepacket will spread out a little, so that it is not exactly any single element of the basis, but (we are assuming that) it remains reasonably localized.) This is a somewhat remarkable property of the phase-space basis which we might call 'basis preservation': the dynamics takes elements of the basis to other elements of the basis.

Fairly clearly, this can only occur exactly for an orthonormal basis in the trivial cases where that basis is an eigenbasis of the Hamiltonian. (If not, the dynamics would have to flip the system discontinuously from one basis vector to another.) In this case, though, the overcompleteness of the basis (and, in most realistic situations, our willingness to settle for a very high but not 100% level of precision) allows basis preservation and nontriviality to coexist.

Because of the property of basis preservation, the various classical histories instantiated by different wavepacket states can coexist. To see this, suppose $|\psi_1(t)\rangle$ and $|\psi_2(t)\rangle$ each instantiate some classical history. The

---

[3] Note for philosophers: I am helping myself here to something that was not actually developed in Ch. 2: namely, an identity criterion for structures. Something like 'two structures are the same when they are instantiated by precisely the same states of the instantiating theory' will probably do, but in practice I am again happy to fall back on the fact that in practice we have no trouble working out when two structures are really the same one differently described, and to leave the task of making this precise to future work in general philosophy of science.

structures which make up those classical histories are, as we have seen, structures in the expectation values of the phase-space POVMs, and so a superposition

$$|\Psi(t)\rangle = \alpha|\psi_1(t)\rangle + \beta|\psi_2(t)\rangle \qquad (3.6)$$

will instantiate both histories simultaneously provided that those structures are not erased by interference between the terms in the superposition.

The particular expectation values in this case are

$$\langle\Psi(t)|\widehat{\Pi}(\mathbf{q},\mathbf{p})|\Psi(t)\rangle = |\alpha|^2\langle\psi_1(t)|\widehat{\Pi}(\mathbf{q},\mathbf{p})|\psi_1(t)\rangle$$
$$+|\beta|^2\langle\psi_2(t)|\widehat{\Pi}(\mathbf{q},\mathbf{p})|\psi_2(t)\rangle$$
$$+2\mathrm{Re}(\alpha^*\beta\langle\psi_1(t)|\widehat{\Pi}(\mathbf{q},\mathbf{p})|\psi_2(t)\rangle) \qquad (3.7)$$

The first two terms are simply the weighted sum of the two expectation values of the original structures. The third term—the interference term—will vanish, to a very good approximation, at all times, because if $|\psi_1(t)\rangle$ and $|\psi_2(t)\rangle$ are instantiating different quasi-classical histories in the way described above, they will be localized at different phase-space points at all times (this is basis preservation in action: a superposition of two orthogonal terms in the basis will forever after remain a superposition of two orthogonal terms in the basis). So we are just left with the first two terms, and with the observation that the expectation values of the phase-space POVMs have the structure of two independent, non-interacting classical worlds.

Notice that it is not merely the linearity of quantum mechanics which allows us to interpret superpositions as instantiating multiple structures.[4] Rather, it is the disappearance of interference terms between the relevant terms in those superpositions. Basis preservation is a sufficient condition for this to occur; as we will shortly see, it is not a necessary condition.

So: in this simple model, we seem to have achieved emergent classicality—and to have achieved it in a way which leads to superpositions representing multiple quasi-classical worlds. Furthermore, nothing we did really relied on the system being a single particle: generalizing to a system with N degrees of freedom, with some Hamiltonian like

---

[4] Notwithstanding the overly simplistic claims of Wallace (2003a).

$$\widehat{H} = \sum_{i=1}^{N} \frac{1}{2m_i} \widehat{P}_i^2 + V(\widehat{Q}_1, \ldots \widehat{Q}_N) \qquad (3.8)$$

is straightforward. (In realistic cases the degrees of freedom will normally be grouped into triples, of course, given the three-dimensional[5] nature of the universe we live in.) Localized wavepackets of this system will now pick out trajectories in a high-dimensional space, and these trajectories will instantiate the dynamics of a classical theory with N degrees of freedom. Superficially, this *seems* to be everything that Everett-interpreted quantum mechanics needs.

We shall see shortly (in section 3.4) that in fact this account has a number of conceptual problems. However, there is a technical problem that is at least as severe: namely, we are relying on the assumption that the wavepackets of isolated macroscopic systems do, indeed, remain in fairly well localized states whose trajectories satisfy classical dynamics. As we shall see, things are not actually that simple.

3.3 Dynamical properties of isolated quantum systems

In this section I want to investigate how initially localized quantum states actually do behave under different Hamiltonians. We can consider this under fairly general conditions: we will assume that the system has N degrees of freedom and that its Hamiltonian is of the form of equation 3.8: that is, the sum of a term in $\widehat{Q}_1, \ldots \widehat{Q}_N$ and of a quadratic term in each $\widehat{P}_i$. For convenience I will just write (q, p) to encode the $2N$ position and momentum coordinates in the system's phase space.

As we saw in Box 1.1 (generalizing to N dimensions), given a set of coherent (wavepacket) states $|q, p\rangle$, each one representing a Gaussian wavepacket of width λ in position space localized around q in position space and p in momentum space, then the set of (improper) operators $(2\pi)^{-N}|q, p\rangle\langle q, p|$ provides a satisfactory phase-space POVM for the system. (Recall that there is a certain amount of arbitrariness in the particular

---

[5] A worry: is it really three-dimensional, given that the theory seems to be about the quantum state and not about entities in space at all? I will postpone a full answer until Ch. 8; for now, it suffices to note that the theory is at least *emergently* three-dimensional, in the sense that the emergent classical dynamics that it instantiates is on three-dimensional space.

choice of POVM, in particular in the choice of λ.) It follows that the function

$$H_\rho(q,p) = \frac{1}{(2\pi)^N} \langle q,p|\rho|q,p \rangle \qquad (3.9)$$

(known as the *Husimi function*) expresses the phase-space structure of the mixed quantum state ρ, and in particular,

$$H_\psi(q,p) = \frac{1}{(2\pi)^N} \langle \psi|q,p \rangle \langle q,p|\psi \rangle \qquad (3.10)$$

expresses the phase-space structure of the pure quantum state $|\psi\rangle$. (The $1/(2\pi)^N$ term is a normalization factor.) It can further be shown that, given the Husimi function, ρ can be recovered (and thus, for pure states, $|\psi\rangle$ can be recovered up to phase).[6]

Because the Husimi function is somewhat cumbersome to track, however, it will be useful to set out an alternative way of representing the phase-space structure of the state: the so-called Wigner function[7]

$$W_\rho(q,p) = \frac{1}{\pi^{N/2}} \int d^N y \, e^{2ipy/\hbar} \rho(q-y, q+y) \qquad (3.11)$$

(where $\rho(x,y)$ is the density operator in the position basis), which is related to the Husimi function by

$$H_\rho(q,p) = \frac{1}{\pi^N} \int d^N q' \, d^N p' \, e^{-(q-q')^2/\lambda^2} e^{-(p-p')^2\lambda^2/\hbar^2} W_\rho(q',p'). \qquad (3.12)$$

(That is, the Husimi function is obtained from the Wigner function by smearing it over a small region of phase space.)

It is sometimes said that the Wigner function 'is not a probability distribution because it is not positive definite'. This is misleading at best. It is indeed the case that the Wigner function is not guaranteed to be non-negative, but the deeper reason why it is not a probability distribution is that (at the risk of being repetitive): If 'phase space' means 'space representing the positions and momenta of all the particles', then there is no phase space in quantum mechanics (except emergently), and the Husimi function, positive definite though it may be, is no more a probability distribution

---

[6] The Husimi function was first introduced in Husimi (1940); see Hillery et al. (1984) for a review of its properties.

[7] The Wigner function was first introduced in Wigner (1932) and explored further by Moyal (1949); see Hillery et al. (1984) for a review of its properties.

on phase space than the Wigner function. The only reason for using these 'phase space' representations of the state at all is that we are interested in the emergent quasi-classical structures within the state, and these structures are most perspicuously identifiable in the phase-space representation.

The Wigner function is computationally somewhat more tractable than the Husimi function (being obtained rather more straightforwardly from the position representation of the state). Its dynamics can be expressed in closed form as

$$\dot{W} = \{H, W\}_{MB} \equiv \frac{2i}{\hbar} \sin\left(\frac{\hbar}{2i}\{\cdot, \cdot\}_{PB}\right) \cdot (H, W), \qquad (3.13)$$

where $\{\cdot, \cdot\}_{PB}$ is the classical Poisson bracket, $\{\cdot, \cdot\}_{MB}$ is known as the *Moyal bracket* (Moyal 1949), and the sine function is (most straightforwardly) understood in terms of its power-series expansion. Less compactly but more illuminatingly, we can expand (3.13) as

$$\dot{W} = \{H, W\}_{PB} + \frac{\hbar^2}{24} \frac{\partial^3 V}{\partial q^3} \frac{\partial^3 W}{\partial p^3} + O(\hbar^4), \qquad (3.14)$$

showing that the quantum dynamics is the classical dynamics plus correction terms in successively higher powers of $\hbar^2$. This seems very reassuring: as $\hbar \to 0$, we revert to classical dynamics. But as Zurek and Paz reminded us in the quotation at the start of this chapter, this formal mathematical limit is not directly physically relevant: what matters for emergent classicality is the behaviour of macroscopic systems for fixed $\hbar$.

The simplest such system is a free particle in one dimension. For this system, the higher-order terms in the Moyal bracket vanish, and classical dynamics holds exactly. The spread of a wavepacket in this situation is then a purely classical phenomenon, in the following sense: if the wavepacket has position spread Δq (and thus momentum spread at least $\sim \hbar/\Delta q$), over a time t the part of the packet with momentum $p + \hbar/\Delta q$ will travel a distance $\hbar t/m\Delta q$ further than the part with momentum p, and so the position spread will increase to $\Delta q + \hbar t/m\Delta q$. Over a time t, then, the minimum size that a packet will have is

$$\Delta q(t) \sim \sqrt{\frac{\hbar t}{m}}. \qquad (3.15)$$

Not only does this decrease to zero as $\hbar \to 0$, it does so satisfactorily fast. An invisibly small dust mote, for instance (ten microns across, say, with a mass of $\sim 10^{-12}$kg) , if evolving freely, could be prepared in a wavepacket state that remained of width $\leq$ 1cm for the age of the Universe; a bowling ball with a mass of $\sim$1 kg, could be similarly prepared in a state that remained of width $\leq 10^{-8}$m.

No real systems are entirely free, of course; but some real systems (sometimes called *regular*) share with free systems the property that phase-space distributions spread out at a rate linear in time. For these systems, (3.15) will remain a fairly good approximation for the minimum achievable spread over time of a classical distribution of area $\sim$ $\hbar$. (I continue to work in one dimension for convenience; the generalization is straightforward.) Furthermore, the classical spread will be a good approximation to the quantum spread as long as the higher terms in the Moyal bracket are small. The first such term, evaluated for a wavepacket of size Δq, will be of order

$$\hbar^2 V'''(q) \times \left(\frac{1}{\Delta p}\right)^3 \sim \hbar^{-1} V'''(q)(\Delta q)^3. \tag{3.16}$$

If we estimate the size of this via the free-particle assumption that $\Delta q \sim \sqrt{\hbar t/m}$, then this is proportional to $\hbar^{1/2}$, and so goes to zero as $\hbar \to 0$. Again, it does it sufficiently quickly that, for systems of micron size or above, quantum corrections are utterly negligible.

So: regular isolated systems do indeed instantiate quasi-classical dynamics if they are above a certain size. Unfortunately, most Hamiltonians do not give rise to regular dynamics. Much more commonly, a system is *chaotic*: phase-space regions in such systems spread out exponentially, not linearly. (Or, more accurately: they spread out exponentially in some directions and contract exponentially in others, so as to conserve phase-space volume.) In such a system, the spread of a classical packet of initial width Δq (and so of a quantum wavepacket of width Δq, as long as classical dynamics remains approximately valid for it) will be of the form[8]

$$\Delta q(t) \simeq e^{t/\tau_L} \Delta q \tag{3.17}$$

---

[8] The results in this section are based on results in Berry and Balzas (1979), Zurek and Paz (1994; 1995a), and Joos (2003).

(the characteristic timescale τ_L is known as the Lyapunov exponent).[9] Since the wavepacket cannot be dramatically narrower than (3.15) on pain of being so delocalized in momentum space that it rapidly spreads out anyway, a crude estimate for the minimum achievable wavepacket spread after time t is

$$\Delta q(t) \sim e^{t/\tau_L} \sqrt{\frac{\hbar t}{m}}. \tag{3.18}$$

Equivalently, we have

$$\ln \Delta q(t) \sim \frac{t}{\tau_L} + \frac{1}{2} \ln\left(\frac{\hbar t}{m}\right) = \left(\frac{t}{\tau_L} + \frac{1}{2} \ln\left(\frac{t}{\tau_L}\right)\right) + \frac{1}{2} \ln\left(\frac{\hbar \tau_L}{m}\right). \tag{3.19}$$

In the regime where $t \gg \tau_L$, then, we have

$$\frac{t}{\tau_L} \sim \ln\left(\Delta q(t) \sqrt{\frac{m}{\hbar \tau_L}}\right). \tag{3.20}$$

If the packet becomes so spread that it samples regions of appreciably different potentials, it certainly will no longer instantiate a classical trajectory, so a criterion for emergent classicality (at least of the form we have so far discussed) is that $\Delta q(t)$ remains below the lengthscale on which this happens. Writing this lengthscale as L, we find that classicality fails once

$$t \geq \tau_L \ln\left(L \sqrt{\frac{m}{\hbar \tau_L}}\right). \tag{3.21}$$

The good news is: the lower bound for the onset of non-classical behaviour given by (3.21) does go to infinity as $\hbar \to 0$. The bad news is: thanks to the logarithm in (3.21), it does so alarmingly slowly. Suppose that our dust mote (mass $\sim 10^{-12}$ kg) is experiencing chaotic dynamics with a Lyapunov timescale of ~ 10 seconds in a region where the potential varies on a scale of ~ 10cm. (These numbers are off the top of my head; the logarithm means that (3.21) is enormously insensitive to the details.) Then classicality fails when

---

[9] In the classical theory of chaos, a system is chaotic if (roughly) infinitesimally close points in phase space diverge exponentially in some directions; the Lyapunov exponent is the timescale of this exponential divergence. See e.g. Cvitanović et al. (2009) for a formal definition.

$$t \geq 10 \text{ s} \times \ln 10^{11.5}. \tag{3.22}$$

The logarithm of $10^{11.5}$ is about 25, so the system will cease to behave classically after about 250 seconds. This is uncomfortably short compared with, say, the age of the Universe. Nor does the problem go away for still larger systems. To borrow an example from Zurek and Paz (1995a), Saturn's moon Hyperion tumbles chaotically in its orbit on a Lyapunov timescale of about 20 days. Hyperion weighs $\sim 10^{20}$ kg and is $\sim 10^{5}$ m in size, so (if treating it as an isolated system were appropriate) its wavefunction would become highly nonclassical once

$$t \geq 20 \text{ days} \times \ln 10^{28} \sim 4 \text{ yrs.} \tag{3.23}$$

Since we are discussing a supposed *many-worlds theory*, one tempting idea is to say: this spreading out of the quantum state is exactly the branching of worlds that we were expecting to find. Whether or not this is conceptually appropriate, though (more on this later, in section 3.4), it fails on technical grounds in this case. For presumably a necessary condition for the idea is that the phase-space distribution defined by the quantum state—localized or no—continues to follow, approximately, the classical dynamics. If not, the various parts of the wavefunction cannot suffice to instantiate dynamically independent worlds. And it turns out that classical dynamics, too, fail for chaotic systems. For consider the correction term (3.16), the leading-order correction to the classical dynamics. This term grows as $1/(\Delta p)^3$. But—thanks to the conservation of phase-space volume—generically we would expect Δp to shrink exponentially as Δq grows. (Chaos generally 'fibrillates' systems, turning compact regions into long, thin ones.) In this case, the correction term will also grow exponentially, and so on a timescale which increases logarithmically with $1/\hbar$, but will in general still be uncomfortably short, we would expect classical dynamics to fail for the system's Wigner function.

To conclude: chaotic, isolated, unitarily evolving quantum systems cannot approximate classical ones on long enough timescales to reproduce classical phenomenology.

3.4 The need for decoherence

Leaving aside for the moment section 3.3's technical problems with chaotic isolated systems, there remain severe *conceptual* problems with the naive recovery of quasi-classicality which was sketched in section 3.2. For a start, notice that we found the emergent structure in the quantum state not by any principled means, but by our pre-existing intuitions that those variables which we call 'position' and 'momentum' would indeed turn out to function like classical position and momentum. We might worry that, in fact, this supposed 'structure' is an artefact of our choosing those variables, and that we might have found similar results in any number of alternative ways.

I think that this is more of a 'niggling doubt' than it is a real worry. As Chapter 2 stressed, emergent properties cannot be deductively found by applying any sort of algorithm to the instantiating theory; for example, the fact that biology is instantiated by molecular physics is something we realized after the development of both sciences and following detailed investigation of (many of) their features, not something we deduced from molecular physics alone. If quasi-classical dynamics are present, then this is a real, objective fact about the system. Nonetheless, it would be more satisfactory if we were able to gain a better understanding of why the structures we seek are instantiated in the phase-space basis.

A much more serious reason to be unsatisfied is that we have assumed, without any justification, that the system we are studying—consisting, recall, of the *macroscopic* degrees of freedom of some isolated system—can indeed be considered as isolated. For a system such as a rigid body, we know (from the translational invariance of the global Hamiltonian) that the centre-of-mass degrees of freedom are dynamically independent of the internal degrees of freedom, but we have no reason to assume that those centre-of-mass degrees of freedom are dynamically isolated from *other* systems. And in more general cases we cannot even neglect the internal degrees of freedom—in a fluid, for instance, the macroscopic coordinates would normally be taken to be spatial averages of fluid density and momentum over small regions, but there is no reason at all to suppose that those coordinates are dynamically independent of the remaining coordinates (no reason except, perhaps, classical intuition—but to invoke *that* would be to beg the question). Indeed, even in the case of the 'rigid body' we do not escape such worries—the very claim that the body is 'rigid' cannot be taken as primitive, but must be regarded as something which ought to be derivable from the underlying physics of its constituents.

A further concern is that, if quantum systems always behave approximately classically, we would not have needed quantum mechanics! Obviously our theory must accommodate situations—such as quantum measurements—where classical mechanics breaks down even at the macroscopic scale. In these situations, we have as yet no solid reason to expect the 'branching' behaviour which the Everett interpretation claims is the correct description of measurement.

To summarize, the main problems with directly reading off quasi-classical structure from the dynamics of isolated macroscopic systems are:

1. It is inaccurate, or at least question-begging, to treat the macroscopic degrees of freedom of a system as dynamically isolated from its residual degrees of freedom.
2. In chaotic systems, it is simply false that the system has any states which behave quasi-classically over sufficiently long timescales to recover the successful predictions of classical physics.
3. In situations like quantum measurements where the dynamics are not even approximately classical, we have no reason to assume that a macroscopic quantum system remains treatable as a collection of non-interacting quasi-classical systems.

As we will see in the remainder of this chapter, all of these problems are satisfactorily solved once *decoherence*—the suppression of interference with respect to a system's macroscopic degrees of freedom via interaction with its internal and external environments—is properly allowed for.[10] Furthermore, this section's 'niggling doubt' is also at least partially assuaged: decoherence provides at least a substantial part of the answer to the question of why it is the quasi-classical degrees of freedom

---

[10] There is a terminological issue here. Some authors (such as Wojciech Zurek, Erich Joos, and H. Dieter Zeh) use 'decoherence' to mean specifically an *environment-induced* process. Others (such as Jonathan Halliwell, James Hartle, and Murray Gell-Mann) use 'decoherence' to mean any process by which interference between quasi-classical histories is suppressed: to them, then, the evolution of the isolated regular system in section 3.2 is also decoherent. Halliwell (2010), in fact, calls this sort of decoherence 'conservation-induced decoherence', and distinguishes it from 'environment-induced decoherence'. In this book, I largely follow the former authors' terminology, writing just 'decoherence' where Halliwell would write 'environment-induced decoherence'; I do, however, follow standard terminology in referring to a *history space* (as discussed in section 3.9 and subsequently) as decoherent in the event that its decoherence functional vanishes.

which instantiate the interesting structures in macroscopic quantum systems.

3.5 Environment-induced decoherence: a simple model

'Decoherence' is the process by which the environment of a system continually interacts with, and becomes entangled with, that system. Its best-known property is the suppression of coherence in coherent superpositions of states in a particular basis picked out by the system–environment interaction—hence the name—but, as we will see, its real significance is much greater. However, suppression of coherence is a convenient way to begin our investigations.[11]

Let us begin by considering a simple model: suppose that we have two one-particle systems, the first much heavier than the other, and that the first system is prepared in a superposition of two localized wavepackets separated from one another by some distance large compared to the packet width. That is: let the first system be in state

$$|\psi\rangle = \alpha|\psi_{q_1}\rangle + \beta|\psi_{q_2}\rangle \tag{3.24}$$

where $|\psi_{q_i}\rangle$ is localized around q_i $(i = 1, 2)$ and suppose for simplicity that $|\psi\rangle$ is stationary on relevant timescales. And suppose that the Hamiltonian of the system contains some interaction term

$$\widehat{H}_{int} = V(\widehat{X} - \widehat{x}) \tag{3.25}$$

where $\widehat{X}$ and $\widehat{x}$ are the position operator of the first and second particles respectively.

If one of α or β is zero, then to a very good approximation this problem reduces to a standard piece of scattering theory: the second particle is scattering off a scattering centre at $x = q_i$, and (again, to a very good approximation) the first particle does not change at all. (See Box 3.1 for a proof of this.)

---

[11] Here and subsequently I draw extensively on the discussions of decoherence by Zurek (1991; 1998; 2003), Joos et al. (2003), and Schlosshauer (2007), and while my models and analyses are in many cases not explicitly lifted from any single source, I claim no particular originality for any of them.

Box 3.1. Scattering of light particles off heavy ones

If two interacting particles have position operators $\widehat{X}_1$ and $\widehat{X}_2$ and Hamiltonian

$$\widehat{H} = \frac{1}{2m_1}\widehat{P}_1^2 + \frac{1}{2m_2}\widehat{P}_2^2 + V(\widehat{X}_2 - \widehat{X}_1), \qquad (3.26)$$

we define the centre-of-mass and relative coordinates by

$$\widehat{R} = \frac{m_1}{M}\widehat{X}_1 + \frac{m_2}{M}\widehat{X}_2; \quad \widehat{r} = \widehat{X}_2 - \widehat{X}_1 \qquad (3.27)$$

where $M = m_1 + m_2$ is the total mass of the system, and the conjugate momenta by

$$\widehat{P} = \widehat{P}_1 + \widehat{P}_2; \quad \widehat{p} = \mu\left(\frac{\widehat{P}_2}{m_2} - \frac{\widehat{P}_1}{m_1}\right) \qquad (3.28)$$

where $\mu = m_1 m_2/(m_1 + m_2)$ is the *reduced mass*. It is then easy to verify that $[\widehat{r}, \widehat{P}] = [\widehat{R}, \widehat{p}] = 0$ and $[\widehat{r}, \widehat{p}] = [\widehat{R}, \widehat{P}] = i\hbar$, and that the Hamiltonian can be rewritten as

$$\widehat{H} = \frac{1}{2M}\widehat{P}^2 + \frac{1}{2\mu}\widehat{p}^2 + V(\widehat{r}). \qquad (3.29)$$

In other words, the system is mathematically equivalent to the tensor product of a free particle with mass M and a particle with mass μ interacting with a scattering centre at the origin of r-space.

We now shift to the position basis. If $\Psi(x_1, x_2; t)$ is the system's wavefunction, we will suppose that at time $t = 0$ it is factorized:

$$\Psi(x_1, x_2; 0) = \psi(x_1)\phi(x_2); \qquad (3.30)$$

in the centre-of-mass coordinates, then, this is

$$\Psi(R, r; 0) = \psi(R - m_2 r/M)\phi(R + m_1 r/M). \qquad (3.31)$$

We now assume that $m_1 \gg m_2$. Then to a very good approximation, $m_2/M = 0$, $m_1/M = 1$, and we have

$$\Psi(R, r; 0) \simeq \psi(R)\phi(R + r). \qquad (3.32)$$

If we further assume that ψ is tightly localized around $R = q$ then we can approximate this as

$$\Psi(R, r; 0) \simeq \psi(R)\phi(q + r) : \qquad (3.33)$$

that is, the wavefunction factorizes. Since there is no interaction between q and R, this remains the case over time: ϕ evolves as if scattering from a centre at $r = -q$, and ψ remains stationary (and, in doing so, justifies our continuing to assume it to be tightly peaked around $R = q$). Reversing the coordinate transformation at the end of the interaction process gives us our result.

So the dynamics is

$$|\psi_{q_i}\rangle \otimes |\phi_0\rangle \longrightarrow |\psi_{q_i}\rangle \otimes |\phi_i^+\rangle \qquad (3.34)$$

where $|\phi_i^+\rangle$ is some post-scattering state: for instance, if $|\phi_0\rangle$ was a plane wave or nearly so, then $|\phi_i^+\rangle$ will be a superposition of a plane wave with an outgoing spherical wave centred on q_i. By the linearity of the Schrödinger equation, then, the general evolution has the form

$$|\psi\rangle \otimes |\phi_0\rangle \longrightarrow \alpha|\psi_{q_1}\rangle \otimes |\phi_1^+\rangle + \beta|\psi_{q_2}\rangle \otimes |\phi_2^+\rangle. \qquad (3.35)$$

That is: in the case where the first particle is in a superposition, but not in the case where it is not, the scattering interaction causes the two particles to become entangled. We might even say (though nothing hangs on this way of talking) that the second particle has measured the position of the first.

The level of entanglement can be quantified by considering the density operator for the first particle in the $|\psi_{q_i}\rangle$ basis. If we idealize it as having exactly two possible position states, $|\psi_{q_1}\rangle$ and $|\psi_{q_1}\rangle$, then tracing over equation 3.35 tells us that the first particle's density operator evolves like

$$\rho_0 = |\alpha|^2 |\psi_{q_1}\rangle\langle\psi_{q_1}| + |\beta|^2 |\psi_{q_2}\rangle\langle\psi_{q_2}| + \alpha^*\beta|\psi_{q_2}\rangle\langle\psi_{q_1}| + \beta^*\alpha|\psi_{q_1}\rangle\langle\psi_{q_2}|$$

$$\implies \rho_+ = |\alpha|^2 |\psi_{q_1}\rangle\langle\psi_{q_1}| + |\beta|^2 |\psi_{q_2}\rangle\langle\psi_{q_2}| + \alpha^*\beta \langle\phi_1^+|\phi_2^+\rangle |\psi_{q_2}\rangle\langle\psi_{q_1}|$$
$$+ \beta^*\alpha \langle\phi_2^+|\phi_1^+\rangle |\psi_{q_1}\rangle\langle\psi_{q_2}| \qquad (3.36)$$

or, in matrix form,

$$\rho_0 = \begin{pmatrix} |\alpha|^2 & \alpha\beta^* \\ \alpha^*\beta & |\beta|^2 \end{pmatrix} \longrightarrow \rho_+ = \begin{pmatrix} |\alpha|^2 & \alpha\beta^* \langle\phi_2^+|\phi_1^+\rangle \\ \alpha^*\beta \langle\phi_1^+|\phi_2^+\rangle & |\beta|^2 \end{pmatrix}. \quad (3.37)$$

The off-diagonal terms provide a measure of the coherence between the two possible positions of the first particle: when they have magnitude equal to $|\alpha^*\beta|$, the first particle is in a pure state and so not at all entangled with the second particle; if they are equal to zero, then the entanglement is maximal, and the quantum measurement algorithm gives the same predictions as it would were the first particle's state a probabilistic mixture of the two positions.

Hence, if the scattering is very weak, or if the wavelength of the incoming particle is large compared with $q_2 - q_1$, then $\langle\phi_2^+|\phi_1^+\rangle \simeq 1$, and the systems become only slightly entangled. At the other extreme, if the incoming particle's wavefunction is highly localized, incident on q_1, and strongly scattered, then $\langle\phi_2^+|\phi_1^+\rangle \simeq 0$, and entanglement is almost maximal.

So: prepare a heavy particle in a macroscopic superposition and expose it to a scattering environment, and that environment will become entangled with the particle, causing the coherence between the terms in the superposition to decay. If the environment consists of short-wavelength particles which interact strongly with the system, the coherence will be completely lost after a single scattering event. Even if the environment is not so constituted, sufficiently many scattering events will still suffice to remove the coherence: it can be shown (Joos et al. 2003: 64–7) that the rate is approximately given by

$$\langle x_1|\rho(t)|x_2\rangle = \langle x_1|\rho(0)|x_2\rangle \exp\left[-\Lambda t(x_1 - x_2)^2\right] \quad (3.38)$$

where

$$\Lambda \sim k^2 F\sigma/\lambda^2, \quad (3.39)$$

where F the incoming particle flux, σ is the interaction cross-section, and λ is the wavelength.

In fact, it is by now well known that in realistic situations, coherence is lost very, very quickly. For a one-micron dust particle, the value of Λ due to the atmosphere is $10^{36}\mathrm{m}^{-2}\mathrm{s}^{-1}$; the value due to sunlight is $10^{21}\mathrm{m}^{-2}\mathrm{s}^{-1}$; even the value due to the cosmic background radiation is $10^6\mathrm{m}^{-2}\mathrm{s}^{-1}$. The rates for larger objects are correspondingly more rapid: Schrödinger's cat, for instance, would endure in a coherent macroscopic superposition for only $\sim 10^{-35}$ seconds before the microwave background radiation—let alone the atmosphere—sufficed to destroy the coherence.

Of course, absent some non-unitary dynamical process of a kind for which we have no evidence, the cat-plus-environment system remains in a superposition of live-cat and dead-cat states. Decoherence, alone, does not solve the measurement problem.

Furthermore, although these examples all involve an *external* environment, there is no need to make this restriction. There is, in fact, every reason to think that the microscopic degrees of freedom of even an isolated system suffice to destroy coherence between macroscopic superpositions of that system's macroscopic degrees of freedom.[12] The upshot, in either case, is that for systems above quite small lengthscales, coherent superpositions of states with macroscopically distinct positions rapidly become entangled with their environment. Conversely, though, if a macroscopic system is prepared in a state highly localized in spatial position, very little entanglement will occur.

3.6 Environment-induced decoherence: further details

So far we have been ignoring the dynamics of the system itself. Qualitatively, though, it is easy to see—at least, for regular systems—how this dynamics will proceed. Systems prepared in superpositions of macroscopically different positions will decohere on timescales much more swift than their characteristic dynamical timescales. Systems prepared in superpositions of macroscopically different *momentums* will quickly evolve into states with macroscopically different positions, and these too will swiftly decohere. But if the system is prepared in a state which is approximately localized in both position and momentum, then this state will undergo very little decoherence, and will simply be able to evolve under the system's own Hamiltonian. Since we already know that that evolution

---

[12] For a concrete model, consider a solid-state system—a crystal, say—which is approximately but not exactly harmonic. The macroscopic degrees of freedom of the system correspond to the long-wavelength phonons; these will be decohered by scattering of the short-wavelength phonons in qualitatively the same way that massive particles are decohered by scattering of light particles. (Systems like this will also, in general, behave quasi-classically even absent the anharmonic terms, for the reasons explained in sections 3.2 and 3.3: they are regular. See Halliwell 1998; 2010 for a detailed analysis.)

takes localized states to localized states—again, for regular systems—then this evolution will continue to be unaffected by decoherence.

It is fairly straightforward to write down purely phenomenological dynamical equations for the density operator of a decohering system: the exponential decay in equation (3.38), in particular, is generated by the equation[13]

$$\dot\rho = -\Lambda[\widehat{X},[\widehat{X},\rho]], \qquad (3.40)$$

where $\widehat{X}$ is the position operator for the decohered particle. which (recalling that the unitary quantum dynamics are given by $\dot\rho = -i[\widehat{H},\rho]$ where $\widehat{H}$ is the particle Hamiltonian) suggests the equation

$$\dot\rho = -i[\widehat{H},\rho] - \Lambda[\widehat{X},[\widehat{X},\rho]]. \qquad (3.41)$$

A microphysical derivation of such an equation would require a specific model for the environment, and a number of such models have been analysed. One of the best-studied is the Caldeira–Leggett model[14] in which a particle interacts linearly with an environment of harmonic oscillators; under appropriate simplifying conditions,[15] this model yields an equation of the form

$$\dot\rho = -i[\widehat{H} + \frac{1}{2}m\Omega^2\widehat{X}^2,\rho] - \eta k_B T\Lambda[\widehat{X},[\widehat{X},\rho]] - i\frac{\eta}{2m}[\widehat{X},\{\widehat{P},\rho\}]$$

$$(3.42)$$

where T is the temperature of the environment, k_B is Boltzmann's constant, and η and Ω parametrize the system–environment interaction.

---

[13] For further discussion of this expression see Joos et al. (2003: 64–75) and references therein.

[14] The Caldeira–Leggett model was first analysed in Caldeira and Leggett (1983); see Schlosshauer (2007: 71–4) for a discussion.

[15] The 'appropriate simplifying conditions' are a nice example of the way theoretical physics works in practice. One of the assumptions is that the system's internal dynamics are harmonic i.e. that the internal potential is quadratic—and this is clearly much too strong to rigorously justify applying the Caldeira–Leggett equation to e.g. chaotic systems. On the other hand, any potential is approximately quadratic as long as the system remains confined to a sufficiently small region of it. So, provided we are entitled to assume that the system is never in a coherent superposition which is large compared with the lengthscales on which the potential deviates from being quadratic, we can derive the equation on the basis of a quadratic potential. And what justifies *this* assumption? Earlier, qualitative arguments, of the form described above. The self-consistency of the whole thing can be seen when it is noted that Caldeira–Leggett dynamics do indeed suppress coherent superpositions on the required lengthscale. Philosophers of science take note: theoretical physics does this sort of thing all the time, and naturalistically inclined philosophers should be fine with this.

Equations derived from different environments have the same general form, consisting of:

1. the system's intrinsic unitary dynamics, given by $\dot{\rho} = -i[\hat{H}, \rho]$ (which, generically, will turn superpositions in momentum into superpositions in position via wavepacket spreading);
2. a decoherence term which suppresses superpositions in the position basis;
3. a dissipation term (in the Caldeira–Leggett equation, this is the last term) corresponding to classical friction;
4. a renormalization term (like the term proportional to Ω^2 in the Caldeira-Leggett equation).

In situations of the sort discussed earlier—a macroscopic system interacting relatively weakly with a microscopic environment—the dissipation and renormalization terms are negligible compared with the other two terms, and the decoherence term suppresses macroscopic superpositions very quickly relative to the dynamical timescale of the unitary term.

In the Wigner-function representation, and ignoring renormalization and dissipation, the Caldeira–Leggett equation (and, as noted, most realistic equations for decoherent systems) takes the form (Zurek and Paz 1995a: 304)

$$\dot{W} = \{H, W\}_{MB} + \Lambda \frac{\partial^2 W}{\partial p^2}. \tag{3.43}$$

It can readily be seen that the decoherence term (the second term in the equation) is a diffusion term: it will cause W to spread out as long as it is sufficiently localized in momentum. For regular systems, this term will normally be negligible for quasiclassical states: such states are and remain sufficiently spread out in momentum space that the diffusion term is almost irrelevant.

Things are interestingly different for chaotic systems. Recall that for such systems, the fact that the system begins in a phase-space-localized state is insufficient to ensure that it remains in such a state. Instead, a state initially localized will begin to spread out in physical space (and so will become squeezed in momentum space)—and, as soon as it starts to spread out, the diffusion term will come into play (i.e. the state will start to become entangled with its environment), so that the pure delocalized state becomes replaced by a mixed state which is an incoherent superposition of localized states. Each of *these* states will spread out under the chaotic dynamics, and so will be decohered in their turn...and so on. At any given time, the density operator of the system will be a weighted sum of localized states, and because of the constant decoherence, each such state will evolve independently of all the others, even though it is constantly splitting into multiple states. So in the case of chaos, 'worlds'—that is, emergent quasi-classical systems—are constantly splitting from one another. And since the system's state is always a mixture of reasonably *localized* wavepackets, the failure of classicality which we predicted for isolated chaotic systems will not occur here.

Notice that the irreversibility induced by decoherence is of a very different character from that which would be induced by the dissipative term: there is no energy loss, no deviation from isolated classical dynamics on long lengthscales, and the process can occur—and occur extremely quickly—in cases where dissipation is negligible. (Consider again Hyperion, for instance: the interplanetary medium decoheres Hyperion essentially instantly, but friction between the medium and Hyperion is dynamically utterly irrelevant.) Nonetheless, decoherence *is* an irreversible process, and so the usual questions arise as to how this is compatible with an underlying reversible dynamics. I postpone this question to Chapter 9; see Schlosshauer (2007: 93–5) for more on the contrast between decoherence and dissipation.

3.7 Decoherent histories

Let us take stock. In section 3.4, I identified three problems with extracting quasi-classical behaviour from macroscopic quantum systems: (i) What justifies our treating the macroscopic degrees of freedom as dynamically isolated from the remainder of the system? (ii) Why do chaotic systems behave quasi-classically given that in isolation they evolve into non-quasi--classical states? (iii) Why even when the dynamics of a system is not even approximately classical—such as in the case of quantum measurement—do macroscopic systems still seem to stay in quasi-classical states?

We can now see that decoherence provides an answer to all three worries.[16] First, it explains why, for the macroscopic degrees of freedom of

---

[16] It is perhaps worth remembering that all three worries concern whether the unitarily evolving quantum state succeeds in instantiating the *structure* of quasi-classical macroscopic systems. Only if the quantum state is taken as representing physical reality does this help us

regular systems, we are justified in ignoring the effects of the environment: the main effect of the environment is to measure the system in the position basis, and this has no effect on the system if it is already in a reasonably localized state.

Secondly, it explains how chaotic systems nonetheless evolve in a classical way, at least at the coarse-grained level: decoherence constantly transforms delocalized states into mixtures of localized states, and so prevents the system ever ending up in a state so delocalized that the dynamics ceases to be approximately classical.

And as for non-classical events like quantum measurement: whatever state they put a system into, if that system's macroscopic degrees of freedom are not fairly localized in position then it will very rapidly become decohered: as such, it will evolve as a collection of non-interacting systems each of which is itself fairly localized in position.

Furthermore, decoherence at least helps to explain why it seems to be only phase-space local states which can instantiate emergent structure. For suppose some state like

$$\alpha|q_1, p_1\rangle + \beta|q_2, p_2\rangle \tag{3.44}$$

is supposed to instantiate a state of some emergent theory. Decoherence will wipe away any information contained in the relative phases: the system will almost immediately move into the mixed state

$$|\alpha|^2|q_1, p_1\rangle\langle q_1, p_1| + |\beta|^2|q_2, p_2\rangle\langle q_2, p_2| \tag{3.45}$$

which is simply a weighted sum of two independently evolving quasi-classical states. So the complete dynamical story of the system is known once we know its quasi-classical dynamics and the relative weights of the quasi-classical histories.

However, our analysis so far—which has been concentrated on the evolution of the system's density operator, and has invariably traced away the environment—makes it somewhat difficult to appreciate how exactly it is that the quantum state has the structure of a collection of quasi-classical *branching* worlds. We may have established that the density operator of such

---

solve the measurement problem proper; and if we do so take it, we need to accept that the 'quasi-classical macroscopic systems' so instantiated are multiply instantiated: that is, we need to accept the many-worlds picture. Decoherence is a necessary part of Everettian quantum theory, not an alternative to it.

systems is diagonalized in a quasi-classical basis, but it is not immediately obvious how to read the branching structure off from this observation.

An example may help to see the difficulty—and how to surmount it. The orbit of the Earth around the Sun is chaotic: over timescales of a few million years it is impossible (using classical physics) to predict where in its orbit the planet may be found.[17] The Earth is also (obviously!) very strongly decohered by its environment. The general considerations of section 3.6 tell us that the system's density operator will evolve, over the same timescales, to be a uniform mixture of states localized at all locations in the orbit, and will thereafter remain in that state indefinitely. That is: if $|\theta\rangle$ is a state of the Earth's centre-of-mass degrees of freedom localized at a particular angular coordinate θ, after a few million years the Earth's centre of mass will have state

$$\rho(t) = \frac{1}{2\pi} \int_0^{2\pi} d\theta \, |\theta\rangle\langle\theta|. \tag{3.46}$$

This stationary state does not look much like what the Everett interpretation predicts: a set of histories of the Earth's orbital position, each one evolving quasi-classically. Nor does it seem to match our own observations of the Earth as in motion.

However, this is an illusion caused by our failure to look at the overall state of the Earth-plus-environment system. The actual structure of this state would be best written as

$$|\Psi(t)\rangle = \int \mathcal{D}\theta \, \Lambda[\theta] |\theta(t)\rangle \otimes |[\theta]\rangle, \tag{3.47}$$

where the integral is over all histories $\theta(\xi)$ of the angular coordinate of the Earth up to time t, and where states $|[\theta]\rangle$, $|[\theta']\rangle$ of the environment are orthogonal if $\theta(\xi)$ and $\theta'(\xi)$ differ significantly for any significant period of time (i.e. any period of time long compared to the decoherence timescale). Each $|[\theta]\rangle$, in other words, encodes a different history of the Earth's location, and this is as we should expect: the position of the Earth at any time leaves an irreversible record in the pattern of light, gravitational waves, and neutrinos radiating outwards from the Solar System at that time. So despite the apparent stationarity of (3.46), actually the system is a superposition of quasi-classical states, each of which is

---

[17] This example is discussed in detail in Zurek and Paz (1995a).

evolving approximately classically but which is branching into multiple approximately-classical states on a long timescale.

For the rest of this chapter, I wish to explore the structure of the quantum state from this more 'historical' perspective. I will begin by getting a little more precise about what it is to say that a system's state is 'branching'.

3.8 Analysing branching structure

What would it mean to say that a quantum state 'has a branching structure'? First, clearly that branching structure would have to be defined by the state *together with* other dynamical structures in the theory: a state, interpreted as a mere vector in a featureless Hilbert space, has no structure at all. Relative to a basis, on the other hand, it is comparatively clear to understand how a state could be branching: if the state evolves from a basis vector to a superposition of such basis vectors, and if each of *those* evolves into a superposition of *different* basis vectors so that no two such superpositions interfere with one another—then we would have branching (relative to that basis, at any rate).

To get rather more precise about this, suppose we have a physical system represented by some Hilbert space $\mathcal{H}$, evolving unitarily under some dynamics $\widehat{U}(t, t_0)$. Instead of restricting ourselves to a basis, we will consider a projection-valued measure $\widehat{P}_1, \ldots \widehat{P}_n$ (that is, a family of disjoint projectors whose sum is the identity but which need not be all of dimension one). At any given time t, and for an initial state $|\psi\rangle$ (at time t_0), the weight of projector $\widehat{P}_j$ is

$$\mathcal{W}_j(t) = \|\widehat{P}_j \, \widehat{U}(t, t_0)|\psi\rangle\|^2 \equiv \langle\psi|\widehat{U}^\dagger(t, t_0)\widehat{P}_j \, \widehat{U}(t, t_0)|\psi\rangle, \qquad (3.48)$$

and the transition weight between $\widehat{P}_j$ at time t and $\widehat{P}_{j'}$ at time t' ($t' > t$) is

$$\begin{aligned} \mathcal{T}(j, t; j', t') &= \frac{\|\widehat{P}_{j'}\widehat{U}(t', t)\widehat{P}_j\widehat{U}(t, t_0)|\psi\rangle\|^2}{\|\widehat{P}_j\widehat{U}(t, t_0)|\psi\rangle\|^2} \\ &= \frac{\langle\psi|\widehat{U}^\dagger(t, t_0)\widehat{P}_j\widehat{U}^\dagger(t', t)\widehat{P}_{j'}\widehat{U}(t', t)\widehat{P}_j\widehat{U}(t, t_0)|\psi\rangle}{\langle\psi|\widehat{U}^\dagger(t, t_0)\widehat{P}_j\widehat{U}(t, t_0)|\psi\rangle}. \end{aligned} \qquad (3.49)$$

For convenience, define $\mathcal{T}(j, t; j', t') = 0$ whenever $\mathcal{W}_j(t) = 0$ (the above definition leaves it undefined).

When quantum mechanics is interpreted instrumentally, of course, the transition weights are supposed to be conditional probabilities and the absolute weights are supposed to be unconditional probabilities; in quantum mechanics interpreted realistically, though, they are just objective properties of the quantum-mechanical Universe.

As we have noted, 'branching' (relative to a given basis) is just the absence of interference. This in turn occurs (between times t and t') when at most one component of the quantum state (in that basis) at time t contributes to the weight of any given component at time t'. In terms of transition weights, this is just to require that no two transition weights of transitions into a given projector are nonzero—that is, to require that

$$\mathcal{T}(j_1, t; j', t') \neq 0, \mathcal{T}(j_2, t; j', t') \neq 0 \implies j_1 = j_2. \qquad (3.50)$$

(To visualize this, think of 'weight' as a fluid, redistributing itself across the projectors over time. (3.50) guarantees that each projector receives weight from exactly one previous projector. Less picturesquely, if (3.50) holds then there is a unique way to connect projectors at later times to projectors at earlier times: each projector's weight may determine the weight of many future projectors but its own weight is determined by exactly one past projector at any given past time.)

The importance of decoherence is: when it occurs, quantum-mechanical systems (approximately) develop a particularly natural branching structure. For decoherence is a process which constantly and (on sub-Poincaré-recurrent timescales) irreversibly entangles the environment with the system so as to suppress interference between terms of the decoherence-preferred basis.[18] (We might say that the environment constantly measures the system and records the result.) If we idealize the dynamics as discrete, then at each branching event, the environment permanently records the pre-branching state, so that at each time the universal state is a superposition of states each of which encodes a complete record of where 'its weight' comes from.

Even if the dynamics is not itself discrete, a branching structure is still readily discernible in decohering systems. We can analyse this for the case where the phase-space description of decoherence is applicable: in that case, we can in full generality write the total state of the system and environment at a given time as

---

[18] For further discussion of this irreversibility, see Ch. 9.

$$|\Psi\rangle = \int dp_0 \, dq_0 \, \alpha(p_0, q_0)|p_0, q_0\rangle \otimes |\phi(p_0, q_0)\rangle \qquad (3.51)$$

Because of decoherence, whatever initial state the system was prepared in, the total state will quickly evolve to one where $\langle\phi(p_0, q_0)|\phi(p'_0, q'_0)\rangle \simeq 0$ for sufficiently separated q'_0, p'_0 and q_0, p_0.

After some further time Δt, the state

$$|p_0, q_0\rangle \otimes |\phi(p_0, q_0)\rangle \qquad (3.52)$$

will evolve to a state of the form

$$|\psi(p_0, q_0)\rangle = \int dp_1 dq_1 \beta_1(p_1, q_1; p_0, q_0)|p_1, q_1\rangle \otimes |\phi(p_1, q_1, p_0, q_0)\rangle.$$

$$(3.53)$$

Again, decoherence ensures that $\langle\phi(p_1, q_1, p_0, q_0)|\phi(p'_1, q'_1, p_0, q_0)\rangle \simeq 0$ for sufficiently separated q'_1, p'_1 and q_1, p_1. But we would also expect, in general, to find that if $\langle\phi(p_0, q_0)|\phi(p'_0, q'_0)\rangle \simeq 0$, then $\langle\phi(p_1, q_1, p_0, q_0)|\phi(p'_1, q'_1, p'_0, q'_0)\rangle \simeq 0$ irrespective of the values of p_1, q_1, p'_1, q'_1. For the information about the system recorded in the original decoherence process will be distributed very widely across the environment (think of our original example of decoherence by particle scattering: the initial particles that caused the decoherence are now a distance $\sim v\Delta t$ from the system). The total state after time Δt is then

$$|\Psi(\Delta t)\rangle \equiv \widehat{U}(\Delta t)|\Psi\rangle$$

$$= \int\int dp_0 \, dq_0 \, dp_1 \, dq_1 \, \beta_1(p_1, q_1; p_0, q_0)\alpha(p_0, q_0)|p_1, q_1\rangle \otimes |\phi(p_1, q_1, p_0, q_0)\rangle$$

$$(3.54)$$

Iterating N times, then (and writing $\mathbf{p}$, $\mathbf{q}$ to symbolize the N-tuples $p_0, \ldots p_N, q_0, \ldots q_N$): after a time $N\Delta t$ the system will have state

$$|\Psi(N\Delta t)\rangle \equiv \widehat{U}(N\Delta t)|\Psi\rangle = \int \cdots \int d\mathbf{p} \, d\mathbf{q} \, C_N(\mathbf{p}, \mathbf{q})|p_N, q_N\rangle \otimes |\phi_N(\mathbf{p}, \mathbf{q})\rangle,$$

$$(3.55)$$

where $\langle\phi_N(\mathbf{p}, \mathbf{q})|\phi_N(\mathbf{p}', \mathbf{q}')\rangle \simeq 0$ if any of the (q_i, p_i) are sufficiently separated from the (q'_i, p'_i). (For the example described by (3.38), for instance, this amounts to requiring that $|q'_i - q_i|$ is much larger than $([\Lambda(t_{i+1} - t_i)]^{-1/2}$.) Each dynamical step can be represented by

$$\widehat{U}(\Delta t)|p_N, q_N\rangle \otimes |\phi_N(\mathbf{p}, \mathbf{q})\rangle$$

$$= \int dp_{N+1} dq_{N+1} B_N(q_{N+1}, p_{N+1}; \mathbf{q}, \mathbf{p})|p_{N+1}, q_{N+1}\rangle \otimes |\phi_{N+1}(p_{N+1} \oplus \mathbf{p}, q_{N+1} \oplus \mathbf{q})\rangle \tag{3.56}$$

where $q \oplus \mathbf{q}$ is the sequence obtained by appending q to the sequence $\mathbf{q}$ (and similarly for $p \oplus \mathbf{p}$).

Informally, it should be clear that a state whose dynamics take this form will have a branching structure relative to the basis of $|p_I, q_I\rangle$ states at each time-step. To make this more rigorous, however, let us choose a partition of phase space into cells Σ_i. Classically, a time-indexed sequence of such cells would represent, at a coarse-grained level, a possible history of the system: 'at time t_1 it was in cell Σ_{i_1}, at time t_2 it was in cell $\Sigma_{i_2}, \ldots$'; more abstractly, we can take a sequence of *indices* of such cells as representing the history. To see that the system's evolution has the same structure as a probabilistic mixture of such histories (for all that its *nature* is very different), we can proceed as follows. Writing $\mathbf{i}$ as shorthand for a sequence $i_1, \ldots i_N$ of such indices, we can define the operators

$$\widehat{\Pi}_{\mathbf{i}}^N = \int_{\Sigma_{i_0}} \cdots \int_{\Sigma_{i_n}} d\mathbf{q} \, d\mathbf{p} \, \widehat{1}_S \otimes |\phi_N(\mathbf{p}, \mathbf{q})\rangle\langle\phi_N(\mathbf{p}, \mathbf{q})| \tag{3.57}$$

(where $\widehat{1}_S$ is the identity operator for the system's Hilbert space). If the cells of the partition are chosen to be sufficiently large (in the case described by (3.38), for instance, if they have spatial width $\gg (\Lambda \Delta t)^{-1/2}$ and an appropriate momentum-space width) then these operators will approximately define a PVM:

$$\widehat{\Pi}_{\mathbf{i}}^N \widehat{\Pi}_{\mathbf{j}}^N \simeq \delta_{\mathbf{i}, \mathbf{j}} \widehat{\Pi}_{\mathbf{i}}^N. \tag{3.58}$$

Moreover, we have

$$\widehat{\Pi}_{\mathbf{i}}^N \widehat{U}(N\Delta t)|\Psi\rangle = \int_{\Sigma_{i_0}} \cdots \int_{\Sigma_{i_N}} d\mathbf{p} \, d\mathbf{q} \, C_N(\mathbf{p}, \mathbf{q})|p_N, q_N\rangle \otimes |\phi_N(\mathbf{p}, \mathbf{q})\rangle \tag{3.59}$$

and from this and (3.56) it can readily be seen that

$$\widehat{\Pi}_{\mathbf{i}'}^{N+1} \widehat{U}(\Delta t) \widehat{\Pi}_{\mathbf{i}}^N \widehat{U}(N\Delta t)|\Psi\rangle \simeq 0 \text{ unless } \mathbf{i} \text{ is the initial segment of } \mathbf{i}'. \tag{3.60}$$

That is: the structure of the quantum state relative to the family of PVMs $\{\widehat{\Pi}_{\mathbf{i}}^N\}$ (for each N) is branching.

Notice that although we have imposed a discrete structure on the system so as to make precise the claim that it branches, there is no intrinsic discreteness in the branching process. Less rigorously, but perhaps more perspicuously, we might rewrite (3.55) as

$$|\Psi(t))\rangle = \int \mathbf{D}[q(\xi)] \ C_t[q(\xi)] \ |p(t + \Delta t), q(t + \Delta t)\rangle \otimes |\phi[q(\xi)]\rangle \quad (3.61)$$

where the integral ranges over all classical trajectories defined up to time t and where $\langle \phi[q(\xi)]|\phi[q'(\xi)]\rangle \simeq 0$ if the trajectories $q(\xi)$ and $q'(\xi)$ are sufficiently different for sufficiently long (if they differ by $\gg (\Lambda \delta t)^{-1/2}$ over a period of $\sim \delta t$ in the case of (3.38), for instance). In this formalism, the state has branching structure because $|p(t), q(t)\rangle \otimes |\phi[q(\xi)]\rangle$ evolves over time Δt to

$$\int \mathbf{D}[q'(\xi)] \ B_{t,t+\Delta t}[q'(\xi)] \ |p(t), q(t)\rangle \otimes |\phi[q(\xi) \oplus q'(\xi)]\rangle \quad (3.62)$$

where the integral ranges over classical trajectories defined between times t and $t + \Delta t$, and where $q(\xi) \oplus q'(\xi)$ is the trajectory given by $q(\xi)$ up till $\xi = t$ and by $q'(\xi)$ thereafter, i.e. for times between t and $t + \Delta t$.

3.9 The decoherent-histories framework

To talk more generally about the relation between branching and decoherence, and to help the reader to connect my discussion to the literature, it will be useful to develop a more sophisticated mathematical description of branching. We will consider a discrete set of times $t_0, \ldots t_N$, and will generalize our earlier description by allowing the PVMs used to define branching to vary from time to time; we will also (purely for mathematical convenience) switch to the Heisenberg picture.[19] Then the spaces on which the branching structure is defined is just a time-indexed family of PVMs $\widehat{P}_j^i$ (with the superscript indicating that the operator is a member

---

[19] Recall that while in the Schrödinger picture the operators that represent dynamical quantities ('observables', to use the usual—but unfortunate—phrase) are time-invariant and the unitary time-evolution operator $U(t)$ is applied to the quantum state, in the Heisenberg picture this is reversed, so that states are time-invariant and operators evolve via $X(t) = U^\dagger(t - t_0)X(t_0)U(t - t_0)$. Since both pictures generate the same expectation values for all observables, the choice of which to use for calculations is just a matter of convenience. (It has been argued, notably by Deutsch and Hayden 2000 and Deutsch 2002, that the Heisenberg picture is foundationally preferable—see Ch. 8 for further discussion—but I adopt it here for simplicity only.)

of the time-t_i PVM and the subscript indexing it within that PVM), and the transition weights are given (for $t_{i'} > t_i$) by

$$\mathcal{T}(j, t_i; j'; t_{i'}) = \frac{\langle \psi | \widehat{P}_j^i \widehat{P}_{j'}^{i'} \widehat{P}_j^i | \psi \rangle}{\langle \psi | \widehat{P}_j^i | \psi \rangle}. \tag{3.63}$$

The branching criterion (3.50) can then be succinctly expressed as

If $\widehat{P}_{j'}^{i'} \widehat{P}_{j_1}^i | \psi \rangle$ and $\widehat{P}_{j'}^{i'} \widehat{P}_{j_2}^i$ are both non-zero, then $j_1 = j_2$.

It is again useful to define a *history* as a sequence of projectors, one from each of the time-indexed PVMs: I call the set of such histories generated from some such sequence of PVMs a *history space*. Since there is a one-to-one relation between such history spaces and the sequences of PVMs from which they are generated, we can identify a history space with the set of PVMs that define it, and write that set as $\{\widehat{P}_j^i\}$, where i ranges across time indices and j labels the projectors within the PVM for time index i.

Since a sequence of projectors can also be viewed as a function from times to projectors, given a history α I write $\widehat{\alpha}(m)$ for the projector associated by the history α with time index m; each $\widehat{\alpha}(m)$ is specified uniquely by giving its index number in the time-t_m PVM, and I write this index number as α_m, so that

$$\widehat{\alpha}(m) = \widehat{P}_{\alpha_m}^m. \tag{3.64}$$

I call a history *realized* if $\mathcal{T}(\alpha_m, t_m; \alpha_{m+1}, t_{m+1}) \neq 0$ for all $m \leq n$. The branching criterion then guarantees that if two realized histories coincide at some time (that is, assign the same projector to that time) then they coincide at all earlier times, and we will say that any set of histories with this property has a *branching structure*. (Note that this property is relative to a given state.) Given two history spaces $\{\widehat{P}_j^i\}$, $\{\widehat{Q}_j^i\}$, $\{\widehat{Q}_j^i\}$ is a *coarse-graining* of $\{\widehat{P}_j^i\}$ if every projector $\widehat{Q}_j^i$ is a sum of projectors $\widehat{P}_k^i$.

Following Gell-Mann and Hartle (1990), we can define the *history operator* $\widehat{C}_\alpha$ of the history α by

$$\widehat{C}_\alpha = \widehat{\alpha}(n) \cdots \widehat{\alpha}(0), \tag{3.65}$$

and the *decoherence functional*, a complex function on pairs of histories (relative to a choice of state $|\psi\rangle$), by

$$\mathcal{D}(\alpha, \beta) = \langle \psi | \widehat{C}_\alpha^\dagger \widehat{C}_\beta | \psi \rangle. \tag{3.66}$$

A history space is said to satisfy the *decoherence condition* or to be *decoherent*[20] if the decoherence functional between any two incompatible histories is zero. (Hence, implicitly a history space is only decoherent relative to a choice of state.)

The significance of all this formalism is summarized in the following theorem (first stated by Griffiths 1993, so far as I know), which tells us that branching entails decoherence and (up to possible coarse-grainings) vice versa:

Branching-Decoherence Theorem. Suppose $\mathcal{P} = \{\widehat{P}_j^i\}$ is a history space and $|\psi\rangle$ is a quantum state. Then:

(i) If $\mathcal{P}$ has branching structure (relative to $|\psi\rangle$) and α is a history then $\widehat{C}_\alpha|\psi\rangle \neq 0$ iff α is realized (with respect to $|\psi\rangle$).

(ii) If the set Hist of all histories α such that $\widehat{C}_\alpha|\psi\rangle \neq 0$ has branching structure (that is, if no two histories in Hist agree on their ith projector but not on all previous projectors), then $\mathcal{P}$ also has branching structure (relative to $|\psi\rangle$), and the realized histories in that branching structure are just the histories in Hist.

(iii) If $\mathcal{P}$ has branching structure (relative to $|\psi\rangle$), $\mathcal{P}$ satisfies the decoherence condition.

(iv) If $\mathcal{P}$ satisfies the decoherence condition, it is a coarse-graining of a (decoherent) history space which has branching structure relative to $|\psi\rangle$.

The proof of the Branching-Decoherence Theorem is straightforward but tedious and is relegated to Appendix 1; however, the basic ideas behind it are easy to understand. The first two parts are just an iteration of the branching criterion to apply to sequences of more than two projectors, and the third part follows straightforwardly from the first two. The key to understanding the fourth part is to notice that decoherence implies that the states

$$|\alpha\rangle = \widehat{C}_\alpha|\psi\rangle \qquad (3.67)$$

are all mutually orthogonal. These states can be thought of as 'record states', each recording the structure of an entire branch. The state of the

---

[20] Sometimes this condition is called *medium* decoherence, following Gell-Mann and Hartle (1990) and in contrast to *weak decoherence*, defined in the next section.

system at a given time, then, is a superposition of all these histories, and the subsequent evolution of the system will not erase these histories; hence, the terms in the superposition cannot interfere with one another, and so the state has a branching structure.

3.10 Decoherence, records, and consistency

From the Everettian perspective, the decoherence functional is a purely technical tool: its significance comes from the branching-decoherence theorem, which tells us that the vanishing of the decoherence function between any two distinct histories is a necessary and sufficient condition for a history space to have a branching structure. An alternative perspective, however—developed by Robert Griffiths (1984; 1996; 2002), Roland Omnés (1988; 1992; 1994), and (from a rather different viewpoint) by Murray Gell-Mann and James Hartle (1990; 1993; 2007)—was historically important and remains frequently discussed in the literature, and is the subject of this section. For clarity, I follow Griffiths in calling this approach a *consistent histories* approach, though these authors', and their commentators', terminology has been somewhat varied. What is 'consistent' is our interpretation of mod-squared amplitude as probability: Griffiths et al., as we will see, interpret the decoherence condition as a sufficient condition for probabilistic talk to make sense in quantum theory.

This framework starts with the idea that quantum mechanics ought somehow to be interpreted as a stochastic theory. Doing this consistently would require the theory to specify a space of histories and some probability measure over those histories. Within quantum mechanics, the obvious mathematical representation of a history is that of the previous section: a string $\widehat{\alpha}_1, \ldots \widehat{\alpha}_N$ of time-indexed projectors (note that for the moment I do not assume that a history is part of some previously specified history *space*). And the obvious probability to assign to a history α is

$$\Pr(\alpha) = \|\widehat{\alpha}_N \cdots \widehat{\alpha}_1 |\psi\rangle\|^2 \tag{3.68}$$

That is, start with the quantum state, sequentially project it out by the projectors, and take the mod-squared amplitude of the resulting state—in the Schrödinger picture it would also be necessary to evolve the state unitarily between sequential projections. Using the history operator $\widehat{C}_\alpha$ and decoherence functional $\mathcal{D}(\alpha, \beta)$ defined in the previous section, we can write this succinctly as

$$\Pr(\alpha) = \langle \psi | \widehat{C}_\alpha^\dagger \widehat{C}_\alpha | \psi \rangle = \mathcal{D}(\alpha, \alpha). \tag{3.69}$$

The problem, of course, is the same problem that besets all attempts to interpret quantum mechanics probabilistically: interference. In this case, the mathematical representation of interference is as a failure of the probability calculus. Suppose, for instance, that α and β are histories with $\widehat{\alpha}(k) = \widehat{\beta}(k)$ for all time indices k except some m, and that $\widehat{\alpha}(m)$ and $\widehat{\beta}(m)$ are orthogonal. If the history γ is defined by

$$\widehat{\gamma}(k) = \widehat{\alpha}(k) = \widehat{\beta(k)} \ (k \neq m)$$

$$\widehat{\gamma}(m) = \widehat{\alpha}(m) + \widehat{\beta}(m) \tag{3.70}$$

then the probability calculus would require that $\Pr(\gamma) = \Pr(\alpha) + \Pr(\beta)$. But this, of course, is generally not the case in quantum theory.

In the consistent-histories approach, this is solved by restricting the set of allowed histories. The starting point here is the *history space* of section 3.9, which was defined (recall) as the set of histories generated from a particular time-indexed family of PVMs. To allow for histories which are sums of other histories (as in the above case), we now permit histories which assign to a time t_i a sum of projectors (rather than just a single projector) in the time-t_i PVM. A history which assigns only one projector in the appropriate PVM to each time is called *atomic*. (In fact, once we generalize history spaces in this way, the notion of atomic histories becomes dispensable, as I explain in Box 3.2, but for expository purposes it is convenient to retain them.)

Box 3.2. Atomless history spaces

Given a Hilbert space, a Boolean algebra of projectors on that Hilbert space is just a set of projectors which contains the identity and is closed under taking countable sums and complements; such an algebra is *atomic* if there is a countable set of projectors such that all elements of the algebra are sums of elements of the set. I specify a *history algebra* $\{\mathcal{S}^i\}$ by assigning to each time index t_i a Boolean algebra $\mathcal{S}^i$ of projectors; the histories in that algebra are sequences of such projectors, and I call the history atomic iff all its Boolean algebras are atomic. The history

(continued)

Box 3.2. Continued

operator and the decoherence functional can be defined as before; the probability of history α is by definition $\mathcal{D}(\alpha, \alpha)$.

Two histories α, β are *overlapping* if for each k, $\widehat{\alpha}(k)\widehat{\beta}(k) \neq 0$. Given a history α in $\{\mathcal{S}^i\}$, a *decomposition* of α is a set of histories specified by giving, for each k, a set of mutually orthogonal projectors $\widehat{P}_i^k \in \mathcal{S}^k$ whose sum is $\alpha(k)$; the histories in the refinement are exactly those histories constructed from projectors in this set.

A history *space* continues to be specified by a time-indexed sequence of sets of projectors; each history space determines an atomic history algebra in the obvious way, and conversely a history space is *contained within* a history algebra if all its histories are histories in the algebra. Given a history algebra, and two history spaces contained within it, the first is a *refinement* of the second iff each projector in each time-t_k projector set in the second space is the sum of projectors in the time-t_k projector set in the first space. (It follows that a history algebra is atomic iff it contains some history space with no proper refinements.)

We can then make the following definitions. Given a history algebra, then with respect to some state $|\psi\rangle$:

- the algebra is *branching* if it contains some history space relative to which $|\psi\rangle$ has branching structure.
- the algebra satisfies *decoherence* iff $\mathcal{D}(\alpha, \beta)$ vanishes whenever α, β are non-overlapping, and *weak decoherence* if the real part of $\mathcal{D}(\alpha, \beta)$ vanishes for non-overlapping α, β.
- the algebra is *consistent* iff for any history α, and any decomposition of that history, the probability of α is the sum of the probabilities of the histories in its decomposition.

It then follows that:

1. A history algebra is decoherent iff it is branching (atomless version of the Branching-Decoherence Theorem)
2. A history algebra is weakly decoherent iff it is consistent

The former is proved in Appendix A; the latter is proved by the method used in section 3.10.

Given histories α and β, I call α a *subhistory* of β iff $\widehat{\alpha}(k)$ is a sub-projector of $\widehat{\beta}(k)$[21] for all k. And $\mathrm{Dec}(\alpha)$, the *decomposition* of α, is then the set of all atomic histories that are subhistories of α: in effect (if a stochastic interpretation is required) the various elements of the decomposition of α are the various ways of filling in those details of a system's history which α itself leaves unspecified.

A succinct way of writing the condition required by the probability calculus is then that for any history α,

$$\Pr(\alpha) = \sum_{\alpha_i \in \mathrm{Dec}(\alpha)} \Pr(\alpha_i), \tag{3.71}$$

or in terms of the history formalism,

$$\langle \psi | \widehat{C}_\alpha^\dagger \widehat{C}_\alpha | \psi \rangle = \sum_{\alpha_i \in \mathrm{Dec}(\alpha)} \langle \psi | \widehat{C}_{\alpha_i}^\dagger \widehat{C}_{\alpha_i} | \psi \rangle. \tag{3.72}$$

We now define a history space as *consistent* if this condition holds; it follows that in general, consistency is relative to the quantum state.

Now, since

$$\widehat{C}_\alpha = \sum_{\alpha_i \in \mathrm{Dec}(\alpha)} \widehat{C}_{\alpha_i} \tag{3.73}$$

we can rewrite the left-hand side of (3.72) as

$$\langle \psi | \widehat{C}_\alpha^\dagger \widehat{C}_\alpha | \psi \rangle = \sum_{\alpha_i,\alpha_j \in \mathrm{Dec}(\alpha)} \langle \psi | \widehat{C}_{\alpha_j}^\dagger \widehat{C}_{\alpha_i} | \psi \rangle = \sum_{\alpha_i,\alpha_j \in \mathrm{Dec}(\alpha)} \mathcal{D}(\alpha_i, \alpha_j) \tag{3.74}$$

and the right-hand side as

$$\sum_{\alpha_i \in \mathrm{Dec}(\alpha)} \mathcal{D}(\alpha_i, \alpha_i). \tag{3.75}$$

It follows that any history space which is decoherent—i.e. which satisfies $\mathcal{D}(\alpha, \beta) = 0$ for $\alpha \neq \beta$—is also consistent. Because $\mathcal{D}(\alpha, \beta) = \mathcal{D}(\beta, \alpha)^*$, a slightly weaker condition—that the real part of $\mathcal{D}(\alpha, \beta)$ vanishes for $\alpha \neq \beta$,—suffices to guarantee that a history space is consistent. For this reason, sometimes it is this condition which is called 'consistent'; another

---

[21] Recall that given projectors $\widehat{P}, \widehat{Q}$, then $\widehat{P}$ is a sub-projector of $\widehat{Q}$ iff the range of $\widehat{P}$ is a subspace of the range of $\widehat{Q}$.

term sometimes used is 'weak decoherence'. However, weak decoherence does not seem to have any dynamical significance (in the way that decoherence proper has been shown to have), and composite systems satisfying weak but not full decoherence have been shown to have various unsatisfactory properties (Diósi 2004). It follows from part (iii) of the branching-decoherence theorem that any branching history space is consistent, and from part (iv) that physically interesting consistent history spaces are coarse-grainings of branching history spaces.

Originally, it was possible to suppose that consistency, or decoherence, or some reasonable strengthening of these conditions, would suffice to pick out a *unique* history space: the measurement problem would thereby have been solved and quantum mechanics could have been interpreted as a stochastic theory. Unfortunately for the consistent-histories program, this turns out not to be the case: Fay Dowker and Adrian Kent demonstrated convincingly (Dowker and Kent 1996; Kent 1996) that an enormous number of consistent history spaces exist and that many of them are pathologically unlike the observed macroworld.

The responses[22] of Griffiths, Omnes, and Gell-Mann and Hartle to this problem differ interestingly. Griffiths and Omnes attempt to hold on to the idea of quantum mechanics as a stochastic theory of a single quasi-classical world, and in doing so end up advocating interpretations of quantum mechanics that offer 'vestiges of reality' as I put it in section 1.6—i.e. which are not just unapologetic instrumentalism—but fall short of conventional scientific realism. Griffiths, for instance, tries to regard different history spaces as different ways of describing the same underlying reality. But while in classical mechanics such multiple descriptions can always be understood as coarse-grainings of a single exhaustive description (a principle which Griffiths dubs the *principle of unicity*), this fails in the consistent-histories setting:

The principle of unicity does not hold: there is not a unique exhaustive description of a physical system or a physical process. Instead, reality is such that it can be described in various alternative, incompatible ways, using descriptions which cannot be combined or compared. (Griffiths 2002: 368)

---

[22] I do not want to make any *historical* claim here as to the influence or otherwise of Dowker and Kent's work on proponents of consistent-histories approaches: my account is intended to capture the *logic* of the situation, rather than its chronology.

Approaches of this kind, of course, fall outside the scope of this book.

Gell-Mann and Hartle, on the other hand, rule out pathological history spaces by requiring histories to be 'quasi-classical', which they define (consistently with my usage in this chapter) as histories

such that the individual histories obey, with high probability, effective classical equations of motion interrupted continually by small fluctuations and occasionally by large ones.

This is not the kind of criterion which can be formalized as a new law of physics: it is a criterion for emergent structure of very much the same kind as I discussed in Chapter 2. Gell-Mann and Hartle's exploration of consistent histories, in other words, can be understood as an exploration of those emergent structures which exist within the unitarily evolving state: that is, it can be understood as an exploration of Everettian quantum mechanics. (And indeed, this is how Hartle, on my reading, does seem to understand it; see Hartle 2010.)

3.11 How many worlds?

We are finally in a position to answer one of the most commonly asked questions about the Everett interpretation:[23] how much branching actually happens? As we have seen, branching is caused by any process which magnifies microscopic superpositions up to the level where decoherence kicks in, and there are basically three kinds of these processes:

1. Deliberate human experiments: Schrödinger's cat, the two-slit experiment, Geiger counters, and the like.
2. 'Natural quantum measurements', such as occur when radiation causes cell mutation.
3. 'Classically chaotic' processes: that is, processes governed by Hamiltonians whose classical analogues are chaotic.

The first is a relatively recent and rare phenomenon, but the other two are ubiquitous. Chaos, in particular, is everywhere, and where there is chaos, there is branching (the weather, for instance, is chaotic, so there will be different weather in different branches). Furthermore, there is no sense in which these phenomena lead to a naturally *discrete* branching process:

---

[23] Other than 'and you believe this stuff?', that is.

as we have seen in studying quantum chaos, while a branching structure can be discerned in such systems, it has no natural 'grain'. To be sure, by choosing a certain discretization of (configuration-)space and time, a discrete branching structure will emerge, but a finer or coarser choice would also give branching. And there is no 'finest' choice of branching structure: as we fine-grain our decoherent history space, we will eventually reach a point where interference between branches ceases to be negligible, but there is no precise point where this occurs. As such, the question 'How many branches are there?' does not, ultimately, make sense.

This may seem paradoxical—certainly, it is not the picture of 'parallel universes' one obtains from science fiction. But as we have seen in Chapter 2, it is commonplace in emergence for there to be some indeterminacy. (Recall: when *exactly* are quasi-particles of a certain kind present?) And nothing prevents us from making statements like:

Tomorrow, the branches in which it is sunny will have combined weight 0.7

—the combined weight of all branches having a certain macroscopic property is very (albeit not precisely) well defined. It is only if we ask: '*How many* branches are there in which it is sunny?', that we are asking a question which has no answer.

Box 3.3. A metaphor for indefinite branch number

1. Imagine a world consisting of a very thin, infinitely long and wide, slab of matter, in which various complex internal processes are occurring—up to and including the presence of intelligent life, if you like. In particular one might imagine various forces acting in the plane of the slab, between one part and another.

2. Now, imagine stacking many thousands of these slabs one atop the other, but without allowing them to interact at all. If this is a 'many-worlds theory', it is a many-worlds theory in the style of the philosopher David Lewis (1986b): none of the worlds are dynamically in contact, and no (putative) inhabitant of any world can gain empirical evidence about any other.

3. Now introduce a weak force normal to the plane of the slabs—a force with an effective range of two or three slabs, perhaps, and a force which is usually very small compared to the intra-slab forces. Then other slabs will be detectable from within a slab but will not normally have much effect on events within a slab. If this is a many-worlds theory, it is a science-fiction-style many-worlds theory (or maybe a Philip Pullman/C. S. Lewis many-worlds theory): there are many worlds, but each world has its own distinct identity and is mostly unaffected by the other worlds.

4. Finally, turn up the inter-slab interaction a lot: let it have an effective range of several thousand slabs, and let it be comparable in strength (over that range) with characteristic short-range interaction strengths within a slab. Now, dynamical processes will not be confined to a slab but will spread over hundreds of adjacent slabs; indeed, *evolutionary* processes will not be confined to a slab, so living creatures will exist spread over many slabs. At this point, the boundary between slabs becomes epiphenomenal. Nonetheless, this theory is *stratified* in an important sense: dynamics still occurs predominantly parallel to the slabs and events hundreds of thousands of slabs away from a given slab are dynamically irrelevant to that slab.[a] One might well, in studying such a system, divide it into layers thick relative to the range of the inter-slab force—and emergent dynamical processes in those layers would be no less real just because the exact choice of layering is arbitrary.

> [a] Obviously there would be ways of constructing the dynamics so that this was not the case: if signals could easily propagate between distant slabs, for instance, the stratification would be lost. But it's only a thought experiment, so we can construct the dynamics how we like.

This bears repeating, as it will be central to some of the arguments of Part II:

> Decoherence causes the Universe to develop an emergent branching structure. The existence of this branching is a robust (albeit emergent) feature of reality; so is the mod-squared amplitude for any *macroscopically described* history. But there is *no* non-arbitrary decomposition of

macroscopically-described histories into 'finest-grained' histories, and *no* non-arbitrary way of counting those histories.

Put another way: asking how many worlds there are is like asking how many experiences you had yesterday, or how many regrets a repentant criminal has had. It makes perfect sense to say that you had many experiences or that he had many regrets; it makes perfect sense to list the most important categories of either; but it is a non-question to ask *how many*.

If this picture of the world seems unintuitive, the metaphor in Box 3.3 may help. Ultimately, though, that a theory of the world is 'unintuitive' is no argument against it, provided it can be cleanly described in mathematical language.

CHAPTER 3. If we apply to quantum mechanics the same principles as we apply right across science, we find that a multiplicity of quasi-classical worlds are emergent from the underlying quantum physics. These worlds are structures instantiated within the quantum state, but they are no less real for all that.

CHAPTER 4. Quantum mechanics is a probabilistic theory; how is this compatible with unitary quantum mechanics' deterministic dynamics?

First Interlude

DAVIES: So the parallel universes are cheap on assumptions but expensive on universes.
DEUTSCH: Exactly right. In physics we always try to make things cheap on assumptions.

Interview between Paul Davies and David Deutsch[1]

I do not know how to refute an incredulous stare.

Attributed to David Lewis[2]

Testability

SCEPTIC: Isn't the Everett interpretation untestable?

AUTHOR: Unitary quantum mechanics is testable, and Everettian quantum mechanics just is unitary quantum mechanics, so tests of the former are tests of the latter. Case in point: it's a basic prediction of unitary quantum mechanics that the superposition principle holds absolutely, so that (if you can eliminate decoherence) you ought to be able to do interference experiments with any system you like. So the recent experiments showing diffusion with buckyballs[3] count as experimental confirmation of the Everett interpretation. Any of the various proposed tests of dynamical collapse theories would either confirm or falsify the Everett interpretation, depending on the result.

SCEPTIC: What makes those tests of the Everett interpretation, and not just tests of quantum mechanics itself?

---

[1] Davies and Brown (1986).

[2] I have been unable, however, to find this (widely quoted) expression in any of Lewis's writings, though it paraphrases remarks in Lewis (1986b).

[3] For a review of these and other recent experimental tests of the superposition principle, see Schlosshauer (2006).

AUTHOR: Because as I say, the Everett interpretation just is quantum mechanics. If you're after some experiment to distinguish Everett-interpreted quantum mechanics from an operational interpretation of quantum mechanics, then sure, I haven't got anything to offer. But there's nothing special about quantum mechanics there. I don't know of an experiment that distinguishes the theory that dinosaurs existed from the 'theory' that dinosaurs didn't exist but the fossils are exactly the way they would have been if dinosaurs had existed.

SCEPTIC: Isn't it still a bit uncomfortable for you that you're arguing that all these other worlds exist but that there's no possible way to observe one of them?

AUTHOR: Not especially. Our best current theory of physics (a) predicts that they exist and (b) explains why we can't normally see them. (And it's just a matter of degree. Various interference experiments—the Michelson interferometer, for instance, or Shor's algorithm—let us see the many-worlds structure of the quantum state pretty directly, as I'll discuss in more detail in Chapter 10.)

In any case, we see this sort of thing a lot in science. We can't directly observe a dinosaur, or a quark, or a quasar, or the interior of the sun, but that doesn't stop us taking them seriously.

SCEPTIC: In most of those cases, it's just happenstance that we can't make the observation. If we were properly situated in space and time, we'd be able to.

AUTHOR: Well, if we were properly situated in the multiverse, we'd be able to see other worlds.

SCEPTIC: How can you know that?

AUTHOR: My best theory of physics tells me so. How do you know that you'd be able to see dinosaurs or quasars if you were appropriately situated?

Extravagance

SCEPTIC: The Everett interpretation wildly violates Ockham's razor.

AUTHOR: Well (as I noted in Chapter 1) Ockham's razor only requires us not to multiply entities *beyond necessity*. Necessary entities are fine.

SCEPTIC: OK, but isn't the extraordinary ontological extravagance of the Everett interpretation a good reason to look very hard indeed for an alternative?

AUTHOR: I don't especially see why. Generally in physics, we try to keep our number of postulates, and the complexity of our theories, as low as possible. But we're not usually that bothered about *how much* there is in the Universe of any given entity we postulate. For instance, we don't tend to assume that cosmological theories are a priori more or less likely to be true according to how many galaxies they postulate.

SCEPTIC: The cases aren't analogous.

AUTHOR: Funny you should say that. One of Galileo's critics objected to his theory on the grounds that it required 'vast useless voids'.[4]

SCEPTIC: It's a nice talking point, but there are still lots of differences between the two cases. For instance—

AUTHOR: —of course you can find differences, ways in which Everett's theory differs from any previous example of posited ontology. Given any two distinct theories, of course there are going to be things that make them distinct. But the bottom line remains: If you say, 'Ockham's razor rules out the Everett interpretation', you're not really making just one more application of a tried and tested philosophical principle. You're appealing to a new principle—that we should reject theories according to which the Universe exceeds some threshold size—and that principle doesn't have any independent motivation.

Uniqueness

SCEPTIC: Why do you keep on insisting that this is just 'quantum mechanics, taken literally'? The Everett interpretation is just one interpretation among many, just one more theory built around the same framework.

AUTHOR: Start by separating out other realist solutions to the measurement problem—the dynamical collapse theories, the hidden-variables and modal 'interpretations', the transactional approach. All those solutions are different from quantum mechanics on the formal level, not

---

[4] Giovanni Agucchi, quoted in Drake (1978); the analogy to the Everett interpretation was made previously by Barrow and Tipler (1986: 496). See also the Epilogue to this book.

just interpretationally, and their advocates are completely relaxed about acknowledging that.

SCEPTIC: Granted.

AUTHOR: Now set aside quantum-logic theories, the Copenhagen interpretation, operationalism, and the like. None of these count as 'taking quantum mechanics literally' in the sense of accepting that the mathematical structure of the quantum state directly represents the structure of physical reality, just as in classical physics.

SCEPTIC: Again, granted.

AUTHOR: That just leaves the Everett interpretation.

SCEPTIC: Not so fast! For a start, there are so many varieties *of* the Everett interpretation. Deutsch's 1985 many-worlds theory,[5] all the many-minds theories[6] ...

AUTHOR: There's a certain problem of language, to be sure. There are approaches to quantum mechanics that do call themselves 'Everettian' and which I agree are distinct from the approach in this book, and the ones you list are among them. But they all add something to the formal theory. Sometimes the thing that's added is explicit extra formalism, as with Deutsch's old theory. Sometimes it's less explicit: in the many-minds theories, it's usually added in the form of some nonfunctionalist theory of mind, and not formulated in mathematical terms. But it's no less added for all that: the fundamental principles of the world have to include, along with the laws of physics, the 'mental laws' that govern the supervenience of mental on physical: it's not something that's derivable from, or emergent within, quantum physics. (I think that kind of approach to the mind is absurd, for wholly unoriginal reasons:[7] but, absurd or not, it adds something to the theory, and it's that something that does the work of solving the measurement problem.)

What I'm calling 'Everettian quantum mechanics', though, doesn't add anything to the formalism. It's just unitary quantum mechanics, on its own. As it happens, I think that's pretty close both to what Everett

---

[5] Deutsch (1985a).

[6] See esp. Albert and Loewer (1988), Lockwood (1989; 1996a; 1996b), and Donald (1990; 1992; 2002).

[7] See e.g. Dennett (2005).

himself meant and to what most contemporary physicists who talk about 'the Everett interpretation' mean—but that's secondary.

SCEPTIC: I'm not actually prepared to grant that it makes sense for your theory just to be 'unitary quantum mechanics, on its own'. Textbook quantum mechanics, with its 'projection postulate' and 'measurement probability rule', is far too vaguely formulated to be treated as a physical theory without some sort of supplement. Any 'interpretation' of quantum physics has to sharpen that theory into something well defined, and the Everett interpretation is no exception. It's a nice talking point that the Everett interpretation is a pure interpretation, but it's just a talking point.[8]

AUTHOR: Actually, any given quantum theory—N-particle nonrelativistic mechanics, say, or the theory of the helium atom, or qubit models of computation, or QED (with some short-distance cutoff)—is as well-defined, mathematically, as any given classical theory. It's got a well-defined state space and a well-defined dynamics on that state space.

Notice that I'm not counting 'the projection postulate' as part of the formalism. I think it's better understood as an ad hoc rule by which (absent a more principled treatment, which Everett provides) we extract empirical content *from* the formalism. It isn't even any more a particularly good reflection of practice in physics: any serious treatment of the quantum physics of measurement uses something much more sophisticated.

SCEPTIC: Even granted that quantum theory is *mathematically* well formulated, that doesn't tell us that it's *physically* coherent without extra assumptions. The theory tells us about the evolution of a complex function on some absurdly high-dimensional space, whereas what we want is an appropriate account of what's going on in physical space. The theory is silent on that—unless you resort to sleight of hand and start calling that absurdly high-dimensional space 'configuration space', despite the lack of anything that could *have* configurations.[9]

AUTHOR: Basically, my answer is going to be: it makes contact with 'what's going on in physical space' because at the emergent level, that complex function instantiates a vast number of dynamically autonomous

---

[8] I'm borrowing this phrase from an anonymous referee report on Wallace and Timpson (2010).

[9] Here the sceptic is channelling Maudlin (2010).

systems functionally isomorphic to an approximately classical world on spacetime. But there's more to say on this—and actually, thinking about the underlying quantum reality as a wavefunction is quite misleading, and obscures the connections between spacetime and quantum physics. I'll come back to this topic in Chapter 8.

Absurdity

SCEPTIC: The Everett interpretation is just one more sceptical hypothesis. John Norton says it very well:

> Many worlds theorists in quantum mechanics protect the possibility of some superpositions of systems at the macroscopic level by asserting that the most basic fact of laboratory experience that experiments have unique outcomes is an illusion. . . .
>
> [According to the Everett interpretation, t]he cat evolves into a superposition of "dead" and "alive". That we do not see this macroscopic superposition–the cat is just dead, say, when we check–would seem to put an end to the supposition of macroscopic linearity. . . .
>
> The desperate stratagem intervenes. We are told that there is another live cat we cannot see, so that the definiteness of its death is an illusion. (Norton 2010)

AUTHOR: The cat doesn't evolve into a superposition of 'dead and alive'. The world evolves into a superposition of a dead cat and an alive cat. There's nothing illusory or indefinite about the dead cat's death: he's dead, Jim (or John). It's just that there's another cat, elsewhere in physical reality, that's alive. I can't rule out that hypothesis by checking the dead cat, any more than I can rule out, by the same means, the hypothesis that the neighbour's cat is alive.

SCEPTIC: Look: the eigenstate–eigenvalue link tells us that the cat isn't definitely alive or dead.[10]

AUTHOR: Possibly, but the eigenstate–eigenvalue link isn't part of the Everett interpretation. It's not really part of quantum theory at all, actually: it's a philosopher's invention.[11]

---

[10] Here the sceptic is no longer following Norton.

[11] For what it's worth, 'eigenstate–eigenvalue link' turns up exactly once in *Physical Review*'s archives ('decoherence' turns up 8,650 times).

SCEPTIC: Okay, forget the cat *after* the measurement. According to the Everett interpretation, it's an illusion that the *process* of measurement gives only one outcome.

AUTHOR: It's actually a bit delicate to work out whether that's true or false in the Everett interpretation, as it rather depends on issues of personal identity. (See Chapter 7.) But we can bypass that to some extent and just notice that nothing in our laboratory experience gives us any evidence that the process of measurement gives only one outcome. It's an additional posit, one based on the false assumption that our world isn't a branching world.

SCEPTIC: Leaving that aside, the basic problem remains: when it comes right down to it, I just find the Everett interpretation unbelievable.

AUTHOR: That's an interesting psychological observation.

SCEPTIC: What I mean is, it's impossible to take it seriously as a scientific theory.

AUTHOR: Well, empirically that's false: plenty of people (me for a start) seem to manage. But I think what you're saying is: it's so weird that we *shouldn't* take it seriously.

SCEPTIC: Basically, yes.

AUTHOR: In that case, I'd like to hear an argument as to why weirdness is a criterion in rejecting scientific theories.

SCEPTIC: But it's just intuitively obvious that the world isn't like this!

AUTHOR: Now I'd like to hear an argument as to why intuitive obviousness is a good guide to the truth in fundamental physics.

...Look, physics is weird. Science in general is weird. Black holes, the relativity of simultaneity, Darwinian explanations of complexity, Keynes' paradox of thrift—none of them are exactly intuitively obvious.

SCEPTIC: It's not comparable. The Everett interpretation is weird on a much deeper level.

AUTHOR: I'm inclined to put that down to unfamiliarity. But for the sake of argument, let's suppose you're right. That still doesn't amount to any kind of argument.

Let me put it this way. The mainstream position in philosophy of science, these days, is some kind of realism: to understand what scientific theories are telling us about the world, we basically should take them literally and believe what they say. And there have been lots of principled attempts—from positivism to constructive empiricism—to take a different view of science, to say that scientific theories *shouldn't* be taken literally, or that we *shouldn't* believe their claims about unobservable objects.

But to make your position work, you'd have to defend an entirely new philosophy of science, one that no one has taken seriously thus far—a philosophy of science according to which we should believe our best theories *until they cross some kind of weirdness threshold*, and then stop believing them. Now, if you can argue—and I mean *argue*, not thump the table or appeal to intuitions—for that kind of philosophy of science, more power to you. But I doubt it can be done.

PART II

Probability in a Branching Universe

4

The Probability Puzzle

Philosopher's Syndrome: mistaking a failure of imagination for an insight into necessity.

Daniel Dennett[1]

4.1 Problem? What problem?

Probability is by general consensus the most serious difficulty in making sense of the Everett interpretation. Yet it is difficult and controversial to say what exactly the problem is, or even why there is a problem at all. The most common argument that probability doesn't make sense in the Everett interpretation goes something like this:

In physics, probability basically enters in one of two ways. Some-times—as in classical statistical mechanics—it represents our ignorance of the microstate of the system. If we knew that microstate, and had sufficient computing power, we could eliminate the probabilities; in practice, they are indispensable. Other times, probability represents the fact that the system is not deterministic. Even if we did know its exact microstate at one instant of time, that wouldn't be sufficient to determine its microstate at later times. (And in non-Everettian solu-tions to the measurement problem, we see both of these: in hidden variable theories, the probabilities represent our lack of knowledge; in dynamical-collapse theories, they represent indeterministic ('stochas-tic') microdynamics.[2])

---

[1] Dennett (1991a: 401).

[2] Actually, this is a bit too simplistic. There is no reason why hidden-variable theories should not also have indeterministic dynamics (Bub 1997 gives many examples of this; the QFT models of Dürr et al. 2004; 2005 are also indeterministic). And conversely, it is possible

Neither of these applies to the Everett interpretation. The Schrö-dinger equation is deterministic, and (although of course knowing the *exact* microstate is unrealistic), normally we know enough about the microstate to calculate what the branching structure will be, including what weights should be attached to each branch.

What is more, in both the ignorance case and the stochastic case the probabilities can be taken as labelling *alternative possibilities*. We can't understand the weights in the Everett interpretation that way: all the branches are actually there!

So whatever those 'weights' may be, they are not probabilities. And since our empirical evidence for quantum mechanics is entirely com-posed of its probabilistic predictions, that makes the Everett interpreta-tion empirically inadequate.

And the most straightforward dismissal of the problem goes like this:

Never mind philosophical quibbles about the *nature* of probability. Formally, mathematically, how does it enter our theories? Well, any such theory, stripped down to its mathematical core, consists of a space of instantaneous configurations, together with a rule saying which paths through that space—which histories of the system—are dynamically possible (satisfy the laws of physics, if you like). In a deterministic theory, that rule is all-or-nothing: some paths are allowed, some are not, and for each initial segment of a path (i.e. for each path specified up to some time *t*) there is at most one path which has that segment as its initial segment.

Stochastic theories relax that rule: for a given initial segment, the theory places a probability measure over all histories which have that segment as their initial segment. Again, never mind what that measure is conceptually: mathematically, it's just a function from sets of histories to positive real numbers, additive over disjoint sets of histories, and such that the set of all histories with the given segment as initial segment gets probability one.[3]

---

to explore models of dynamical collapse where the collapse occurs through coupling to some source of statistical-mechanical noise, which in principle could be deterministic.

[3] The reader familiar with measure theory will recognize that I gloss certain technical details here—details complicated further by the fact that the space of histories is in general infinite-dimensional. The details are inessential for my purposes, however.

Now, at the *fundamental* level, the Everett interpretation is deterministic, not stochastic. The configuration space is Hilbert space, and the Schrödinger equation picks out exactly one dynamically allowed trajectory through every point in the space. But we have already seen that there is an *emergent* branching structure realised by the underlying unitary dynamics. In that emergent theory, the configuration space can be taken to be the space of instantaneous decoherence-selected projectors discussed in Chapter 3. The emergent dynamics assign a weight to each history, and thus a relative weight to each history relative to each of its initial segments. And because of decoherence, these weights obey the axioms of a probability measure (at least within the degree of precision at which the emergent description itself is valid).

So, mathematically, formally, the branching structure of the Everett interpretation *is* a stochastic dynamical theory. And nothing else need be said.

For the advocate of the second argument, there is no probability 'problem'. Decoherence provides the link between our precisely formulated, deterministic, microdynamics and our emergent stochastic macrodynamics. The microscopic theory is precisely formulated, extends cleanly to relativistic dynamics, and can be understood as a literal description of the world; the macroscopic theory shows how this literal description reproduces the empirical data; anything else is useless metaphysics.

I have a good deal of sympathy with this position; still, it must be admitted that prima facie the advocate of the first argument has a point, and deserves a reply. But what the second argument shows us is that the reply will have to be of a conceptual, even a philosophical, nature. Given decoherence, there are no *technical* problems with probability in the Everett interpretation.

Furthermore, once we recognize that the problems with probability are conceptual rather than formal, we need to be alert to the possibility of a double standard. For ordinary, non-Everettian probability is also highly controversial from a philosophical standpoint, for all that it is mathematically well-behaved. Familiarity is said to breed contempt, but it can also breed complacency, and in fact we will see that many issues that seem to be problems specifically with the Everettian concept of probability will turn out to be problems with probability itself.

However, this does create difficulties for an author. To give a really thorough account of probability in the Everett interpretation requires a fairly extensive discussion of the foundational and conceptual problems of probability and of statistical inference, independently of quantum mechanics. But although some may need an account this thorough, others may be misled into seeing probability *in the Everett interpretation* as so complex and knotty as to need tens of pages of discussion, when in fact the complexity and the knots are within the general theory of probability and are if anything simpler in the particular context of the Everett interpretation.

To address this issue, I have tried in Part II of this book to provide several accelerated routes, for readers with various kinds of concern about probability which fall short of a desire to be fully immersed in the philosophy of probability. Readers who simply agree with the second viewpoint espoused above may wish to skip the rest of Part II entirely; readers tempted by that viewpoint, but concerned either about the lack of 'uncertainty' and 'possibility' in the Everett interpretation or about the idea that each branch ought to have equal probability (probably the two most commonly raised objections) should read at least sections 4.2 and 4.3 of this chapter, in which these objections are addressed.

For those with deeper concerns, in the current chapter I survey the main approaches that have been taken to making sense of probability, and argue that in fact, they work just as well—and in some cases better—if the Everett interpretation is true. Readers with limited interest in the conceptual foundations of probability *per se* but with residual doubts about Everettian probability in particular will, I hope, feel reassured by the chapter, and may wish to skip the remaining chapters of Part II.

In those remaining chapters, I develop a specific theory of probability in the Everett interpretation, providing a good deal more detail than is possible in this chapter, and indeed a good deal more detail than is usually provided for any theory of classical probability. Indeed, one of my main themes in these chapters is to show that far from being a special problem for the Everett interpretation, probability can actually be put on a far firmer footing if Everettian quantum mechanics is true. I have two audiences in mind for that part of the book, then: those whose concern about Everettian probability is so great that it can only be assuaged by a detailed theory, and those whose interest is the philosophy of probability itself.

So much for the general structure of Part II. In this chapter I will consider (in section 4.4) just how probability is actually used (especially in science). I will then consider two general strategies for understanding probability: the strategy of deriving probabilities from frequencies, and the strategy of deriving probabilities from rational constraints on how we quantify our uncertainty and our preferences between actions. (A third 'strategy', that of taking probability as a primitive term about which nothing useful can be said beyond the axioms of the probability calculus, has already been considered: as we have seen, the Everettian is just fine with that strategy.)

I will argue that the Everettian is just as well off as the non-Everettian as far as the frequency-based strategy is concerned, but that in fact there are severe problems (both conceptual and technical) in getting the strategy to work. The other strategy, too, has severe problems, but I will argue that they are in fact *less* severe given the Everett interpretation. I will conclude that part of the discussion, and the chapter, by sketching a derivation of the Born probability rule within the second strategy; much of the remainder of Part II consists of a careful filling out of that sketch.

But before any of this work begins, I wish to confront directly what are perhaps the two most common detailed objections to Everettian accounts of possibility. They are in a sense complementary to one another, beginning as they do from very different standpoints: the first doubts the very coherence of probability; the second accepts its coherence but objects to its actual values being what they are.[4] In the next two sections, I expound and criticize both.

4.2 Against intuition-based objections

The first objection was briefly mentioned in the previous section, and can be phrased rhetorically thus:

> The very notion of probability requires (Alternative possibilities only one of which is actualized/Genuine uncertainty as to what the outcome is) [delete as applicable]. This cannot be found in the Everett interpretation, which is deterministic and in which all outcomes of measurement

---

[4] Amusingly, this does not seem to prevent them being espoused simultaneously by the same person.

processes occur. So there can be no probabilities in Everettian quantum mechanics.

The response, too, can be put rhetorically, and more succinctly:

What is your evidence for believing that probability requires this?

It is, to be sure, extremely *intuitively plausible* that probability requires these things. But it is hard to see why this matters. Intuition, absent supporting arguments, is no guide to truth, as twentieth-century physics has made clear. (And if there are supporting arguments, the intuition itself is no longer required.)

But perhaps what is intended here is an appeal not to intuitions but to language. And indeed, we *could* choose to use 'probability' to refer only to something which parametrises our uncertainty, or which requires that only one of the outcomes really occurs, or any such constraint, and then perhaps there will indeed be no probabilities in the Everett interpretation. But we are equally free to use 'probability' to refer only to something which only occurs in daylight, in which case there will be no probabilities at night. We can use words however we like, ultimately.

On the other hand, when we talk about 'probability' in science it does seem as if the term really does pick out some important, non-arbitrary feature of the world. But what evidence do we have that this non-arbitrary concept really does require alternative possibilities, or that only one outcome occurs, or whatever other Everett-incompatible restriction is under consideration? If this argument is to have any weight, it must rest on some aspect of the restriction which actually does explanatory work in science.

Notice that it will not do to say that we have lots of evidence that probability works in non-branching contexts but no evidence that it does in branching ones. If the Everett interpretation is true, it has been true forever, and much of our evidence actually pertains to branching situations.

Now, if we had a successful reductive analysis of probability (i.e. a successful definition, in terms of other previously understood concepts, of something with the formal properties required by probability) and that analysis required that we lived in a non-branching Universe, this would count against the Everett interpretation. But as we will see, if anything the reverse turns out to be true.

(Incidentally, none of this is to concede that we cannot make sense of uncertainty, or of alternative possibilities, in the Everett interpretation. In fact, we can make sense of them just fine, as Chapter 7 will argue. But we do not have to make sense of them in order to make sense of Everettian probability.[5])

4.3 Against naive branch counting

The second objection I want to consider is of a very different character, and might be put like this:

> The problem is not so much that probability does not make sense in the Everett interpretation: it is that the numbers come out wrong. There is an extremely natural, obvious probability rule to use: all the branches are equally real, so just give all equal probability. If you're about to carry out an experiment with N outcomes, each one should have probability $1/N$ regardless of the weights of the branches.
>
> Of course that rule is wildly at variance with empirical data. The only way to restore empirical accuracy is to weight the probability of each branch with its mod-squared amplitude. But there is no justification for doing so: the weight of a branch is empirically inaccessible, after all. The only justified rule is the equal-probability rule, and if that rule is correct, the Everett interpretation is ruled out by the data.

This is a very old objection to the Everett interpretation: the first time I am aware of it in print is in Graham (1973). But it is constantly reinvented in subtly different forms. Sometimes it is given an anthropic gloss ('There are N copies of me, and I could be any one of them'), sometimes a decision-theoretic gloss ('If I make this measurement, I will branch into N versions of myself: they are all equally important to me, so I should give them all probability $1/N$').

Fortunately for the Everettian, it doesn't work. In fact, there are two independent problems with this rule (which we might call the branch-counting rule).

The first is purely logical, and would apply in any universe where branching is ubiquitous: namely, the rule is inconsistent over time.

---

[5] In the interests of scholarship, I should note that this marks a departure from my position in Wallace (2006a).

Consider, for instance, this simple example. At time t_0, there is only one branch. At time t_1 it splits into two branches, which we call A and B, and my future self in branch A gets given a fine red hat. At time t_2, A splits into two branches $A1$ and $A2$ (and the version of me with the hat likewise splits); B remains unsplit.

What probability should I give, at time t_0, to getting the hat? Well, at t_1 there will be two branches, in one of which I get the hat; so the branch-counting rule says that the probability of getting it is $1/2$. At t_2 there will be three branches, in two of which I get the hat; so the branch-counting rule says that the probability of getting it is $2/3$. Which is right?

Well, couldn't the probability change over time? Not without breaking the standard rules of the probability calculus. For those rules say that the probability Pr(Hat at t_2) at t_0 of me getting the hat at t_2 ought to be

$$\text{Pr(Hat at } t_2 | \text{Hat at } t_1) \times \text{Pr(Hat at } t_1)$$

$$+ \text{Pr(Hat at } t_2 | \text{No hat at } t_1) \times \text{Pr(No hat at } t_1). \tag{4.1}$$

But having got the hat, I never lose it: my future selves at t_1 will keep the hat iff they already have it. So this formula says that Pr(Hat at t_2) $= 1/2$, not $2/3$.

Now, it is open to a sceptic to worry about whether rules like this are actually appropriate in the Everett interpretation. Maybe Everettian probabilities ought to be updated differently from non-Everettian ones. I think this worry is misplaced, actually (I will have much to say about updating and its rationale later in Part II), but it is difficult to reply to except in the context of some particular theory of probabilities.

Fortunately, the second problem with the branch-counting rule is not at all dependent on the details of our theory of probability. This problem is simply that *there is actually no such thing as the number of branches*. I argued for this in detail in section 3.11, but in brief: the branching structure is given by decoherence, and decoherence does not deliver a structure with a well-defined notion of branch count. Very small changes in how the decoherence basis is defined, or the fineness of grain that is chosen for that basis, will lead to wild swings in the branch count. Insofar as a particular mathematical formalism for decoherence does deliver something that looks like a branch count (and many do not), that something is a mathematical artefact of no physical significance. And of course, if there is no such thing as branch count then there can be no branch count rule.

So much for branch counting. Of course, even to discuss it I have been assuming that we can make sense of probability in the Everett interpretation, and it is to that question that I now return. I begin, as promised, with a discussion of the role probability plays in our conceptual scheme, and particularly in our scientific theories.

4.4 The role of probability

When unsure what something is, it often pays to ask what it does. What do we actually use probability for? What role does it play in scientific practice and in human activity more generally?

A little thought reveals that there are broadly two sorts of thing which we use probability for, two ways in which it interfaces with the rest of our activity in and out of science. (Here, roughly speaking, I follow Papineau 1996, from whom my terminology is drawn.) First, we make inferences from non-probabilistic data to probabilities (the *inferential link*). In particular, we use relative frequency as a guide to probability. If we've rolled a die six thousand times and each side has come up a thousand times (plus or minus a few dozen), we conclude, tentatively, that the die is fair—which is to say that each side has probability 1/6 of coming up. If 30 of the 1,000 subjects who try a given drug develop zebra stripes, we conclude, tentatively, that the probability of that particular side effect is about 3%. Exactly how tentative we are depends (amongst other things) on how large our sample is. In physics, sample sizes are sometimes enormous, of the order of Avogadro's number, and so our confidence is very high that, for instance, the time it takes for half of a sample of a radioisotope to decay is a very good estimate of the time after which each nucleus in the sample has a 50% chance of having decayed. But we are not always so fortunate (in astrophysics, for instance, there is little prospect for experiment, only observation, and the sample size for a given astrophysical phenomenon may be quite small).

Once such inferences are made, we normally treat those probabilities just as we would other measurements. If a given theory makes a certain prediction about the probability of a given atomic decay, we test that theory by measuring the probability, and we make that measurement via relative frequencies. The more samples of the atom that we have available to observe, the more confident we are that the relative frequency is an accurate estimate of the true probability, and so the more problematic it is if that estimate diverges from our theory's predictions.

We also see such inferences indirectly, in estimates of how solidly supported a given theory is by the data. Standard deviations, in effect, are attempts to estimate how probable it is that the real result is indeed what the data suggest: when a deviation from a theory is declared significant at the three sigma level, what is meant is that, if the theory really was true, the probability of the deviation would be about 0.3%. The more sophisticated methods of Bayesian statistics (of which more later) attempt to develop these simple tests into a general analysis of how probable we should regard a theory as being in the light of given data.

Secondly, we use known probabilities as a guide to decision-making (the *decision-theoretic link*). Within our everyday lives, obvious examples of this are gambling and insurance. In the former case, our assessment of the good and bad bets is based on the probability of our winning the bet (weighted by the reward we receive); in the latter, we choose which risks to insure against by estimating how likely the risk is to occur and comparing that likelihood to the loss we would incur if it did. Governments do essentially the same thing when making safety assessments: the standard measure of safety in transport, for instance, is the number of deaths per passenger mile, which is in effect the probability per mile that a given passenger will die. (When we are told that flying is safer than driving, we are not being told that actuarially, fewer deaths per passenger mile will occur in planes than in cars. We are being told that we, personally, are less likely to die if we make a given journey by plane than by car: that is, we are being given probabilistic information.)

Very roughly, there are two kinds of strategies for understanding probability (leaving out the third strategy, which just treats it as primitive and not in need of understanding). The first treats the inferential link between probability and relative frequency as definitional, effectively identifying probability with relative frequency or some close cousin, and treats the decision-theoretic link as derivative. The second, conversely, treats probability as definitionally linked to rational decision-making, and tries to derive its connection with observed relative frequencies.

I will come on to the second strategy in due course, but I begin with the first: frequentism.

4.5 Frequentism

If there is a definitional link between probability and relative frequency, it is easy to see that this link cannot be too simple. I have just tossed a die ten times; I rolled six twice; this does not mean that the probability of a six is 1/5. (I invite any reader who disagrees to join me in a game of chance; I still have the die!) Probability is not relative frequency *simpliciter*.

Of course, no advocate of frequentism has ever believed that it was. The strategy, rather, is to extract the probabilities from the *long-run* relative frequencies: from the relative frequencies obtained after a great many repetitions of the experiment.

There are broadly three ways in which this might be done:

1. Probabilities are to be identified with the limiting relative frequencies as the number of repetitions tends to infinity.
2. Probabilities are to be identified with the actual relative frequencies after an actual infinity of repetitions.
3. Probabilities are to be extracted somehow from a large but finite number of repetitions.

Of these, the first is probably the most familiar. It is, at any rate, often how the textbooks introduce probability: the probability of rolling a six is equal to the limiting relative frequency of sixes as the number of throws approaches infinity.

But it is not at all clear what this even means. The die is not going to be thrown arbitrarily many times; even if it were, it will have abraded away to nothing long before the quintillionth throw. Even if we leave aside such practical scruples, there is a problem of principle. For it is well known that the relative frequency of sixes on an (indestructible!) fair die is not *certain* to tend towards 1/6 as the number of throws tends to infinity. The best that can be proven is that the *probability* of the relative frequency diverging by any given amount from 1/6 tends to zero as the number of throws tends to infinity. (This is one form of the Law of Large Numbers; see any textbook on probability theory for a discussion.)

If we are using relative frequency to *measure* probability, this is reassuring: the more repetitions of the experiment that we perform, the less likely it is that the probabilities are not accurately measured by the relative frequencies. If we are using relative frequency to *define* probability, on the other hand, it is disastrous: if probability *is* limiting relative frequency, what can it possibly mean to say that the long-run relative frequency approaches the probabilities with high probability?

The second frequentist way of defining probability attempts to chart a technical route around this, however (one suggested originally by von Mises (1957); see Howson and Urbach (1993: 320–37) for a clear modern account). Instead of considering arbitrarily long finite sequences, von Mises goes straight to infinite sequences. To claim that the die has probability 1/6 of landing heads is, on von Mises' formulation, to claim that if the die were to be tossed infinitely many times then (i) exactly 1/6 of the tosses would give heads, and (ii) the infinite sequence of tosses would otherwise be random (for the technical sense of 'random' here, see Howson and Urbach 1993).

Both technically and conceptually, though, this may create as many problems as it solves. For one thing, if throwing a die arbitrarily many times is problematic, throwing it an actual infinity of times seems doubly so. For another, while any finite set of die throws has a well-defined frequency of sixes, there is no well-defined notion of what it means to divide one infinite number by another. So what does it mean to speak of the fraction of sixes in an infinite set of throws? It can be understood in a limiting sense, of course, as the limit of the fraction as the number of throws increases without limit (as von Mises himself does), but it is a well-known result in analysis[6] that this limit depends on the order in which the infinite set is arranged.

So neither the strategy of letting the number of repetitions approach infinity, nor the strategy of letting that number actually be infinity, looks terribly promising. At least at first sight, the strategy of considering a large but finite number of repetitions looks equally problematic, though. A non-exhaustive list of its problems:

- it requires probabilities to be rational numbers (despite quantum theory allowing arbitrary real-valued probabilities);
- it raises the question of how many repetitions is sufficient;
- it seems to provide no room for the idea of statistical deviations getting smaller but not dropping exactly to zero;

---

[6] See e.g. Apostol (1974: 187–9).

- it courts paradox: if (say) the required number of repetitions of the die roll is N and we have actually rolled the die $\gg N$ times, there will be a great many N-element subsets of the set of all rolls where the relative frequency of '6' is far from $1/6$.

So any such strategy will have to be a bit more subtle and indirect. One fairly straightforward example might be called 'typicality': one takes the fundamental law to be something like

The observed outcomes of the measurement must be typical given that the probability is p.

In particular, if there are enough repetitions, sequences where the relative frequency is not $\sim p$ will be atypical; hence, the relative frequency ought to be a reasonable guide to the probabilities if the number of repetitions is large.[7] Again, though, such proposals are both technically and conceptually problematic. Technically, one runs up against the fact that *every* sequence of outcomes is 'atypical' in some sense; conceptually, what would be the justification of such a principle? What does 'atypical' mean here, if not just 'improbable'?

A much more sophisticated strategy to extract probability from finite frequencies—the 'best-system analysis'—was proposed by the philosopher David Lewis (1973: 72–7; 1983b; 1994). It aims to offer an analysis of *all* laws of physics—deterministic and probabilistic—and takes as its starting point the idea that the laws have no status beyond the actual physical contents of the Universe: that is, it would not even make sense to suppose that the laws could change without some concrete fact about the Universe changing.[8] Rather, what makes a law of nature correct is that it is that way of describing the actual Universe which best combines simplicity with strength.

For instance, suppose that the universe was as Newton envisaged it. We could describe the material contents of the universe by just saying, for each atom, exactly where that atom is at a given time; such a description would

---

[7] I have most often encountered this suggestion in conversation: it is frequently(!) appealed to as a response to the sorts of worries I have listed with frequentism. For proposals of this kind in the literature, see Dürr, Goldstein, and Zanghi (1992) and Allori et al. (2011).

[8] This kind of approach to laws is called 'Humean' after the philosopher David Hume; it is a moot point whether Hume was actually a Humean in this sense.

be very *strong* (would give a very detailed description of the Universe) but not at all simple. Stating that all the particles obey Newton's laws, on the other hand, is less strong (it underspecifies the initial conditions), but the increase in simplicity is huge, and so in this case (so goes the theory), Newton's laws are a better candidate for being the laws of physics than is the atom-by-atom description.

Now consider probability. If our Universe contains, say, a certain decay process which occurs only in circumstance Q, if circumstance Q has occurred N times, and if on kN of those occurrences the decay in fact took place, it *might* be that there is nothing reasonably simple that can be said about the process *except* that it occurs a fraction k of the time. And so, according to the Best System Analysis, it would be a law that it occurs a fraction k of the time, and no more detailed (in particular, no deterministic) law could be stated for the decay process.

Indeed (goes the theory), it might be (because, for instance, the decay process was some complicated composite reaction derivable from much simpler underlying principles) that the simplest set of laws would give a probability only approximately equal to k. This would be a slight decrease in the strength of the law, to be traded off against a great increase in simplicity. And so we would obtain a theory of probability in which (i) facts about probability depend only on non-probabilistic facts about the world, (ii) only finite sequences of experiments need to be considered, and yet (iii) the frequencies observed in those sequences of experiments need not agree precisely with the probabilities given by the laws of physics. In effect, to say 'each system of type Q has probability k of decaying' is to say, in a highly elliptical way, 'approximately a proportion k of systems of type Q decay, and no simple structure can be discerned beyond this relative frequency'. Whether this actually works at a technical level is controversial, though, and the suggestion that facts about laws supervene only on the physical contents of the world is equally controversial.

The bottom line, then, is that the frequentist program of extracting probabilities from frequencies is, at best, still a work in progress. In fact, though, even if one of the approaches I have discussed—or some other approach—could indeed extract well-defined numerical probabilities from long-term frequencies, it is far from clear that those 'probabilities' are fit to play the role that probability is supposed to play. I explain why this is in section 4.7; first, though, we should consider what the effect on the frequentist program would be if Everettian quantum theory were correct.

4.6 The Everettian frequentist

Early approaches to understanding Everettian probability were usually frequentist in character, and these approaches have been widely criticized as circular and question-begging. I believe that this criticism is quite correct—but only because it is essentially a restatement of the traditional criticisms of frequentism itself.

Without any pretence at historical completeness, let me discuss three key 'frequentist-style' approaches to Everettian probability. All are based on the same observation: that *if*, after sufficiently many repeats of an experiment, all branches display approximately the correct statistics (i.e. the statistics predicted by 'conventional' quantum mechanics), then experimenters would be able to reliably infer what the mod-squared amplitudes are from those statistics.

The first approach goes back to Everett himself: in his original paper (1957) he proved that if a measurement is repeated arbitrarily often, the combined mod-squared amplitude of all branches on which the relative frequencies are not approximately correct will tend to zero. And of course this is circular: it proves not that mod-squared amplitude equals relative frequency, but only that mod-squared amplitude equals relative frequency with high mod-squared amplitude.

Substitute 'probability' for 'mod-squared amplitude', though, and the circularity should sound familiar; indeed, Everett's theorem (as is well known) is just the Law of Large Numbers transcribed into quantum mechanics. So the circularity in Everett's argument is just the circularity in the simplest form of frequentism, disguised by unfamiliar language. That simplest form of frequentism may indeed be hopeless, but so far Everettian quantum mechanics has neither helped nor hindered it.

The second approach is due to Farhi, Goldstone, and Gutmann (1989): instead of considering arbitrarily many runs of an experiment, consider an actual infinity of runs. This requires some delicate mathematics (infinitely many measurements of even a two-state system requires a non-separable Hilbert space[9]), but it can be done, and (so Farhi et al. claim) the result is that the quantum state is a superposition of states in *each* of which the

---

[9] If a given system has an N-dimensional Hilbert space, a composite system consisting of M copies of that system has an N^M-dimensional Hilbert space. Taking $N = 2$ and $M = \aleph_0$ (countable infinity), we get that the dimension of the composite system is $2^{\aleph_0}$, which is well known to be uncountable.

relative frequency equals the probability predicted by 'orthodox' quantum mechanics.

We can criticize Farhi et al.'s result on conceptual and technical grounds. Conceptually speaking, it raises by now familiar worries about an actual infinity of measurements. And an important technical objection was voiced by Caves and Schack (2005): Farhi et al.'s construction of the Hilbert space corresponding to infinitely many measurements is highly non-unique, and other constructions seem equally possible but would lead to different relative frequencies.

Again, there is nothing new here. The conceptual point is familiar; on the technical side, Caves and Schack are making (in effect) the quantum analogue of my earlier criticism of von Mises' use of actual infinities, and the ambiguity in the definition of relative frequency for an infinite set of measurements is formally analogous to the ambiguity in the construction of the Hilbert space identified by Caves and Schack.[10]

So the limiting-case and actual-infinity strategies for frequentism have direct analogues in Everettian quantum theory: they work as well (or otherwise) there as in the classical case. What of the strategies which attempt to extract probabilities from finitely many repetitions of an experiment?

The typicality strategy has also been proposed in the Everett interpretation, by Geroch (1984). His principle of *preclusion* simply states, as a premise, that it is reasonable to treat certain regions of Hilbert space as 'precluded'—i.e. to assume that we are not in such regions—because they have very small weight, and that these regions include ones with anomalous statistics (recall that we already know that such regions indeed have very small weight). I am not at all sure what positive justification could be given for the preclusion principle, but it does seem formally analogous to—and as (un)justified as—the 'typicality principle' alluded to above. More recently, Allori et al. (2011) explicitly invoke typicality as a route to probability in the Everett interpretation.

But of course, if all analogies were perfect, life would be boring. There is not any direct analogy in the Everett interpretation for Lewis's best-systems analysis. (Of course, we can apply that analysis directly to the underlying quantum dynamics, and presumably it would give us back out the deterministic Schrödinger equation, but this is certainly not what

---

[10] In case it isn't obvious, this is no criticism of Caves and Schack; in any case, they too regard classical frequentism as hopeless.

we're after.) There is, however, another strategy which tries to extract probabilities from finite repetitions of an experiment and which (though it has certain aspects in common with the best-systems approach) is available to the Everettian frequentist but not open to his one-world counterpart.

Recall that Everett proved that the combined weight of the anomalous branches is very small after a large finite number of repetitions of an experiment, and tends to zero as the number of repetitions tends to infinity. But this assumes that the branching structure is given exactly, by some particular choice of basis. As was described in detail in Chapter 3, though, in fact the structure is a consequence of decoherence, the branches are only approximately orthogonal, and indeed the theory only approximately specifies them. This suggests that in fact, at the emergent level appropriate to a branching description there may *be* no deviant low-weight worlds. If their weight is small compared to the level of noise in the decoherence process, they will simply be an artefact caused by false precision, and will not have the robustness that Chapter 2 argued was an essential requirement for emergent structure. As such (we might hope) *all* of the branches will exhibit approximately correct statistics.

An approach along these lines, going by the name of 'mangled worlds', has been developed by Robin Hanson (2003; 2006). In Hanson's terminology, low-weight worlds are 'mangled'—that is, disrupted or destroyed—by interaction with larger-weight worlds.

The proposal is somewhat tricky to make precise sense of. In particular, it is important to remember that the set of worlds with any *given* statistically typical collection of measurement outcomes may not have any higher weight than the set of worlds with some statistically atypical set of outcomes.[11] For instance, suppose that 10^{12} particles are prepared in equally weighted superpositions of spin-up and spin-down along some axis, and then their spin is measured along that axis. There will be $2^{10^{12}}$ different sets of outcomes from this measurement, and the worlds corresponding to each set will have weight $2^{-10^{12}}$. Collectively, the set of all worlds with statistically typical results is vastly larger than its complement, but each individual set of statistically typical measurements still has tiny weight.

Nonetheless, it is an extremely interesting observation that the evolution of the quantum state is extremely close in Hilbert space norm to the

---

[11] I am grateful to Huw Price, who first pointed this out to me.

evolution of a state in which all branches display non-anomalous statistics. It seems to be a live possibility that a more careful analysis of the decoherence process would indeed show that the set of worlds with anomalous statistics is not a genuinely emergent feature of the unitary dynamics. There might, at most, be isolated instantaneous states displaying anomalous statistics; linking them together dynamically might be unjustified by the details of decoherence.

Ultimately, at a technical level the success of Hanson's proposal depends on whether decoherence really delivers the goods, so to speak. The level of noise in the decoherence process is really very small indeed, and to the best of my knowledge neither Hanson nor anyone else has carried out a decoherence-based analysis in a physically realistic situation to show whether the low-weight branch dynamics really is washed away by the residual interactions with other branches. More work is doubtless needed on this interesting approach.

4.7 Frequentism and the short run

So technically speaking, there are Everettian versions of most classical approaches to developing frequentism, and there are interesting research programs on both sides which do not generalize to the other side; in both the nonbranching and the Everettian case it is at best very unclear at present whether any of these approaches can really succeed.

But as I noted above, there are actually other, more serious, conceptual worries for frequentism, which beset Everettian and non-Everettian alike. Any attempt to define probability in terms of the arbitrarily long (or infinite, or merely very long) run falls foul of the banal fact that we don't actually carry out experiments arbitrarily many times: we carry them out a few thousand times, or a few hundred times, or in many cases just once.

The point can be put in a great many closely related ways: for instance,

- Many probabilistic events occur only once; yet we feel happy associating probabilities to them. To take a frivolous example: imagine connecting a planet-destroying bomb to a Geiger counter and putting the counter next to a radioactive source of given mass for a few minutes; would you rather that source was uranium-238 (half-life 4.6 billion years), carbon-14 (half-life 5,700 years), or sodium-24 (half-life 15 hours)? If probability is long-run relative frequency, it's not clear that you should care: *this* experiment certainly isn't going to be carried out a great many times. To take a more serious example: consider the inhomogeneities in the cosmic microwave background mapped by the COBE satellite. Probabilistic methods are routinely used to study them, but it is hard to think of a less repeatable experiment than the Big Bang.

- Even when it does seem to make sense of arbitrarily many repetitions of an experiment, often we only care about the first few. Suppose you are gambling and you are informed (by God, presumably) that a die is fair, which is to say that in the infinite long run it will land six one sixth of the time. At what odds would you bet on a six on the next throw? On fewer than 500 sixes on the next 600 throws? Frequentism appears silent on these questions: the first throw, or the first 600 or 6 million throws, have no effect at all on the infinite long-run relative frequencies, and there are as many sequences (i.e. an infinite number of them) on which the first 600 throws are sixes as there are sequences on which ~100 are.

- By considering arbitrarily long sequences, we seem to have broken the connection between probability and experimental practice that motivated frequentism in the first place. If we repeat an experiment 1,000 times and get a certain result 42% of the time, according to frequentism this tells us nothing at all about the probability of that result. Suppose the probability is actually 99% for the result; this is perfectly compatible with the first 1,000 runs giving the result only 42% of the time.

In the latter two cases, the temptation is to say that it is extremely unlikely that the anomalous sequences occur, and of course it is indeed unlikely—but not if 'likely' means 'probable' and 'probable' means 'high relative frequency in the long run'. Just as in the first case, the lesson seems to be that probability may be measurable by means of frequencies, but it cannot be identified with them.

Once again, however strong or weak these objections, there does not seem to be anything particularly different about them when transposed to the Everett interpretation. In this case as with the more technical objections I discussed previously, taking Everett into account leaves the frequentist program looking about as strong or weak as it did before. As often occurs in discussions of the Everett interpretation, what at first sight look like Everett-specific problems turn out to be general philo-sophical problems thrown into sharp relief by the unfamiliar setting. The reader happy with frequentism, therefore, should have no concerns about Everettian probability. The reader who agrees with me that frequentism is hopeless should read on.

4.8 Personal probability: the rationalist approach

Recall the earlier discussion on the role of probability: we measure proba-bility through relative frequencies; we use probability as a guide to rational action (whether that 'action' is accepting a bet or accepting a theory as confirmed). Frequentism, in *identifying* probability with relative frequency, fails (amongst other deficiencies) to reproduce its role in guiding action; an obvious alternative strategy to try is to regard probability's role in rational action as definitional and try to recover its connection with frequencies.

This strategy is sometimes called a *Bayesian* approach to probability, but that term is used in many subtly different ways and can be misleading. To avoid confusion, I will coin a new term, and call this strategy *rationalist*. The rationalist strategy will be our concern for the remainder of this chap-ter; I will argue that it, too, runs into severe difficulties in giving a complete account of probability in physics, but that the Everett interpretation in fact greatly alleviates its difficulties; in fact, the positive theory of probability which I will present in Chapters 5 and 6 will be rationalist in character.

It is easiest to see how the rationalist approach to probability works in the rather stylized context of betting for small sums. Betting odds give us a context to measure an agent's assessment of probabilities: if someone is willing to bet on the die landing six at six-to-one odds but not at worse odds, we can assume his assessment of the probability of the die landing six is $1/6$.

According to the rationalist approach, the *criterion* for the probability of some event x being P is that a rational agent is willing to bet on it at $1/P$-to-one odds. That is, that the agent would be willing to pay λP dollars—but no more—in exchange for receiving λ dollars in the event that x obtains. To the rationalist, the betting preferences are not indications of what the agent thinks the probabilities are: the betting preferences define what the probabilities are for that agent. It follows immediately—at least at first sight—that this notion of probability is an agent-relative affair: we cannot speak of the probability of x period, only of the probability of x for agent A.

Furthermore, we can argue that these betting-defined probabilities had better satisfy the probability calculus. For instance, suppose that an agent assigns probability x to some event occurring, and y to that event not occurring. By definition, this means that he will pay λx dollars for a bet on the event occurring that has payoff λ dollars, and μy dollars for a bet on the event *not* occurring that has payoff μ dollars. If he accepts *both* bets, then, his return will be

- $\lambda(1 - x) - \mu y$ if the event occurs;
- $-\lambda x + \mu(1 - y)$ if the event does not occur.

Suppose that $y \neq 1 - x$; then we can choose λ and μ such that both of these returns are negative (we might need to let λ and/or μ be negative: paying $-\lambda$ for a bet is effectively the same as selling the bet to someone else for λ). So an agent who does not assign probabilities that add up to one is committed to accepting bets according to which he will certainly lose money.

Arguments of this form are known in the literature as *Dutch book arguments*: a Dutch book is a set of bets such that anyone accepting the bet will always lose money, whatever outcome actually occurs. And it is fairly straightforward to prove[12] that any agent whose personal probabilities do not satisfy the usual axioms of probability theory is vulnerable to a Dutch book.

The Dutch book argument has much room for improvement. (For one thing, talk of literal dollars had better be metaphor or loose talk, unless we want our probabilities to take only finitely many values and to depend on the existence of a monetary economy!) Indeed, a sizeable section of the field of *decision theory* is concerned with exactly how arguments of this kind can be made more powerful and convincing,[13] and I will return to this topic *in extenso* in Chapter 5 and in Appendix B. But they show in outline how constraints on rational action can require that an

[12] Classic versions of the proof can be found in Ramsey (1926) and in de Finetti (1931; 1937); for a modern account, see Talbott (2008).

[13] A particularly good presentation of the weaknesses of the Dutch book argument, and the potential of decision theory to alleviate these weaknesses, is given in Kaplan (1996: ch. 1).

agent's preferences determine a unique probability measure on outcomes of uncertain processes.

That being the case, it seems apposite to ask how well the rationalist strategy works in the case of Everettian quantum mechanics. On the surface, it seems that it cannot work at all: how can it make sense to bet on whether or not some outcome occurs when *all* outcomes occur? When we consider concretely how such a bet would go, however, this turns out to be too simplistic.

Suppose, for instance, that a quantum system is prepared in some superposition of N states in a given basis, and that in five minutes' time it is to be measured in that basis. In the non-Everettian context, there are then N interestingly different possible outcomes (although there are of course vastly more than N microscopically distinct possible results) and we can ask, of any given agent, what probability he puts on each outcome. We have seen that we can operationally make sense of this question by looking at which bets on outcomes the agent is willing to accept, and that if he does not assign probabilities in accordance with the probability calculus then there will be some bet which he is committed to accepting but which will lose him money whatever outcome occurs.

And what is it for an agent to bet on an outcome? It is to pay a certain amount of money, up front, in exchange for a promise that another certain amount will be returned to the agent if a given outcome obtains. It will help (though it is not essential) if we imagine automating the process: if the agent makes a given bet, an automaton is set in motion which waits until the measurement is made, looks at the result of the measurement, and provides the payment if that result is appropriate.

What happens in this situation if the Everett interpretation is true? Instead of there being N interestingly different possibilities, there are N sets of interestingly different branches, one for each distinct outcome of the measurement (although there are of course vastly more than N—indeed, an undefinable number—of branches). But this does not in any way prevent our bet-implementing automaton from functioning. Being a physical system, it too will be split into indefinitely many copies, one in each branch, and the ones in the branches where the outcome is appropriate will pay out. So a bet, in this sense, is just as meaningful in the Everettian as in the non-Everettian context. Its metaphysical status may be different: in buying a bet, an agent is purchasing rewards for some but not all future versions of himself, rather than purchasing the possibility of a reward for his single future self. But operationally, the process is the same.

And this means that the rationalist definition of probability, too, works just as well in the Everettian as in the non-Everettian context. To say that an agent assigns probability P to an outcome is to say that she is committed to paying λP dollars in exchange for her successors receiving λ dollars in all branches in which the outcome occurs. And just as in the non-Everettian case, if the agent's assignations of probabilities to outcomes do not conform to the probability calculus, she will be vulnerable to Dutch books: she will be committed to accepting collections of bets which guarantee that all of her future selves lose money.

Why should she care? Well, an agent's future self *is* her future self just by virtue of the causal, structural, dynamical relations between it and the agent's past self. There is (I assume) no indivisible, immaterial soul which passes through my life and magically makes me a single being: what makes the stages of me at different times all *me* is that they are appropriately related. And it seems, at least, that an Everettian agent's future selves stand in all the same relations to her as a non-Everettian agent's future selves stand to her. It's hard to see why, in the Everett case, I should regard my future self as any the less *me*—why I should not treat his goals and desires, his hopes and dreams, as my own—just because I actually have multiple such selves.

(At this point, readers may worry that a theory according to which, in some sense, 'all possible outcomes occur' has no room for an agent to freely choose between possibilities. I address this worry in Box 4.1.)

Box 4.1. The Everett interpretation and free will

It is common for people encountering the Everett interpretation to worry that it makes nonsense of any idea of rationality and free choice—let alone moral responsibility (Egan 2002, for instance, is an exploration of this worry in fiction). The worry is that since all possible choices I might make will be made on *some* branch, in what sense is my choice really free?

To see that this is mistaken, consider an agent choosing whether or not to bet on some quantum event, and if so, what odds to accept. Each *choice* is a choice of betting strategy: if the agent chooses a given strategy, the consequence of this choice is a certain pattern of outcomes

(continued)

Box 4.1. Continued

across those branches in the agent's future. A given choice, for instance, may lead to those of her successors who are in branches where a quantum measurement gives a spin-up result becoming rich and those successors in other branches becoming poor; a different choice may have the opposite result. In each case, different choices are represented not by different branches but by different distributions of outcomes across all branches.

Now, to be sure: the agent, ultimately, is a physical system, and her actions are determined physically. And if the neurological processes in her brain which underpin her decision-making are quantum-mechanically random (a neurological question about which I know nothing whatever), then there will be some branches in which she makes one choice, some branches in which she makes another (in which case, futurewards of each of those branches will be an entire pattern of branching which represents that choice).

Does this in any way undermine the fact that the choice (or the multiple choices, if the agent's decision-making process is quantum-mechanically random) is freely made? Only if there is a clash between the idea of a choice being *free* and the idea of it being the result of some mechanistic physical process (deterministic or otherwise).

This, of course, is one more case of an already existing philosophical problem just being thrown into sharp relief by the unfamiliar setting. In this case, the problem is the old question of free will, generally described as a clash between free will and determinism but better understood as a clash between free will and *mechanism*. The Everett interpretation, so far as I can see, has nothing to add to that debate; nor is free will any *more* threatened in Everettian quantum mechanics.

For the record, I don't myself think there is any incompatibility between freedom and mechanism; see Dennett (1984; 2003) for arguments to this effect, and Kane (2002) for a variety of other views. The reader who disagrees, however, need not reject the appeals to rationality made elsewhere in this book: nothing is lost by recasting them in terms of the optimal design for a (however non-free) control system designed to make maximally rational decisions when presented with alternatives.

So, the very idea of defining probability in terms of betting preferences seems to work just as well in the Everett interpretation as in non-branching situations.

However, in the Everettian case we want more than this: we do not just want the agent to assign *some* probabilities to the branches, we want those probabilities to be given by the Born probability rule. So far, although I have argued (in section 4.3) why the superficially obvious strategy of giving all branches equal weight cannot be right, I have provided no general argument as to why the Born Rule should be correct. But (again) this is not a problem peculiar to the Everett interpretation, as we will see.

4.9 Objective probability

If probabilities are personal things, reflecting an agent's own preferences and judgements, then it is hard to see how we could be right or wrong about those probabilities, or how they can be measured in the physicist's laboratory. But in scientific contexts at least, both of these seem commonplace. One can erroneously believe a loaded die to be fair (and thus erroneously believe that the probability of it showing '6' is one sixth); one can measure the cross-section of a reaction or the half-life of an isotope.

Indeed, one can make judgements about—even, in principle, make bets on—questions of what the real probabilities are. It seems reasonable enough to say, for instance, that the die is probably fair, or that the decay rate is probably less than such-and-such, and on the surface at least, these are effectively statements about the probabilities *of* probabilities.

Put another way: as well as the *personal probabilities* (sometimes called *subjective probabilities* or *credences*) which I discussed in the previous section, there also appear to be *objective probabilities* (sometimes called *chances*), which do not vary from agent to agent, and which are the things scientists are talking about when they make statements about the probabilities of reactions and the like.[14]

---

[14] A note on terminology: 'chance' and 'credence' have become terms of art in the philosophical literature since their introduction by Lewis (1980). 'Chance' is used by Lewis to denote only those objective probabilities associated with stochastic processes. For Lewis, the chance of a die showing six, assuming dice to be deterministic, is either zero or one. ('To the question of how chance can be reconciled with determinism . . . my answer is: *it can't be done*': ibid. 118; emphasis in original.)

What is the connection between these two notions? There are basically three positions that could be taken:

1. Objective probabilities do not really exist: there are only personal probabilities.
2. Objective and personal probabilities both exist, but are really the same thing: the apparent 'subjectivity' of personal probability is illusory.[15]
3. Objective and personal probabilities both exist, and are distinct.

The first of these positions (sometimes called 'subjectivism') seems committed to the idea that (say) the half-life of uranium-235 is not an objective fact about the Universe but no more than some collective agreement amongst scientists (that there *is* a collective agreement as to the half-life is not in dispute, of course: what is in dispute is whether or not there is something objective about which they are agreeing).

Defenders of subjectivism appeal to a variety of results to the effect that agents with very different initial probability functions who conditionalize on the same data will expect to converge on the same probability function. But this is of limited comfort: conversely, given any finite amount of data, one can find two initial probability functions which will still differ significantly when conditionalized on that data.

More relevantly for my purposes, subjectivism seems committed either to a rejection of realism about science or to a radical revision of our extant science, which certainly seems to be formulated on the assumption that the probabilities—whatever their nature—do have an objective status of some kind. There is nothing illicit about attempting such revisions, or about denying scientific realism (though I am sceptical about the merits of either), but in any case they lie outside the scope of this book, as I noted in Chapter 1.[16]

The second position requires that we close the apparent gap between the two notions of probability, which in turn requires us to reduce the apparent subjectivity of personal probability. For this reason, the position

[15] I chose to refer to 'personal probabilities' rather than the more common 'subjective probabilities' to avoid ruling this possibility out by fiat.

[16] Both de Finetti (1931; 1937) and Savage (1972) endorsed subjectivism. For contemporary advocacy of subjectivism in quantum mechanics, see Caves, Fuchs, and Schack (2002), Fuchs (2002), and Schack (2010).

is sometimes called 'objective Bayesianism'; in physics, it is associated in particular with Edwin Jaynes.[17] Objective Bayesians take the view that there are rational constraints on agents' personal probabilities over and above conforming to the probability calculus—constraints so strong, in fact, that any two rational agents with the same information will have the same personal probability functions.

'Objective' probability for the objective Bayesian, then, is the probability function that a rational agent will adopt if he is in possession of all the physically relevant facts. Indeed, it would be possible to adopt this position and hold that *only* for agents in possession of all the physically relevant facts is it coherent to talk about probability at all (although Jaynes himself would not have held this position).

(Of course, this does raise the question of just what the 'physically relevant facts' really are. In the case of a fair die, for instance, if the die is deterministic then sufficiently good knowledge of the physical state of the die will fix its probability of showing six to one or zero. I postpone this question until section 4.11.)

The third position might be called *Lewisian dualism*: 'Lewisian' for the philosopher David Lewis, one of its strongest proponents,[18] and 'dualism' because of its commitment to two kinds of probability: personal and objective. We have a good understanding (says the Lewisian dualist) of personal probability via Dutch book arguments and the like; it is much less clear that we know what objective probability is. But we know that we use it as a guide to action: that is, if we believe that the objective probability of an event is $\sim q$, we will bet on it at odds of $\sim 1/q : 1$. This suggests a link between personal and objective probability: roughly speaking, if we know what the objective probability of something is, we should set our personal probability of that something equal to the objective probability.

On its own, though, this is too weak to really get at the role objective probability plays: in practice, we are never 100% confident that we really know what the objective probabilities are. A strengthened rule that really does suffice to link them is available, though: Lewis's so-called *Principal Principle* (Lewis 1980), which states, in my terminology, that

---

[17] See e.g. Jaynes (1957a; 1957b; 1968); for examples of the application of Jaynesian methods in physics, see Baierlein (1971), Grandy (1987), and Grandy and Schick (1990) and references therein.

[18] See esp. Lewis (1980).

For any real number x, a rational agent's personal probability of an event E conditional on the objective probability of E being x, and on any other background information, is also x.

(There is an additional restriction—that the background information is not *inadmissible*—but I postpone discussion of this for now.)

It is easy to see that this principle does indeed capture the way we use objective probability in decision-making. Suppose, for instance, that I am 90% sure that a die is fair, and 10% sure that it is loaded so as always to show six. Then the Principal Principle states that my personal probability in it showing six *conditional on it being fair* is $1/6$, and my personal probability in it showing six *conditional on it being loaded* is 1. By the normal rules of conditional probability, the intuitively correct answer follows: my personal probability in the die landing six is $(0.9 \times 1/6 + 0.1 \times 1 =) 0.25$.

Slightly less obviously, the Principal Principle also captures the idea that we use frequencies to measure probabilities. For instance, suppose that some random process is to be carried out N times, and that I am totally sure that the objective probability of the process giving some particular result is the same on each repetition.

Now, in the language of probability theory, suppose that C denotes my current beliefs prior to the experiment being performed and that Pr is my personal probability function (so that my personal probability in a hypothesis H at the moment is $\Pr(H|C)$). In particular, let X_p be the hypothesis that the objective probability on each repetition of the process giving the particular result is p, and let Y_i be the proposition that the ith repetition of the process gives the result. The Principal Principle, then, says that $\Pr(Y_i|X_pC) = p$.

Now let K_M denote the proposition that the experiment reproduces the particular result on M runs out of N. Simple combinatorics tells us that

$$\Pr(K_M|X_pC) = {}^NC_M p^M (1-p)^{N-M}. \tag{4.2}$$

For large N, we can approximate this as

$$\Pr(K_M|X_pC) \simeq \frac{1}{\sqrt{2\pi Np(1-p)}} \exp - \left[\frac{N}{2p(1-p)} (M/N - p)^2 \right] \tag{4.3}$$

(this is the well-known normal approximation to the binomial distribution), which will be very small for large N unless $M \simeq Np$.

By the normal updating rule for probabilities,

$$\Pr(X_p|K_M C) = \frac{\Pr(K_M|X_p C)\Pr(X_p|C)}{\Pr(K_M|C)}, \tag{4.4}$$

so as N gets large, the personal probability in X_p given K_M will become very small unless $p \sim M/N$. That is, as more and more experiments are carried out, any agent conforming to the Principal Principle will become more and more confident that the objective probability is close to the observed relative frequency.

So in fact, the Principal Principle together with the rationalist concept of probability allows us to make precise sense of section 4.4's inferential and decision-theoretic links between objective probability and the rest of our concepts.

Indeed, in a sense we can take this to be a definition of objective probability. That is, if some physical theory T enables us to define some magnitude Q for events, then Q is objective probability just if anyone believing T is rationally required to constrain his personal probabilities to equal Q. More formally, Q is objective probability iff for any event E, if T together with (admissible) background information B entails that $Q(E) = p$, then $\Pr(E|BT) = p$ (where again Pr is any experimenter's personal probability function).

In other words, the Principal Principle can be used to provide what philosophers call a *functional* definition of objective probability: it defines objective probability to be *whatever thing* fits the 'objective probability' slot in the Principal Principle. In the same way, I could define the queen (or king) of England as *whatever person* has been crowned at Westminster, is acknowledged as head of state by the government, signs Acts of Parliament, etc.; equally, I could define electric charge as *whatever property* leads to the generation of certain electric fields, accelerates in a certain way the presence of such fields, etc.[19]

Notice that this actually makes the gap between Lewisian dualism and objective Bayesian (our third and second positions) narrower than it may have seemed. In both cases, the objective probability of an event is p iff an agent who knows all the relevant information rationally judges its

---

[19] Philosophers of language will recognize that there are actually two somewhat different entities here: (i) definite descriptions, and (ii) natural kind-terms where the natural kind is identified by virtue of a definite description.

probability to be *p*; the two positions differ only on the correct way to analyse agents who lack some of the relevant information.

The next question to ask, then, is: 'what could make it true that in a given situation, agents knowing all the relevant information are rationally compelled to set the probability equal to *p*?' In other words: what could objective probability be?

4.10 The nature of objective probability

It is tempting to say that this is a non-question: that once we have our *functional* definition in hand, no more needs be said about what objective probability is. After all, asked to *define* charge—or length, or time, or mass—ultimately there is little that we can say except to explain how those concepts are used and measured, how they are represented mathematically in our theories, and how they interact with other entities in our theories. That is, there is nothing more to do other than to give a functional definition of them. Is probability any different? If not, it is no more mysterious than charge or length.

This kind of position is often called *realism* about probability ('realism', because probabilities are real features of the world, perfectly objective and irreducible to anything else), and sometimes called a 'propensity interpretation', a term coined by Karl Popper (1959).[20] I prefer to call it *primitivism* (following Wallace 2006a), to denote the fact that probability is a primitive (i.e. undefinable) term in much the same way as length or charge is.

Is it reasonable? Possibly; but there are some severe difficulties, of which I will focus on two (they are related): first, objective probability seems to enter into our theories in a fundamentally different way from other physical magnitudes; secondly, the functional definition itself has a fundamentally different character from other such definitions.

The first (and probably the lesser) difficulty can be seen by looking at the mathematical formulation of stochastic theories. Charges, masses, lengths and the like seem to be properties of things in the actual universe;

---

[20] It should be noted that the characteristic of realist approaches such as Popper's (and see Mellor 1971 or Maudlin 2007: 50–77 for other examples) is that they treat probability as irreducible to anything non probabilistic, not that they treat it as identified functionally. (I maintain that they *should* so treat it!)

mathematically, they are represented by features of the mathematical object (a set of functions on a differentiable manifold, say) that we use to model the actual universe. Probability, on the other hand, does not appear to be a property of the actual universe at all: whatever the probability was that the coin landed heads at time t, the plain fact of the matter is that it either did or did not. Probabilities appear to be properties of the set of all physically possible worlds; mathematically, they are represented not as a feature of the mathematical object representing the actual world, but as a measure over the set of all such objects.[21] As such, it is rather difficult to see what role that measure is really playing in the theory. It could, for instance, be changed without any effect on the actual world, which makes it difficult to see why it should be linked to observed frequencies and used to guide our actions.

(This argument may be in danger of proving too much, however. For just as the probabilities seem to 'float free' of the actual world, so do the dynamical laws—deterministic or stochastic. A change in the laws of physics which left the actual world as dynamically possible would be just as unobservable as a change in the probability function. 'Humeans'—philosophers who regard laws as having no status beyond the actual physical content of the Universe—are likely to have serious problems with primitivism (cf. Lewis 1994), but it is less clear that these worries persist for non-Humeans.)

The second, and to my mind more severe, problem for primitivism can be seen from the the Principal Principle itself. Just as taking charge as a primitive term requires us to accept Maxwell's equations (or their quantum generalizations) as fundamental principles of physics, so taking (objective) probability as a primitive term requires us to accept the Principal Principle as a fundamental, irreducible principle. But Maxwell's equations are dynamical laws, so to accept them as primitive is to accept that there is no further explanation to give as to why certain sequences of events are dynamically not allowed. The Principal Principle, on the other hand, is ultimately a constraint on *rational action*: to say that the Principal Principle is true is to say something about what patterns of preferences are to be regarded as rational, and to say that it is primitive is to say that there is no further explanation to give as to why doing certain things would be

---

[21] For further exploration of this point, see van Fraassen (1980: ch. 6).

irrational. It has seemed to many (me included) that there is something bizarre, even unacceptable, about the idea of a *principle of rationality* which has no deeper analysis, but which just has to be taken as an empirically determined truth.

What alternative is there? The alternative is to regard the functional definition of objective probability as more like the functional definition of the queen of England (or of money, or of intelligence) than the functional definition of charge, mass, or length. That is, to find some alternative characterization of objective probability, independent of the Principal Principle, and then prove that the Principal Principle is true for that alternatively characterized notion.

The obvious question, then, is: what could that 'alternative characterization' actually be? In fact, we have already seen one such program: frequentism. Frequentists do not postulate probability as primitive: they attempt (however unsuccessfully) to derive probability from the long-run relative frequencies in the actual world. (Notice that if this were doable, it would obviate at least the first objection to primitivism: facts about probability would supervene on facts about the actual world, not all possible worlds.)

From the rationalist perspective, then, what the frequentist must do is (i) construct, from the long-run frequencies in the actual world, a probability measure, and (ii) argue that if a rational agent knew that probability measure, they would be rationally compelled to use it to determine their short-run betting preferences. To a first approximation, indeed, we can merge these two steps into one: the frequentist must argue, approximately speaking, that if a rational agent knew the long-run relative frequency of X then they would be rationally constrained to set their personal probability in X equal to that long-run relative frequency. (See Howson and Urbach 1993 for an example of this kind of rationalist frequentism.)

We have seen, however, that both (i) and (ii) cause severe problems for the frequentist. As for (i), deriving probabilities from the long-run data is at best a research program, at worst an impossibility (as argued in section 4.5). As for (ii), I argued in section 4.7 that frequentism has so far proven quite unable to explain why any measure derived from the results of indefinitely many repetitions of an experiment should constrain our activities in the short run. So frequentism does not at present appear to offer much of an improvement on primitivism.

4.11 Objective probability from symmetry?

If long-run frequencies fail to provide a physical basis for objective probability, still that does not prove that no such basis can be found. There
is in fact an alternative tradition, dating back to Laplace's original work
on probability in gambling, which attempts to provide such a basis from a
very different starting point.

A good way to understand this tradition is to return to the 'objective
Bayesian' strategy discussed in section 4.9. Objective Bayesians maintain
that two agents in possession of the same information must assign the same
personal probability distribution, and this in turn requires constraints on
that personal probability distribution over and above the axioms of the
probability calculus.

The most common proposed constraint is the so-called 'principle of
indifference':[22] an agent's probability distribution should not differentiate
between possibilities that the agent has no reason to differentiate between.
Applying this to a die, for instance, is supposed to give the result that the
agent has no reason to distinguish one side from another, so each must be
given equal probability.

In its simplest form, this principle is fairly obviously hopeless. For one
thing, having 'no reason to differentiate' is ill-defined. It isn't that the faces
of the die are indistinguishable (the numbers on the faces distinguish them
perfectly well); it's just that we don't think the numbers affect the . . . what?
The probabilities? That would be question-begging.

For another thing, a classical die does not have six possible states; it
has continuum infinity possible states. We need some justification for
regarding the infinite set of states corresponding to 'six' being uppermost
as, collectively, equally likely to the infinite set of states corresponding to
'one' being uppermost.

In both of these cases, the natural response is to look at the dynamics.
What justifies disregarding the numbers on the faces is that they are
dynamically irrelevant to a very good approximation—flipping the die to
a new orientation is a dynamical symmetry (something which would not
be the case if, say, the pips on the die were weighted with lead). And
what justifies regarding the 'six' states as collectively equiprobable with

---

[22] To the best of my knowledge, this term was coined by Keynes (1921), but the idea
goes back to Laplace. For some critical comments, see Howson and Urbach (1993: 51–61).

the 'one' states is the fact that this dynamical symmetry maps the former onto the latter.

So this suggests, as a rationality principle, that our initial probability distribution should not distinguish between states of affairs related by a symmetry transformation—a principle which at least seems plausible. We should note, though, that the principle has already moved some way from the kind of indifference principle which could be applied by an agent to a system about which he knows little. It is a nontrivial matter to determine the dynamics and the symmetries of a system; there is nothing *irrational* about believing a die to be fair even if unbeknownst to me it is loaded. Rather, the 'ideally rational' probabilities presumably have to be those which would be assigned by an agent who knows all the relevant dynamical information. (So in fact, although the objective Bayesian program is pedagogically useful in motivating this symmetry principle, the principle is also available to Lewisian dualists, since according to the Principal Principle, the objective probability of P is the personal probability that an agent would have if they possessed all the admissible information.)

But, plausible or not, does this symmetry-based rationality principle deliver—does it provide an analysis of probability less unsatisfactory than frequentism? Unfortunately, the prospects do not look good, at least in the nonbranching context.

For consider: the supposition is that *if* we knew all the relevant information, we would rationally judge a (fair) die to have probability 1/6 of coming up six. But if we really knew *all* the information, then—since the die is deterministic—we would know precisely how it would land, and so would give six either probability zero or probability one. The symmetry is broken by the actual physical state of the die. Indeed, it could not be otherwise, since only one outcome occurs, so *something* must break the symmetry.

To make this vivid, take a die—as fair as you like—and hold it a millimetre above the table with the six face uppermost. At what odds would you bet on getting six?

To see dynamically why in *most* situations dice are random, imagine colouring in in black those regions of phase space where, if the die begins in that region, it will land on six. The dynamics of the die mean that the phase space will be very stripey—symmetry considerations tell us that one sixth of it in Liouville measure will be coloured black, but in general any reasonably sized region will be much larger than the width of the stripes.

Only in special circumstances, such as those where the die has only a millimetre to fall, will the stripes be wide enough that we can easily tell that its initial state is in or out of a region.[23]

So, *if* we regard equal-volume regions of the phase space as equiprobable (within a given set of known macroscopic constraints), we will indeed predict that the we will roll a six with probability 1/6. But that only raises the question of why we should regard such regions as equiprobable. (Note that we don't just need a reason to regard *each point* of phase space as equiprobable—phase space has continuum many points,[24] so we need a justification for the particular measure we are using.)

Some ingenious attempts have been made (see, in particular, Jaynes 1973) to find a justification of this equiprobability principle on grounds of pure rationality—and it is certainly the most *elegant* choice—but overall, the prospect of justifying the concept of a fair die in pure rationalist probability theory looks rather dim.

Things are not notably better in indeterministic physics. If the dynamical laws are stochastic—as in a dynamical-collapse theory, for instance—then they already embed information about what the probabilities are, and it would seem question-begging to use the symmetries of those laws as a derivation of the probabilities. If instead we just leave the stochastic part of the dynamics unspecified (if, say, we consider quantum mechanics with the Schrödinger evolution interrupted periodically by an unspecified collapse mechanism), we leave open the possibility that the stochastic processes break the symmetries of the deterministic laws.

Whether we are considering a stochastic or a deterministic process, the problem is ultimately the same. We are attempting to use a dynamical symmetry between two possible outcomes to argue that the outcomes are equally likely. But since only one outcome actually occurs, something must break the symmetry—be it the actual microconditions of the system, or the actual process that occurs in a stochastic situation. Either way, we have to build probabilistic assumptions into that symmetry-breaking

[23] I am not aware of any explicit (analytic or numerical) investigation of the physics of *dice throws* that goes beyond the qualitative comments I make here; for an interesting analysis of coin tosses, though, see Vulovic and Prange (1986).

[24] As does Hilbert space. *Contra* too many undergraduate textbooks on statistical mechanics, moving to quantum mechanics does not alleviate the problem.

process, and in doing so we effectively abandon the goal of explicating probability.

The reader may have noticed, though, that this is not a problem if the Everett interpretation is true. With this in mind, it is time to return to Everettian quantum mechanics and see how the concept of objective probability fares there.

4.12 Objective probability and the Everett interpretation

We have already seen, that the notion of personal probability, defined through betting preferences, is as coherent in the Everett interpretation as in non-Everettian physics. This means that nothing prevents the Everettian, too, from using the Principal Principle as a functional definition of probability. We have also seen that the one candidate so far discussed for what could fill the functional definition in the non-Everettian case—long-run relative frequencies—is not obviously any less viable if the Everett interpretation is true. But we have also seen that it isn't terribly viable in any case.

However, for advocates of Everettian quantum mechanics, there is an alternative to frequentism available. Rather than attempting to extract single-case probabilities from long-run relative frequencies, they can declare those probabilities to be equal to the relative weights of branches. Asked why this is true, they can (though, we will see, they do not have to) simply shrug and respond that they are no more obliged to explain why this physical magnitude is probability than why another physical magnitude is length: in both cases (they can say) it suffices to confirm that the magnitude has the right formal properties.

In a sense, this would just be a variant of primitivism: the Everettian would postulate that a certain mathematically definable quantity in their theory represents probability, but provide no account of why. But it actually has two notable advantages over primitivism as I described it above.

First, I criticized primitivism on metaphysical grounds, by attacking the analogy between taking charge or length as primitive and taking probability as primitive. Charge and length, I observed, are features of the actual physical Universe and are represented in our theories by features of that mathematical object we use to represent the actual world; probabilities do not seem to have that status, and are mathematically represented by a measure over all dynamically possible worlds.

This disanalogy does not hold in the Everett interpretation. The 'actual physical Universe' is the multiply branching structure generated by unitary evolution under the Schrödinger equation. The branch weights are physical features of this structure; they are represented mathematically within the quantum wavefunction. The Everett interpretation supplies a objective physical correlate for probabilities, but one which—unlike long-run frequencies—is associated with the outcomes of individual experiments.

Furthermore, the Everett interpretation resolves a difficulty with objective probability which caused particular problems in my account of the symmetry-based theory, but which is really a problem for any nonbranching account of objective probability. Recall that, on a rationalist theory of probability, an agent should set his personal probability equal to the objective probability if he knows 'all physically relevant information'. If the agent knows too little information, this is unproblematic: that situation occurs all the time. But what if the agent knows *too much* information: what if, for instance, he knows the actual result that is going to obtain? In that situation, clearly it is irrational not to set one's personal probability in that result to unity, irrespective of its objective probability.

In fact, the Principal Principle (in the Lewis-based form in which I stated it in section 4.9) already makes allowance for this possibility: recall that it applies only to an agent who conditionalizes his personal probability on the objective probabilities and some other *admissible* information. The 'admissible' caveat is there precisely to block out the possibility of knowing too much.

How is knowing too much possible? It is fair to say that, in the case of genuinely stochastic dynamics, it is difficult at best: Lewis himself appeals to rather fanciful examples like time travel or prophecy. But if the dynamics are deterministic, of course, it would suffice to know the current microstate of the system in question. (For this reason, Lewis in fact rules out the existence of objective probabilities in circumstances where there is no stochasticity. Physics cannot be so sanguine, though, as statistical mechanics makes clear; nor can the de Broglie–Bohm theory, of course.)

We have already seen, in the case of symmetry-based theories of probability, that finding a noncircular statement of how much information is admissible is less than trivial. And actually, even in the case of frequentism, things are somewhat awkward: our agent is presumed to have sufficient knowledge of the future to know all about what the relative frequencies will be, but insufficient knowledge to actually know what the actual results are going to be (to some extent, to be fair, this is part of the general philosophical problem of induction).

There is no analogous problem for the Everettian. There simply are *no facts* which could even in principle be available to an Everettian agent so that the agent knows his unique future, because *there is no such unique future*.[25]

Of course, the question remains: why on earth *should* Everettian agents condition their preferences between bets to conform with this mysterious 'branch weight' thing? And of course, Everettians who follow the primitivist line have no answer to give. To them, in effect, this requirement—the Principal Principle—is just a basic law of physics.

And to be sure, taking the Principal Principle as primitive is unattractive: why should it be a *law of physics* that I should care about branches in proportion to their weights? But we have already seen that this problem bedevils non-branching physics too: how can we accept *any* rationality principle as a law of physics? Or put the other way around, the suggestion that Everettians should indeed take this rationality principle as basic and unexplained (as Vaidman 2002 and Greaves 2004 suggest) seems on as solid ground as the analogous nonbranching suggestion.

Of course, deriving the Principle would be far more satisfactory, and if there were a satisfactory derivation of it in a non-branching context which does not generalize to the Everett context, this would count against the Everett interpretation. But we have seen that finding such derivations is a difficult matter, and is not notably *more* difficult if the Everett interpretation is true. Furthermore, presumably the argument holds both ways around: given how unsatisfactory it is to take the Principle as a bare postulate, if there is an *Everett-specific* derivation of the Principle, this is a significant advantage of the Everett interpretation.[26]

---

[25] In Ch. 7 (following joint work with Simon Saunders; cf. Saunders and Wallace 2008), I will suggest a possible metaphysics for worlds in which they are best thought of as differentiating one from another rather than splitting. On this theory, it is rather that each agent does have a unique future, but it is in principle impossible for him to possess reliable knowledge of that future.

[26] This does raise the question: advantage over what? As I stressed in Ch. 1, it is a serious mistake to see the interpretation of quantum mechanics as a game where we contrast several equally empirically viable interpretations of an underlying formalism and weigh up their strengths and weaknesses. A more careful statement of the point might be: if an

And in fact, such a derivation can be given: in the final section, I will sketch a proof that once the single-outcome assumption is dropped, arguments based on symmetry suffice to prove that the objective probabilities can be given only by the Born probability rule.

4.13 Deriving Everettian probability

As I have noted, in Everettian quantum mechanics there is no need for anything to break the symmetry of a particular process in order to give one outcome rather than another—for all outcomes occur. Or, put another way, in Everettian quantum mechanics not just the laws, but the actual microstate of the system are invariant under a symmetry transformation, as could not be the case if only one outcome was to occur.

For instance, suppose that a spin-half system is to be prepared in a certain superposition of spin up ($| \uparrow \rangle$) and spin down ($| \downarrow \rangle$) (on a certain axis), and then its spin is to be measured along that axis. If the measurement device, before measurement, has state $|ready\rangle$, the dynamics of measurement can be written as

$$| \uparrow \rangle \otimes |ready\rangle \longrightarrow |up\rangle$$

$$| \downarrow \rangle \otimes |ready\rangle \longrightarrow |down\rangle, \tag{4.5}$$

where $|up\rangle$ and $|down\rangle$ are joint states of the spin-half system and the measurement device, recording 'up' and 'down' outcomes respectively. (We make no assumption that the measurement is non-disturbing; few measurements are.)

But 'up' and 'down' are just labels. We could perfectly well (at least at a certain level of idealization) relabel them: that is, perform the transformation $|up\rangle \to |down\rangle$, $|down\rangle \to |up\rangle$ on the system after the measurement. In this case, of course, the dynamics of measurement have the form

$$| \uparrow \rangle \otimes |ready\rangle \longrightarrow |down\rangle$$

$$| \downarrow \rangle \otimes |ready\rangle \longrightarrow |up\rangle. \tag{4.6}$$

Everett-specific derivation of the Principal Principle can be given, then the Everett interpretation solves an important philosophical problem which could not be solved under the false assumption that we do not live in a branching Universe.

And (in deterministic non-branching physics, in stochastic physics, and in the Everett interpretation) the probability that the result is 'up' given that the first measurement protocol is used, $\Pr(U|P1)$, equals the probability $\Pr(D|P2)$ that the result is 'down' given that the second measurement protocol is used. (If this is not obvious, recall that the probabilities are ultimately to be quantified in terms of betting preferences, and that the second protocol is performed by performing the first protocol and then swapping the labels. So a bet on 'down' in the second protocol gives a reward to exactly those future selves (if any) of the agent who would have got the reward if the first protocol had been carried out and the bet had been 'up'.)

Now, we specialize to the case of the Everett interpretation, and suppose the system's initial state is $\alpha|\uparrow\rangle + \beta|\downarrow\rangle$. Then if the first measurement protocol is used, the post-measurement state is

$$\alpha|\text{up}\rangle + \beta|\text{down}\rangle, \tag{4.7}$$

and if the second protocol is used, the post-measurement state is instead

$$\beta|\text{up}\rangle + \alpha|\text{down}\rangle. \tag{4.8}$$

In the particular case where $\alpha = \beta$, the post-measurement state is *unaffected* by which protocol is used. Given the reasonable assumption that the probabilities had better depend only on the actual state of the Universe, it follows that the probabilities in this particular case are the same whichever protocol is used: $\Pr(D|P1) = \Pr(D|P2)$. Putting this together with our earlier result, we get $\Pr(U|P1) = \Pr(D|P1)$. Purely by considerations of the symmetries of the quantum state, we derive that equally weighted superpositions are equally likely on measurement to give each outcome.

We can generalize this argument into a full proof of the Born probability rule. Suppose that $\mathcal{H}$ is a Hilbert space (which we assume infinite dimensional); given any collection $\mathcal{F}$ of subspaces of $\mathcal{H}$, define $\vee\mathcal{F}$ as the closure of the span of $\cup\mathcal{F}$, $\wedge\mathcal{F}$ as the closure of $\cap\mathcal{F}$; given any subspace E, define $\neg E$ as the orthogonal complement of E. Write $E \wedge F$ as shorthand for $\wedge\{E, F\}$ and similarly for $E \vee F$. Now let $\mathcal{E}$ be a collection of subspaces of $\mathcal{H}$ which forms a *complete Boolean algebra*: that is, it is closed under the application of these three actions to arbitrary countable subsets of $\mathcal{E}$ (or elements of $\mathcal{E}$ in the case of $\neg$), it contains $\emptyset$, and $\wedge$ and $\vee$ are distributive over one another.

We write Π_E for the projector onto the subspace E, and define $\mathcal{U}$ as the set of unitary maps from elements of $\mathcal{E}$ into $\mathcal{H}$.

(I intend $\mathcal{E}$ to be a set of projectors defining a quasi-classical history space and $\mathcal{U}$ to be a set of physically performable operations available to experimenters in that space, so that if $U \in \mathcal{U}$ has domain E, it is performable by an experimenter whose branch has state in E. To do full duty $\mathcal{U}$ would need additional constraints to rule out recoherence of branches; for simplicity, I leave these unspecified, though they are included in Chapter 5's more detailed model.)

Then suppose that we can make sense of probabilities in this setting (via betting preferences or otherwise), so that for each state ψ wholly contained within some element E of $\mathcal{E}$, and each $U \in \mathcal{U}$ has domain E, we have a function $\text{Pr}_\psi(\cdot|U)$ from $\mathcal{E}$ to the real numbers. We make the following four assumptions about this function:

- We assume that $\text{Pr}_\psi(\cdot|U)$ satisfies the axioms of the probability calculus (that is: Pr is positive, additive, and $\text{Pr}_\psi(\emptyset|U) = 0$ and $\text{Pr}_\psi(\mathcal{H}|U) = 1$).

- We assume that the probabilities are updated in the normal way when new information is received, by conditionalizing on that information. This amounts to the requirement that

$$\text{Pr}_{(\Pi_E U \psi)}(F|(V|_E)) = \frac{\text{Pr}_\psi(F|VU)}{\text{Pr}_\psi(E|U)} \qquad (4.9)$$

(where $V|_E$ indicates the restriction of $V \in \mathcal{U}$ to E) provided that $\Pi_E U \psi \neq 0$.

- We assume that the probabilities supervene on the actual physical state of the universe, or more precisely, of that part influenced by the action performed, so that if $U\psi = U'\psi'$ then $\text{Pr}_\psi(\cdot|U) = \text{Pr}_{\psi'}(\cdot|U')$.

- Finally, we assume that if something is certain then it has probability one, which amounts to the assumption that $\text{Pr}_\psi(E|U) = 1$ whenever $U\psi \in E$.

This suffices to prove that Pr is given by the Born probability rule:

$$\text{Pr}_\psi(E|U) = \mathcal{W}_\psi(E|U) \equiv \langle \psi | U^\dagger \Pi_E U | \psi \rangle. \qquad (4.10)$$

We prove this in three steps. First we prove that when $\mathcal{W}_\psi(E_1|U_1) = \mathcal{W}_\psi(E_2|U_2)$, $\text{Pr}_\psi(E_1|U_1) = \text{Pr}_\psi(E_2|U_2)$, then we prove that when

$\mathcal{W}_\psi(E_1|U_1) \geq \mathcal{W}_\psi(E_2|U_2)$, $\mathrm{Pr}_\psi(E_1|U_1) \geq \mathrm{Pr}_\psi(E_2|U_2)$, and lastly we prove that the two functions $\mathcal{W}_\psi$ and Pr_ψ are the same.

Step 1. Suppose that

$$\mathcal{W}_\psi(E_1|U_1) = \mathcal{W}_\psi(E_2|U_2). \tag{4.11}$$

for unitary maps U_1, U_2 and projectors E_1, E_2. Then we can choose orthogonal elements F, G of $\mathcal{E}$, and unitary maps V_1, V_2, W_1, W_2 in $\mathcal{U}$, such that:

  (i) V_i has E_i as its domain and some subspace of F as its range.
 (ii) W_i has $\neg E_i$ as its domain and some subspace of G as its range.
(iii) $V_1 \Pi_{E_1} U_1 \psi = V_2 \Pi_{E_2} U_2 \psi$ and $W_1(1 - \Pi_{E_1}) U_1 \psi = W_2(1 - \Pi_{E_2}) U_2 \psi$. (This last condition requires (4.11).)

If we define $X_i = (V_i \Pi_{E_i} + W_i(1 - \Pi_{E_i}))$, and apply our assumption about how probabilities are updated, we get

$$\mathrm{Pr}_{(\Pi_{E_i} U_i \psi)}(F|V_i) = \frac{\mathrm{Pr}_\psi(F|X_i U_i)}{\mathrm{Pr}_\psi(E_i|U_i)}. \tag{4.12}$$

Since $V_i \Pi_{E_i} U_i \psi \in F$, this simplifies to

$$\mathrm{Pr}_\psi(E_i|U_i) = \mathrm{Pr}_\psi(F|X_i U_i). \tag{4.13}$$

But since $X_1 U_1 \psi = X_2 U_2 \psi$, we have $\mathrm{Pr}_\psi(F|X_1 U_1) = \mathrm{Pr}_\psi(F|X_2 U_2)$ and so $\mathrm{Pr}_\psi(E_1|U_1) = \mathrm{Pr}_\psi(E_2|U_2)$.

Step 2. Now suppose that actually

$$\mathcal{W}_\psi(E_1|U_1) > \mathcal{W}_\psi(E_2|U_2). \tag{4.14}$$

Then we can find $S, T \in \mathcal{E}$, and an operator $Y \in \mathcal{U}$, such that the range of Y is in $S \vee T$, the domain of Y is E_1, and

$$\mathcal{W}_\psi(S|YU_1) = \mathcal{W}_\psi(E_2|U_2). \tag{4.15}$$

Our previous result tells us that

$$\mathrm{Pr}_\psi(S|YU_1) = \mathrm{Pr}_\psi(E_2|U_2) \tag{4.16}$$

and the probability axioms tell us that

$$\mathrm{Pr}_\psi(S|YU_1) \leq \mathrm{Pr}_\psi(S \vee T|YU_1). \tag{4.17}$$

But

$$\mathcal{W}_\psi(S \vee T | YU_1) = \mathcal{W}_\psi(E_1 | U_1), \qquad (4.18)$$

so

$$\mathrm{Pr}_\psi(S \vee T | YU_1) = \mathrm{Pr}_\psi(E_1 | U_1); \qquad (4.19)$$

putting all this together, $\mathrm{Pr}_\psi(E_1 | U_1) \geq \mathrm{Pr}_\psi(E_2 | U_2)$.

Step 3. Suppose that $E_1 \ldots E_N$ are a set of mutually orthogonal subspaces whose disjunction is $\mathcal{H}$ and that $\mathcal{W}_\psi(E_i | U) = 1/N$ for all i. By the above result $\mathrm{Pr}_i(E_i | U)$ is independent of i; since it also satisfies the probability calculus it must be equal to $1/N$. Note that we will always be able to find N such projectors for any N.

Now assume F is an element of $\mathcal{E}$ such that $\mathcal{W}_\psi(F | V)$ is rational—equal to M/N, say. Then if

$$G = \vee_{i \leq M} E_i, \qquad (4.20)$$

then

$$\mathcal{W}_\psi(F | V) = \mathcal{W}_\psi(G | U) \qquad (4.21)$$

and by the probability axioms

$$\mathrm{Pr}_\psi(G | U) = \sum_{i \leq M} \mathrm{Pr}_\psi(E_i | U) = M/N, \qquad (4.22)$$

so $\mathrm{Pr}_\psi(F | V) = M/N$.

So $\mathrm{Pr}_\psi(\cdot | U)$ is an increasing function of $\mathcal{W}_\psi(\cdot | U)$:

$$\mathrm{Pr}_\psi(E | U) = f[\mathcal{W}_\psi(E | U)] \qquad (4.23)$$

where $f : [0, 1] \to [0, 1]$ is increasing. We also know that $f(M/N) = M/N$. So for any $\alpha \in [0, 1]$, let $\{\alpha_i\}$ be an increasing sequence of rational numbers converging on α. Since $f(\alpha) \geq \alpha_i$ for all i, it follows that $f(\alpha) \geq \alpha$. Repeating this argument with a decreasing sequence establishes that $f(\alpha) = \alpha$, and we're done.

The argument just given establishes that *if* probability basically makes sense, and has the usual qualitative features, in unitary quantum mechanics, then quantitatively it is given by the Born rule. My task for the next two chapters is to develop it into a fully rigorous argument that branch weight fills the functional role of objective probability, without any need to assume anything *probabilistic* at the outset. However, what I hope to have shown in this chapter is that carrying out this task is interesting more for the insight it might give us into the nature of probability than because the Everett interpretation cannot be defended without it. We have seen that classical probability is more controversial than is often recognized, that the intuitive idea that probability requires a single nonbranching Universe actually plays no role in most approaches to the foundations of probability, and that the problems that those approaches face are almost invariably no worse—and often actually less bad—when the Everett interpretation is taken into account. It would have been sufficient for the Everett interpretation to do *as well* as nonbranching physics in making sense of probability; the fact that it can do *better* may be remarkable, but it is ultimately a bonus rather than a necessity.

CHAPTER 4. It may be unintuitive to regard the relative branch weights as probabilities. But no matter what approach we take to the interpretation of probability in general, we find that the approach works at least as well—and sometimes better—in Everettian quantum mechanics as in nonbranching theories.

CHAPTER 5. I have already given a sketch of a derivation of the Born probability rule within the general framework of a rationalist approach to probability. How can this be developed into a fully rigorous derivation of the Born Rule?

5

Symmetry, Rationality, and the Born Rule

Thus we see that quantum theory permits what philosophy would hitherto have regarded as a formal impossibility, akin to "deriving an ought from an is", namely deriving a probability statement from a factual statement. This could be called deriving a "tends to" from a "does".

David Deutsch[1]

5.1 A positive theory of Everettian probability

In this chapter I begin the work of developing a specific, positive account of probability in the Everett interpretation. My starting point will be Chapter 4's analysis of probability in terms of betting preferences; as promised in section 4.13, I will show how such an analysis can be developed into a full-blown account of Everettian probability.[2]

In developing such an account, it can be useful to break the probability problem into subproblems. Recall from Chapter 4 that probability connects to our other concepts in two ways: through its role as a guide to action (Papineau's decision-theoretic link), and through our use of evidence to measure probabilities (Papineau's inferential link). This suggests the following terminology (noted in Chapter 1, and first introduced in Greaves 2007):

---

[1] Deutsch (1999).

[2] The first argument along these lines was given by Deutsch (1999); a preliminary version of the argument here appeared in Wallace (2007). See section 6.10 for more details of the brief but complex history.

The practical problem. Justify why quantum-mechanical branch weight plays the role in decision-making that 'ordinary' probability is supposed to play.

The epistemic problem. Justify why quantum-mechanical branch weight plays the role in inference, including the testing of theories, that 'ordinary' probability is supposed to play. In particular, justify why we can measure branch weight by long-run relative frequencies, and justify why the experimental evidence generally taken to support quantum mechanics does indeed support it when the Everett interpretation is taken into account.

(Before considering how these problems might be solved, it is worth reiterating the basic point of Chapter 4 (cf. also Papineau 1996): namely, that we actually have no idea at all how to solve the analogous problems for non-Everettian probability. The fact that we can actually solve them in the Everett interpretation is a bonus, not a requirement.)

In outline, my solution goes as follows. I use considerations of rationality to make sense of probability in terms of agents' preferences between bets, an approach to probability which makes just as much sense in the Everett interpretation as in non-branching contexts (a claim which I defended in section 4.8, and will discuss further in this chapter).

In the bulk of this chapter, by considering the preferences of a rational agent who knows the current quantum state of his branch, I will prove that such an agent acts exactly as if he assigned probabilities to the outcomes of future events in accordance with the Born Rule. Clearly, if this is possible then it is at least substantial progress towards solving the practical problem. In Chapter 6, I extend this result to agents with less knowledge (of what the quantum state is, and/or of whether the Everett interpretation is correct). This will complete the solution of the practical problem, and also solve the epistemic problem.

Prima facie, there might also exist a third problem:

The residual problem. Establish not only that quantum-mechanical branch weight plays the same roles in decision-making and inference that probability is supposed to, but that it actually is probability.

But—as I argued in section 4.2—there is no residual problem. If we can solve the practical and epistemic problems, then quantum-mechanical branch weight will do everything for us that probability was supposed to;

any other reason to deny that it *is* probability will be either purely semantic, or will involve appeal to intuitions which have no epistemic value. Nonetheless, it is interesting to note that one of the most natural such intuitions is the belief that probability must have something to do with uncertainty and with alternative possibilities. In Chapter 7, I argue that in fact this intuition is actually satisfied in Everettian quantum mechanics (see also Saunders 1996; 1998; 2010 and Saunders and Wallace 2008).

This chapter, then, has a very delineated goal: to prove, rigorously and from general principles of rationality, that a rational agent, believing that (Everett-interpreted) quantum mechanics correctly gives the structure and dynamics of the world and that the quantum state of his branch is $|\psi\rangle$, will act for all intents and purposes as if he ascribed probabilities in accordance with the Born Rule, as applies to $|\psi\rangle$. That is, if U is a unitary transformation which might be applied to the agent's branch, and Π is one of the decoherence-defined projectors that pick out branches, then the agent will act to all intents and purposes as if Π has a probability and that probability is $\langle\psi|U^\dagger\Pi U|\psi\rangle$.

At its mathematical core, the argument I will present is not really decision-theoretic at all: it rests on the same symmetry considerations as the proof of the Born Rule which I presented in section 4.13. The reason for appealing to decision theory, as I have been at pains to stress, is simply that it provides us with a sharp, unambiguous notion of probability applicable to a context—that of a branching universe—in which it has been questioned whether probability makes sense at all. If it can be proved, though, that it is rational to act *as if* Everettian branch weights are probabilities, the question of whether they are 'really' probabilities decreases in importance and becomes largely one of semantics.

From a purely logical perspective, the correct way to present this material would be to give a detailed account of how decision theory works in the classical context, and then consider to what extent that account applies to the Everett interpretation. This would, however, involve a very substantial digression: decision theory is a somewhat involved and often technical topic, and most of its technical results are of only limited relevance to the matter at hand, not least because quantum mechanics actually offers a significantly friendlier arena in which to formulate decision theory than does classical mechanics. To complicate matters further, the classical decision theory to which I appeal is in some respects (notably its formalization of an agent's strategy when considering actions made at different times) decidedly nonstandard.

For this reason, although I *do* provide a detailed account of classical decision theory, I have relegated that account to Appendix B. In this chapter, I will give only a very abbreviated account, aimed at providing just as much general decision theory as is required for my quantum-mechanical arguments. Readers seeking a really comprehensive treatment may wish to begin with Appendix B, which might be thought of as 'Chapter 4.5' (I hope in particular that philosophers interested in decision theory might find something of interest in Appendix B quite apart from its relevance to quantum mechanics.) But Chapters 5 and 6 do not assume familiarity with that appendix, and I use its results only sparingly in the more formal parts of Chapter 6. (As such, there is a small amount of repetition between Appendix B and Chapters 5 and 6.)

In this chapter, after giving an outline review of decision theory I will set out the assumptions required to prove the Born Rule, which will be of two kinds: some are related to purely physical facts about the setup of the decision-theoretic problem confronting our agent, some are related to the assumption that the agent is rational. (In the terminology of Appendix B, these are the richness and the rationality axioms, respectively.) I will then show how, from these assumptions, the Born Rule can be established.

I will eventually state my assumptions, and prove my result, in full mathematical rigour. In foundational work, though, rigour can be a double-edged sword: it provides precision, and (I hope) instils confidence that the conclusion does follow from the premises, but it can feel pedantic and obscure clarity. I approach this problem by stating and proving my result twice: once in a relatively informal way, and again in formal mathematical language. (I borrow this strategy from Wallace 2010b, on which this chapter is largely based.)

I finish this chapter by looking at some of the many alternative strategies for Everettian agents which have been proposed in the recent literature. I will show how in each case they are in violation of some physical or decision-theoretic principle; in doing so I hope to provide further insight into those principles and into the assumptions of my proof.

5.2 Preamble: the decision-theoretic approach

Suppose a coin is to be tossed in five minutes' time; and suppose that an agent bets $5 (at even odds) that it will land heads. There are two interestingly different possible results: (i) the measurement gives result 'up' and the agent gets \$5; (ii) the measurement gives result 'down' and the agent loses \$5. If the result is heads, the agent will be pleased about the bet; if it is tails she will be less delighted, though (if she is of an appropriate character) she may well still regard the bet as having been the right choice given her information before the coin toss. (There are of course vastly more than two microscopically distinct possible results; the division into two sets is based on the pragmatic interests of the agent.)

In deciding whether to accept this bet as opposed to any number of other bets, the agent has to weigh the cost to her if the result is 'down' against the benefit if it is 'up'. Decision theory gives a precise answer to the question of how she should carry out this weighting: she should assign some utility (some real number) $\mathcal{V}(+\$5)$ to receiving \$5, some other utility $\mathcal{V}(-\$5)$ to losing \$5, and some third utility $\mathcal{V}(\$0)$ to neither getting nor losing it; and she should assign a probability $\Pr(H)$ to heads; and she should take the bet only if

$$\Pr(H) \times \mathcal{V}(+\$5) + (1 - \Pr(H)) \times \mathcal{V}(-\$5) > \mathcal{V}(\$0). \qquad (5.1)$$

More generally, decision theory mandates that an agent should assign a utility to each payoff, and a probability to each outcome, and that faced with *any* decision, the agent should choose that option which maximizes expected utility with respect to those assignments.

As I explain *in extenso* in Appendix B, it is a mistake to think of the utility and probability functions as independent functions which jointly tell an agent how to act. Rather, each can be defined in terms of each other and of the agent's preferences. We saw in our discussion of Dutch book arguments in section 4.8 that we can measure probabilities in terms of betting odds (thus, in effect, in terms of the utility we put on a bet on the event in question actually occurring). But equally, we can measure utilities in terms of probabilities: to say that an agent regards the utility of X (relative to the utility of getting nothing) as K times the utility of Y is to say that this agent is indifferent between a bet with (as she judges) a probability $1/K$ of getting X and a 'bet' which in fact just delivers Y with certainty. In either case, the principle of maximizing expected utility (MEU) is a consequence of this definition (together with other qualitative assumptions about agents' behaviour), and not an independent postulate.

In fact, the second way is significantly to be preferred. For one thing, if probabilities are defined in terms of utilities, this leaves the question of how utilities are to be defined; and here no non-circular answer is available: we can define a purely qualitative ordering on possible payoffs of bets in terms of an agent's preferences between those bets, but the set of outcomes is essentially structureless and there are indefinitely many ways to represent that qualitative ordering with a utility function. A purely qualitative ordering can also be defined on the outcomes of bets: we can say that an agent regards E as more likely than F if she prefers a bet on E to a bet (with the same stakes) on F. But in contrast to the case of payoffs, the space of outcomes has a fairly rich structure (that of a Boolean algebra, in fact), since for any two outcomes E and F we can consider their union $E \cup F$ ('either E or F happens'), their intersection $E \cap F$ ('both E and F happen') and their negations $\neg E$ and $\neg F$ ('E/F doesn't happen'). With a sufficiently rich structure of outcomes, and with some additional qualitative assumptions about the agent's preferences, we can prove that this qualitative ordering is uniquely represented by some uniquely determined probability measure. We can then use this probability measure to construct utilities such that the agent's preferences are represented by the MEU rule with respect to the probability measure and those utilities.

The overall process, then, is one where we begin with some assumptions about an agent's preferences between pairs of bets ('rationality axioms') and about the structure of the set of bets ('richness axioms'), and derive from that the existence and uniqueness of a probability measure and a utility function which jointly describe those preferences. (Or near-uniqueness, at any rate: a moment's thought will show that the utility functions U and $aU + b$ ($a > 0$) deliver exactly the same preferences.)

I will adopt largely the same strategy in this chapter. I begin with the idea of a bet on the result of an outcome of a measurement of a quantum state: to take such a bet is to commit to a process whereby the measurement is made and then, conditional on the outcome of that measurement, a reward is provided. In general such bets lead to branching, with different rewards being given to different sets of the agent's future selves. But regardless, the result of the bet is to cause good, bad, or indifferent things to happen to the agent in the future—just as in classical decision theory, except that there is more than one future version of the agent. As such, the agent has all the reasons to care about the interests of those multiple future selves as non-Everettian agents (if there were any) would have had to care about their single future self.

Since different actions lead to different costs and benefits for these future selves, the agent needs to choose which act she prefers to carry out.

This defines a preference order between bets. By considering in particular that agent's preferences between bets in which a reward of fixed value is provided on different sets of branches, I construct a qualitative preference order over those sets of branches; I show that the order is uniquely represented by a probability measure; for completeness I construct a utility function, though at this stage we are just doing conventional decision theory.

(Notice, incidentally, that a given choice by the agent is represented by a given branching structure and *not* by a given branch within that structure. This should obviate one worry sometimes voiced with Everettian decision theory: that an agent will make all possible choices, one per branch, and so the idea of 'deciding' between choices is meaningless; see Box 4.1 for more on this point.)

There is, however, one crucial difference. The representation theorems of ordinary decision theory show that an agent's preferences are represented by *some* probability measure, but they put no constraint at all on which one. This quantum representation theorem will tell us what that probability measure is: it is rationally required to be the probability measure defined by the Born Rule.

5.3 The quantum decision problem

The situation I wish to consider is the following. A quantum state is to be prepared in some superposition; the system is measured in some basis; a bet is made by the agent on the outcome of that measurement. Our agent knows (we assume) that unitary quantum mechanics is correct; she is also assumed to know the universal quantum state, or at least the state of her branch. (The latter is an unrealistic but convenient assumption; in practice, however, it suffices for the agent to know the mod-squared amplitudes for each outcome of a measurement.) Her preferences can be represented by an ordering relation on these bets.

Since (in Everettian quantum mechanics, at any rate) preparations, measurements, and payments made to agents are all physical processes, there is a certain simplification available: any preparation-followed-by-measurement-followed-by-payments can be represented by a single unitary transformation. So our agent's rational preference is actually representable by an ordering on unitary transformations.

We should acknowledge that not all unitary transformations represent something physically possible.[3] In particular, transformations which lead to *recoherence*—i.e. to Everett branches merging—are certainly not performable by any agent localized to a specific branch. Nonetheless we will consider a fairly wide set of transformations to be available—exactly how wide is something that the axioms will spell out. (It might be worth recalling at this stage that decision theory is concerned with the preferences an agent would have when confronted with a particular decision—her dispositional preferences, in philosophers' language—and not just with what actually happens. It is most unlikely that I will be offered a choice between the presidency of the World Bank and the deputy leadership of the Al-Qaeda terror group, but I have a definite preference between the two. As such, the assumption of a reasonably wide set of transformations seems legitimate enough.)

We should also acknowledge in our decision-theoretic setup that decoherence imposes a certain structure on the Hilbert space. We can represent this by a resolution of the identity on the Hilbert space: that is, by a decomposition of the space into subspaces, with each subspace corresponding to a possible macrostate. The choice of macrostates is largely fixed by decoherence, although the precise fineness of the grain of the decomposition is underspecified. (In the model, of course, it will be precisely specified, but this just illustrates that the model is artificially precise.) We call a macrostate a *possible outcome* of an act (for a particular agent) if performing that act causes the agent's branch to evolve into a state overlapping nontrivially with that macrostate (i.e. causes some of the agent's future selves to be in that macrostate); we call an macrostate *available* to an agent if there is an available act with that macrostate as a possible outcome.

Part of the point of the decomposition into macrostates is that an agent can be assumed not to care exactly what the microstate is within a given macrostate (if she does care, we have defined the macrostates too coarsely). But in fact, usually an agent will also be indifferent between a great many macrostates: for instance, if offered a million dollars, I am indifferent as to

---

[3] Doesn't *only one* unitary transformation represent something physically possible? Doesn't the Hamiltonian of the universe uniquely determine which transformation is performed? If this is a problem, it is not specific to Everett: it is the ancient debate of free will vs. determinism. Again, see Box 4.1 for further discussion.

the colour of the cheque.[4] It will be useful to consider a coarse-graining of the macrostate subspaces into *reward subspaces*, such that an agent's only preference is to which reward subspace she is in.

For mathematical reasons it will be convenient to work both with the set of macrostates and with the Boolean algebra $\mathcal{E}$ of arbitrary disjunctions of macrostates,[5] which (following standard practice in decision theory) we call the *event space*. The formal development of the theory will not actually require the assumption that the event space can be constructed from a set of macrostates (though it does not rule out that assumption). Indeed, since the fineness of grain of branches is indeed underspecified, the branch structure might be best idealized in some particular situation by a model in which the algebra is not constructed this way.[6] (For instance, if the Hilbert space is $L^2(R^N) \otimes \mathcal{H}_E$, where $\mathcal{H}_E$ represents some subsystem of environmental degrees of freedom, then we might wish to take

$$\mathcal{E} = \{\Sigma_E : E \text{ is an open subset of } R^N\} \tag{5.2}$$

where

$$\Sigma_E = \text{Span}\{f \otimes v : f \text{ has support in E}\} \tag{5.3}$$

which cannot be generated from macrostates (unless we are willing to relax rigour and consider eigenstates of position).

For simplicity, we will refer to the set of unitary transformations over which an agent's preference order is defined as *acts*. A different set may be relevant for different physical states of the universe, so we will have cause to speak of the *acts available at* a macrostate π. (In view of the previous paragraph's comment, we might do better to talk of the acts which are *contemplatable* at π; I avoid this terminology mostly because it's cumbersome.)

In fact, it will be simpler to talk of which acts are available at a given event (not just a macrostate)—informally an act available at an event $E = \pi_1 \vee \pi_2 \cdots \pi_N$ is the conditional act 'if the macrostate is actually π_i, perform U_i'. This makes it much more straightforward to talk about the composition of acts: if U is available at an event E, and V is available at

---

[4] The reader who doubts this claim is encouraged to test it empirically.

[5] Recall that the disjunction $E \vee F$ of two subspaces of a Hilbert space is the closure of the span of their union.

[6] Cf. the discussion of atomless history spaces in Box 3.2.

the smallest event containing the range of U, for instance, then VU ought to be the act of performing U and V sequentially and so also should be available at E. In the formal development we will state explicit rules to ensure that these and similar compositions are available; for now we take it as tacit that they are.

We now need to represent the agent's preferences between acts. Since those preferences may well depend on the state, we write it as follows: if the agent prefers (at ψ) act U to act U', we write

$$U \succ^\psi U'. \tag{5.4}$$

To be meaningful, of course, this requires that U and U' are both available at ψ's macrostate. So $\succ^\psi$ is to be a two-place relation on the set of acts available at that macrostate. In the event formalism we use later, we will require $\succ^\psi$ to be a two-place relation on the acts available at each event which contains ψ.

So much for the setup; now for the axioms. I begin with the axioms of richness, which concern which acts are available to the agent (how rich the structure of the set of acts is) and which are not connected to a particular agent's preference order.

The richness axioms, then, are

Reward availability. All rewards are available to the agent at any macrostate: that is, the set of available acts always includes ones which give all of the agent's future selves the reward.

Branching availability. Given any set of positive real numbers $p_1, \ldots p_n$ summing to unity, an agent can always choose some act which has n different macrostates as possible outcomes, and (independent of the quantum state[7]) gives weight p_i to the ith outcome.

Erasure. Given a pair of states $\psi \in E$ and $\varphi \in F$ in the same reward, there is an act U available at E and an act V available at F such that $U\psi = V\varphi$.

Problem continuity. For each event E, the set of acts available at E is an open subset of the set of unitary transformations from E to $\mathcal{H}$.

---

[7] This caveat is not actually required for the results of this chapter, but will be convenient in Ch. 6.

These should mostly be uncontroversial. Branching availability and reward availability are consequences of the relatively stylized decision problem we are considering, where measurements are being made and payments are being provided; they reflect the facts (respectively) that quantum systems can be prepared in arbitrary states and that envelopes of cash can always be given to people.

Erasure is slightly more complicated. It effectively guarantees that an agent can just forget any facts about his situation that don't concern things he cares about (i.e. by definition: that don't concern where in the reward space he is). In thinking about it, it helps to assume that any reward space has an 'erasure subspace' available (whose states correspond to the agent throwing the preparation system away after receiving the payoff but without recording the actual result of the measurement, say). An 'erasure act' is then an act which takes the quantum state of the agent's branch into the erasure subspace; the agent is (by construction) indifferent to performing any erasure act, and since he lacks the fine control to know which act he is performing, all erasures should be counted as available if any are. It follows that, since for any two such agents all erasures are available, in particular there will be two erasures available satisfying the axiom.

I postpone a discussion of problem continuity until the axioms of rationality have been introduced.

5.4 The dictates of rationality

Moving on to the rationality axioms, they come in two groups. The first two axioms are very general principles of rationality, as relevant in the classical as in the quantum context.[8]

Ordering. The relation $\succeq^\psi$ is a total ordering in the following sense: $\succeq^\psi$ is transitive (if $U \succeq^\psi V$ and $V \succeq^\psi W$ then $U \succeq^\psi W$), reflexive ($U \succeq^\psi U$) and connected (either $U \succeq^\psi V$ or $V \succeq^\psi U$).

Diachronic consistency. If U is available at ψ, and (for each i) if in the ith branch after U is performed there are acts V_i, V_i' available, and (again for each i) if the agent's future self in the ith branch will prefer V_i to V_i', then the agent prefers performing U followed by the V_is to performing U followed by the V_i's.

---

[8] My justification of them is essentially repeated in section B.8 of Appendix B.

Ordering is utterly familiar (indeed, built into our use of the $\succ^{\psi}$ symbol) and hopefully uncontroversial. But it is worth stressing that the *reason* it is uncontroversial is not (just!) that it would be unintuitive for an agent's preferences to violate ordering, but because it isn't even possible, in general, for an agent to formulate and act upon a coherent set of preferences violating ordering.

Of course, in stylized and artificial special cases, it might be. If an agent knows that she will be offered three acts chosen from a set of ten, she can arbitrarily pick one element from each three-element subset, and elect to choose that one. But of course, real decision problems aren't that cleanly specified: the precise number of acts available is vague or just indeterminate and the cognitive cost of trying to pin down the size of that set is prohibitive (even when the very act of trying to pin it down does not change the problem out of recognition). Excluding stylized and occasional exceptions, then, ordering is *constitutive* of rationality, not just intuitively necessary for it.[9]

I have stressed this because, in fact, very much the same defence can be offered of the less familiar diachronic consistency principle, which in effect rules out the possibility of a conflict of interest between an agent and his future selves. In philosophy examples one often speaks of a (classical) agent as if he were a continuum of independent entities, one for each time, each having his own preference ordering. But of course actual decision-making takes place over time. An agent's actions take time to carry out; his desires and goals take time to be realized. If his preferences do not remain consistent over this timescale, deliberative action is not possible at all. Indeed, since almost any action I decide upon at a given instant is enacted largely through a series of actions performed at later instants, it is not really even coherent to think of agents *acting* unless their behaviour is basically diachronically consistent.

This is not to rule out *localized* violations of diachronic consistency even outside the Everettian context. If I tell my friend not to let me order another glass of wine after my second, I acknowledge that my desires

[9] Readers familiar with the philosophy of mind will notice the parallels with debates there about the normativity or otherwise of ascriptions of rationality; my position presupposes, to some extent, that ascriptions of rationality have a significant normative component (as defended famously by Davidson 1973, Lewis 1974, and Dennett 1987b, and opposed by e.g. Fodor (1985; 1987); a more recent exchange is Wedgwood (2007) and Rey (2007).

at that point will conflict with my desires now. But notice that such situations

(a) are generally not taken to be rational;
(b) are indeed analysed as situations of conflict, where my present self acts to prevent my future self having access to his preferred choice;
(c) are localized, taking place against a general assumption of diachronic consistency in myself and others (as when I assume that my friend's future self will indeed act on her agreement not to let me order the wine, or that the morning after the night before, I'll be glad that she did).

Similarly: in a branching universe, to accept a conflict of interest between my pre-branch and post-branch selves is to cease to see them as the same person. If branching were an isolated occurrence, this might be possible: it is arguably callous to make a copy of myself and send him off to do a dangerous or disagreeable task—and, crucially for the point, to take actions designed to prevent him shirking that task—but it is not *irrational*.[10] But *Everettian* branching is ubiquitous: agents branch all the time (trillions of times per second at least, though really any count is arbitrary). In the presence of *widespread, generic* violation of diachronic consistency, agency in the Everett universe is not possible; so, any *rational* strategy must be a diachronically consistent strategy.[11]

Incidentally, the very idea of composing acts to make further acts also presupposes diachronic consistency: only if an agent can think of future decisions she will make as *her decisions*, so that she can meaningfully make those decisions (for all that there is always some possibility that she will change her mind), does it make sense to consider composite acts.

The remaining rationality axioms are more specific to the Everettian context. Their precise statements get a bit more technical, so I phrase them fairly loosely here; as always, see section 5.6 for details and for reassurance that there isn't sleight of hand going on.

---

[10] See the first part of Egan (1994) for a science-fictional exploration of the idea—but notice that its plausibility relies on the copy's actions being causally relevant to the original, something not possible in the Everettian universe.

[11] I explore the principle of diachronic consistency further, independent of the Everett interpretation, in Wallace (2010a).

Macrostate indifference. An agent doesn't care what the microstate is provided it's within a particular macrostate.

Branching indifference. An agent doesn't care about branching *per se*: if a certain operation leaves his future selves in N different macrostates but doesn't change any of their rewards, he is indifferent as to whether or not the operation is performed.

State supervenience. An agent's preferences between acts depend only on what physical state they actually leave his branch in: that is, if $U\psi = U'\psi'$ and $V\psi = V'\psi'$, then an agent whose prefers U to V given that the initial state is ψ should also prefer U' to V' given that the initial state is ψ'—$U \succ^\psi V$ iff $U' \succ^{\psi'} V'$.

Solution continuity. If for some state ψ, $U \succ_\psi U'$, then sufficiently small permutations of U and U' will not change this.

Macrostate indifference is hopefully uncontroversial: it's built into the definition of macrostates, in fact. (The point being that an agent can have no practical control as to what state she gets, within a particular macrostate, on familiar statistical-mechanics and decoherence grounds, and that we are interested in an agents' preferences only insofar as they show up in her actual dispositions to action.)

Solution continuity and branching indifference—and indeed problem continuity—can be understood in the same way, in terms of the limitations of any physically realisable agent. Any discontinuous preference order would require an agent to make arbitrarily precise distinctions between different acts, something which is not physically possible. Any preference order which could not be extended to allow for arbitrarily small changes in the acts being considered would have the same requirement. And a preference order which is not indifferent to branching *per se* would in practice be impossible to act on: branching is uncontrollable and ever-present in an Everettian universe.[12]

The other way to understand these assumptions is as prohibitions on strategies that just exploit artefacts of our model. The branching structure—including the well-defined number of branches associated with any act—is derived from the set of macrostates, which is in turn derived from decoherence. But as I argued in section 3.11, this structure has a

---

[12] The main source of branching, as discussed in section 3.11, is probably classically chaotic systems.

significant degree of arbitrariness associated with it, primarily in terms of the coarseness of the grain of the macrostates. In the actual physics there is no division of the dynamics into discrete branching events followed by evolution of individual branches: branching, rather, is continuous. But if branching is always going on, and cannot be quantified in a non-arbitrary manner, then no strategy can be formulated which is other than indifferent to the presence of branching.

A quick defence of state supervenience would be: the agent's preferences supervene on the actual state of the branch; transformations which differ only in how they would affect non-actual quantum states do not differ in any relevant respect. However, this brings out a tacit assumption in the formalism: the idea that acts can be represented by *single* unitary transformations rather than by *sequences* of unitary transformations. Why regard a sequence of measurements as decision-theoretically equivalent to a single measurement just because the same unitary transformation is enacted by both?

One possible defence is that the agent is playing a sequence of games which result in rewards that he spends only after the sequence is done. In this case, what does he care about what happens during the brief period in which the games are being played (when having or not having rewards makes no difference to his status)—should he not care only about the state of the universe after the payouts are all made?

However, appealing to intuition in this way—especially when discussing the Everett interpretation—is unsatisfactory. A far better defence is to observe that caring about the final state only is the diachronic equivalent of branch indifference, and can be defended in the same way. There is no 'real' branching structure beyond a certain fineness of grain, so the details of that structure can only be included in terms of their coarse-grained consequences.

Put another way: we could have defined our decision theory in terms of preferences, not over final states, but over decoherent history spaces. But if we had done so, we would have needed both synchronic and diachronic indifference assumptions: indifference both to the fineness of grain of the history projectors at each time, and to the size of the temporal gaps between history projectors. Translated back into our setting, where we consider sequences of decisions made only over very short periods of time, the former assumption entails branch indifference and the latter entails that acts can be represented by single unitary transformations.

5.5 A quantum representation theorem

I can now prove, in succession, four results, the first three of which are (trivially) entailed by the fourth.

Equivalence lemma. If two acts assign the same weight to each reward, the agent must be indifferent between them.

Nullity lemma. An agent is indifferent to a possible outcome of an act iff that act has weight zero.

Dominance lemma. Suppose that two acts each only have two possible rewards r_1, r_2 as outcomes, with $r_1 \succ r_2$[13] and that the first act assigns a higher weight to r_1 than the second act does. Then the first act must be preferred to the second.

Born Rule theorem. There is a utility function on the set of rewards, unique up to positive affine transformations, such that one act is preferred to another iff its expected utility, calculated with respect to this utility function and to the quantum-mechanical weights of each reward, is higher.

Since all these results are proved *formally* in section 5.7, my purpose in this section is explanation and not persuasion: I wish simply to show the general shape of the proof.

The equivalence lemma is best illustrated by examples (here, I basically elaborate the example given in section 4.13). For a simple case, suppose we have two acts (A and B, say): in each, a system is prepared in a linear superposition $\alpha|+\rangle + \beta|-\rangle$ and then measured in the $\{|+\rangle, |-\rangle\}$ basis. On act A, a reward is then given if the result is '+'; on B, the same reward is given on '−' instead. The resultant states are

$$A: \quad \alpha|+\rangle \otimes |\text{reward}\rangle + \beta|-\rangle \otimes |\text{no reward}\rangle; \qquad (5.5)$$

$$B: \quad \alpha|+\rangle \otimes |\text{no reward}\rangle + \beta|-\rangle \otimes |\text{reward}\rangle. \qquad (5.6)$$

By erasure, there will exist acts available to the agent's future self in the reward branch (for both A and B) which erase the result of what was measured, leaving only the reward. Performing these transformations, and

---

[13] That is: with an act which returns some microstate in r_1 with certainty preferred to one which returns some microstate in r_2 with certainty; that this determines a well-defined ordering over rewards follows from macrostate indifference and branching indifference.

the equivalent erasures in the no-reward branch, leaves

A-plus-erasure: $\alpha|0\rangle\otimes|\text{reward}\rangle + \beta|0'\rangle\otimes|\text{no reward}\rangle$;   (5.7)

B-plus-erasure: $\beta|0\rangle\otimes|\text{reward}\rangle + \alpha|0'\rangle\otimes|\text{no reward}\rangle$.   (5.8)

Now, by branching indifference, the agent's future selves are indifferent to whether this erasure is or is not performed. (Branching indifference is needed because we have no guarantee that erasures are nonbranching; if we did, microstate indifference would suffice.) So by diachronic consistency, the original agent is indifferent between A and A-plus-erasure, and between B and B-plus-erasure.

But now: if $\alpha = \beta$, then A-plus-erasure and B-plus-erasure leave the system in the same quantum state. So by state supervenience, the agent is indifferent between them. Since we know from ordering that preferences are transitive, the agent must also be indifferent between A and B. Indeed, we actually require only that $|\alpha| = |\beta|$, for phase differences too can be erased.

For a slightly more complicated case, suppose game C involves a 2-state system being prepared in state

$$\sqrt{2/3}|+\rangle + \sqrt{1/3}|-\rangle$$

and a reward being given on '+', and game D involves a three-state system being prepared in state

$$\sqrt{1/3}(|+\rangle + |0\rangle + |-\rangle)$$

and a reward being given on '+' and on '0'. The resultant states are then

$$C : \sqrt{2/3}|+\rangle\otimes|\text{reward}\rangle + \sqrt{1/3}|-\rangle\otimes|\text{no reward}\rangle; \quad (5.9)$$

$$D : \sqrt{1/3}|+\rangle\otimes|\text{reward}\rangle + \sqrt{1/3}|0\rangle\otimes|\text{reward}\rangle + \sqrt{1/3}|-\rangle\otimes|\text{no reward}\rangle.$$
$$(5.10)$$

But by erasure, there is an act available for the future self of the agent in the 'reward' branch of game C which creates two equally weighted branches:

$$|+\rangle\otimes|\text{reward}\rangle \longrightarrow \sqrt{1/2}|X\rangle\otimes|\text{reward}\rangle + \sqrt{1/2}|Y\rangle\otimes|\text{reward}\rangle. \quad (5.11)$$

Since by branching indifference the agent's future self is indifferent to performing this act or not, by diachronic consistency the original agent is indifferent between C and C-plus-branching. But the state produced by C-plus-branching is

C-plus-branching: $\sqrt{1/3}|X\rangle \otimes |\text{reward}\rangle + \sqrt{1/3}|Y\rangle \otimes |\text{reward}\rangle + \sqrt{1/3}|-\rangle \otimes |\text{noreward}\rangle.$

$$(5.12)$$

By a generalization of our earlier argument, the agent is indifferent between C-plus-branching and D, and so between C and D.

By arguments of this kind, the equivalence lemma can be proved for any act with finitely many outcomes. The null and dominance lemmas are easy further steps, using the second clause of diachronic consistency.

We are now nearly done: the remainder of the proof is actually a standard decision-theoretic method for constructing utilities. Pick two rewards R and S with $R \succ S$, and assign R utility 1 and S utility 0. For any reward T satisfying $R \succeq T \succeq S$, there is a unique number $U(T)$ such that the agent is indifferent between getting T with certainty, and getting R on a branch of weight $U(T)$ and S otherwise. (We need to appeal to the continuity axioms to establish this and rule out the possibility of rewards whose utilities differ only infinitesimally.)

Now consider an act which leads to rewards R, S, T with weights w(R), w(S) and w(T) respectively. The agent's future selves in the T branch are indifferent between doing nothing and performing an act that delivers R with weight $U(T)$ and S otherwise. Applying diachronic consistency once more, the original agent is indifferent between the original act and an act which delivers R with weight $w(R) + w(T)U(T)$ and S with weight $w(S) + (1 - U(T))w(T)$. Note that the utilities of these acts are the same: in this particular case, the agent is indifferent between two acts iff they have the same utility. Generalizing the argument, and applying the dominance lemma, tells us that one act is preferred to another iff its utility is higher.

The continuity axioms play only a limited role in these arguments. They serve to rule out situations where two rewards are infinitesimally, or infinitely, different in value; they are also required to handle the generalization to acts which have infinitely many rewards as possible outcomes.

5.6 Formal statement of the axioms

As promised, in this section and the next I lay out the formal version of my decision theory and its associated proofs. The reader who is happy to take on trust my mathematics—and my reassurances that there has been no sleight of hand—is welcome to skip to section 5.8.

A *quantum decision problem* is specified by:

- A separable Hilbert space $\mathcal{H}$. Given a countable set $\mathcal{S}$ of subspaces of $\mathcal{H}$, I write $\vee\mathcal{S}$ (the *disjunction* of $\mathcal{S}$) for the closure of the span of $\cup\mathcal{S}$, and $\wedge\mathcal{S}$ (the *conjunction* of $\mathcal{S}$) for the closure of $\cap\mathcal{S}$; Given subspaces E and F, I define $E \vee F = \vee\{E, F\}$ and likewise for $\wedge$, and I write Π_E for the projector onto E.

- A complete Boolean algebra $\mathcal{E}$ of subspaces of $\mathcal{H}$, the *event space*. (So $\mathcal{E}$ contains $\mathcal{H}$ and is closed under $\vee$, $\wedge$, and taking the complement.) I define a *partition* of an event E to be a set of mutually orthogonal events whose conjunction is E.

- A subset $\mathcal{M}$ of $\mathcal{E}$, the *macrostates*, such that for any event E, there is a partition of E by macrostates.

- For each $E \in \mathcal{E}$, a set $\mathcal{U}_E$ of unitary operators from E into $\mathcal{H}$, which we call the set of *acts available at* E. We write $\mathcal{O}_U$ for the smallest event containing the range of the act U^{14} and require that the choice of available acts satisfies:

    1. *Restriction*: If $E, F \in \mathcal{E}$ and $F \subset E$, then if U is available at E then the unitary map $U|_F$, defined by $U|_F\psi = U\psi$ whenever $\psi \in F$, is available at F.

    2. *Composition*: If U is available at E, and V is available at $\mathcal{O}_U$, then VU is available at E.

    3. *Indolence*: For any event E, if there are any acts available at E then the identity $\mathbf{1}_E$ is available at E. (More precisely, the embedding map of E into $\mathcal{H}$ is available at E.)

    4. *Continuation*: If U is available at some E, then there is some act available at $\mathcal{O}_U$.

    5. *Irreversibility*: If E and F are orthogonal and U is available at $E \vee F$, $\mathcal{O}_{U|_E} \wedge \mathcal{O}_{U|_F} = \emptyset$.

- A partition $\mathcal{R}$ of $\mathcal{E}$ (i.e. a set of mutually orthogonal elements of $\mathcal{E}$ whose disjunction is $\mathcal{H}$), the set of *rewards*. These represent payoffs an agent could get.

---

[14] We can define $\mathcal{O}_U$ explicitly as the conjunction of all events containing the range of U; this suffices to show that $\mathcal{O}_U$ is well-defined.

The simplest choice of macrostates and event space is to pick some particular set of orthogonal subspaces of $\mathcal{H}$ whose disjunction is $\mathcal{H}$, take this as $\mathcal{M}$, and take $\mathcal{E}$ to be the set of all disjunctions of subsets of $\mathcal{M}$; this is the sense of 'macrostate' and 'event' used in the informal version of the proof. However, we could equally well take $\mathcal{E}$ to be an arbitrary Boolean algebra of subspaces and define $\mathcal{E} = \mathcal{M}$. (As I noted above, this sort of formalization may be more appropriate for decision problems with a less natural discrete structure.)

Rays within $\mathcal{H}$, as usual, are called states. I adopt the usual convention of representing a ray by any vector within it and of blurring the distinction between the two; I do not require that vectors representing states be normalized (this is just for notational convenience) though I often tacitly assume that they are nonzero. If $\mathcal{B}(E, \mathcal{H})$ is the set of unitary maps from E into $\mathcal{H}$, it can naturally be regarded as a subset of $\mathcal{B}(\mathcal{H}, \mathcal{H})$ by identifying U with $U\Pi_E$; as such, $\mathcal{B}(E, \mathcal{H})$ inherits the norm topology.

I introduce a few derived concepts. The *weight* $\mathcal{W}_\psi(E|U)$ of an event E with respect to a (nonzero) state ψ and an act U is defined by

$$\mathcal{W}_\psi(E|U) = \|\Pi_E U|\psi\rangle\|^2 / \||\psi\rangle\|^2 = \langle\psi|U^\dagger \Pi_E U|\psi\rangle / \langle\psi|\psi\rangle. \quad (5.13)$$

A *reward function* is any function w from $\mathcal{R}$ to $[0, 1]$ such that $\sum_{r\in\mathcal{R}} w(r) = 1$. Any pair of a state $\psi \in E$ and an act U available at E determine a reward function

$$R_{\psi, U}(r) = \mathcal{W}_\psi(r|U) \quad (5.14)$$

which I call the *characteristic reward function* of U and ψ.

A set $\mathcal{F}$ of events is *available* if its elements are mutually orthogonal and there is at least one act available at $\vee\mathcal{F}$. (An event is available iff its singleton set is available.)

Finally, if $\mathcal{S}$ is any set of rewards, I say that an act has rewards in $\mathcal{S}$ iff its range is a subset of $\vee\mathcal{S}$. If u is a real function of $\mathcal{S}$ and U is an act whose rewards are in $\mathcal{S}$, the *expected utility* of U with respect to a state ψ (and, tacitly, with respect to u) is

$$\text{EU}_\psi(U) = \sum_{r\in\mathcal{S}} \mathcal{W}_\psi(r|U)u(r) \equiv \sum_{r\in\mathcal{S}} R_{\psi, U}(r)u(r). \quad (5.15)$$

Stating the richness axioms is a little fiddly, because of the need to make sure not only that certain acts (erasures, branchings, etc.) are available everywhere, but to make sure that they are available on multiple branches concurrently. To state them in a concise way, I make the following definitions. First, if $\mathcal{P} = \{p_1, p_2, \ldots\}$ is a (countable or finite) set of positive real numbers whose sum is unity and M is a macrostate in reward r, then a $\mathcal{P}$-branching of M is some act U available at M such that $\mathcal{O}_U \subset r$ and such that there is a partition $\mathcal{M} = \{M_1, M_2, \ldots\}$ of $\mathcal{O}_U$ by macrostates with $\mathcal{W}_\psi(M_i|U) = p_i$ for any $\psi \in M$. (Informally, a $\mathcal{P}$-branching is an act which splits the agent's branch into many branches, each having the same weight as an element of $\mathcal{P}$, but without changing the rewards that the agent gets.)

Secondly, if M and M' are macrostates with $M \subset r$ and $M' \subset r$ for some reward r, and ψ, ψ' are states in M, M' respectively, then an *erasure* of ψ and ψ' is a pair of acts U, U' available at M and M' respectively, such that $\mathcal{O}_U$ and $\mathcal{O}_{U'}$ are both subsets of r and $U\psi = U'\psi'$.

And thirdly, if $\mathcal{F}$ is an available set of events, an *act function* $\mathcal{U}$ for that set is a function which assigns to each $F \in \mathcal{F}$ an act $\mathcal{U}(F)$ available at F. An act function is *compatible* if

$$\sum_{F \in \mathcal{F}} \mathcal{U}(F) \Pi_F \qquad (5.16)$$

is available at $\vee \mathcal{F}$.

The richness axioms are now stateable:

Reward availability. Suppose that $\mathcal{F}$ is an available set of macrostates and f is a function from $\mathcal{F}$ into rewards.

Then there is a compatible act function $\mathcal{U}$ for $\mathcal{F}$ with $\mathcal{U}(F) \subset f(F)$ for all $F \in \mathcal{F}$.

Branching availability. Suppose that $\mathcal{F}$ is an available set of macrostates and for each $F \in \mathcal{F}$, ψ_F is a nonzero state in F and $\mathcal{P}_F$ is a (finite or countable) set of positive real numbers summing to unity.

Then there is a compatible act function $\mathcal{U}$ for $\mathcal{F}$ such that for each $F \in \mathcal{F}$, $\mathcal{U}(F)$ is a $\mathcal{P}_F$-branching of ψ_F.

Erasure. Suppose that $\{r_1, r_2, \ldots\}$ is a (finite or countable) set of rewards, that $\mathcal{M} = \{M_1, M_2, \ldots\}$ and $\mathcal{N} = \{N_1, N_2, \ldots\}$ are two available sets of macrostates with $M_i \subset r_i$ and $N_i \subset r_i$, and that for each i, $\psi_i \in M_i$ and $\varphi_i \in N_i$ are nonzero states.

Then there are compatible act functions $\mathcal{U}$ for $\mathcal{M}$ and $\mathcal{V}$ for $\mathcal{N}$ such that, for each i, $(\mathcal{U}(M_i), \mathcal{V}(N_i))$ is an erasure of ψ_i and φ_i.

Problem continuity. For every available E, the set of acts available at E is an open subset (in operator norm topology) of the set of unitary maps from E to $\mathcal{H}$.[15]

Notice that reward availability and branching availability together entail that for any reward function and any $\psi \in E$, there is an act U available at E such that ψ and U have that reward function as their characteristic reward function.

We now define a *state-dependent solution* to a decision problem as specified by an assignment to every available macrostate E, and every state $\psi \in E$, of a two-place relation $\succ^\psi$ on the acts available at E. (Strictly our notation should include E but for simplicity, its value will always be tacit.)

We call an event N *null* for a given state ψ and act U iff, whenever acts V_1 and V_2 are identical on the complement of N, $V_1 U \sim^\psi V_2 U$. (So an event is null if the agent doesn't care what happens to his future selves, if any, in the branch defined by that event. We will shortly see that, as expected, an event is null iff there are in fact no such future selves.) It is easy to see that any finite union of null sets is null, as is any subset of a null set.

We can now state the rationality axioms:

Ordering. For every ψ for which it is defined, $\succ^\psi$ is a total ordering. That is: it is transitive, asymmetric, and the relation $\sim^\psi$, defined by $E \sim^\psi F$ iff neither $E \succ^\psi F$ nor $F \succ^\psi E$, is an equivalence relation. (As usual, we write '$E \succeq^\psi F$' as an abbreviation for 'either $E \succ^\psi F$ or $E \sim^\psi F$'.)

Diachronic consistency. Suppose U is available at E, and V_1 and V_2 are available at $\mathcal{O}_U$. Then:

(i) If there is some partition $\mathcal{P}$ of $\mathcal{O}_U$ into macrostates such that $V_1|_E \succeq^{\Pi_E U \psi} V_2|_E$ for every element E of the partition not null with respect to ψ and U, then $V_1 U \succeq^\psi V_2 U$.

---

[15] The operator norm topology on the set of linear maps between normed spaces V and W is defined by the norm $\|U\| = \sup\{\|Ux\| : \|x\| = 1\}$. The set of unitary maps from E to $\mathcal{H}$ is a subset of the set of all maps between those two spaces, and inherits the latter's topology.

(ii) If in addition, $V_1|_E \succ^{\Pi_E U\psi} V_2|_E$ for at least one such E, then $V_1 U \succ^\psi V_2 U$.

Macrostate indifference. If:

- U, V are acts available at M;
- U', V' are acts available at M';
- $\mathcal{O}_U \subset M_1 \wedge r_1$ and $\mathcal{O}_{U'} \subset M_1 \wedge r_1$ for some macrostate M_1 and reward r_1;
- $\mathcal{O}_V \subset M_2 \wedge r_2$ and $\mathcal{O}_{V'} \subset M_2 \wedge r_2$ for some macrostate M_2 and reward r_2

then for any ψ, ψ' with $\psi \in M$ and $\psi' \in M'$, $U \succeq^\psi V$ iff $U' \succeq^{\psi'} V'$.

Branching indifference. If:

- r is a reward;
- M is a macrostate with $M \subset r$;
- U is available at M;
- $\psi \in M$ and $U\psi \in r$

then $U \sim^\psi 1_M$.

State supervenience. If:

- $\psi \in E$ and $\psi' \in E'$ for macrostates E, E';
- U and V are available at E, and U' and V' are available at E';
- $U\psi = U'\psi'$ and $V\psi = V'\psi'$

then $U \succ^\psi V$ iff $'U \succ^{\psi'} V'$.

Solution continuity. If E is a macrostate and $\psi \in E$, U, U' are available at E, and $U \succ^\psi U'$, then in the space of unitary maps from E into $\mathcal{H}$ there are neighbourhoods (in norm topology) $\mathcal{N}, \mathcal{N}'$ of U, U' respectively such that any act in $\mathcal{N}$ available at E is preferred (at ψ) to any act in $\mathcal{N}'$ available at E.

(Are these axioms guaranteed consistent? Yes: Box 5.1 provides a concrete model for them.)

Given a solution to a quantum decision problem, we can use it to define a preference ordering on rewards: for any two rewards, $r_1 \succ r_2$ iff there is some macrostate E, some state $\psi \in E$, and acts U_1, U_2 available at E such that $\mathcal{O}_{U_i} \subset r_i$ and $U_1 \succ^\psi U_2$. Provided that the problem is reward-available and the solution is macrostate-indifferent and branching-indifferent, this preference order is a total ordering on $\mathcal{R}$. If r and s are rewards with

Box 5.1. Consistency of the axioms

Hopefully the axioms I have specified look reasonable: realistic decision problems, appropriately idealized, will satisfy them. To show that they are consistent in a mathematical sense, though, requires an actual model, and I give one here.

Let $\mathcal{H}_R$ be a two-dimensional Hilbert space with an orthogonal basis $\{|+\rangle, |-\rangle\}$; for each $N > 0$ let $\mathcal{H}_N$ be an N-dimensional Hilbert space with an orthonormal basis $\{|N, 1\rangle, |N, 2\rangle, \ldots |N, N\rangle\}$.

Now: take the Hilbert space of our decision problem to be

$$\mathcal{H} = \mathcal{H}_R \otimes \left(\oplus_{I=1}^{\infty} \mathcal{H}_I \right), \tag{5.17}$$

so that a complete basis of states is

$$|\pm\rangle \otimes |N, M\rangle \ \ (M \le N), \tag{5.18}$$

and take the macrostates to consist of all the one-dimensional subspaces spanned by each of these states, and the events to be all disjunctions of macrostates. The available events are all those which are contained in some fixed $\mathcal{H}_R \otimes \mathcal{H}_N$, and the acts available at an available event contained in $\mathcal{H}_R \otimes \mathcal{H}_N$ are all unitary maps from $\mathcal{H}_R \otimes \mathcal{H}_N$ to $\mathcal{H}_R \otimes \mathcal{H}_{N'}$, with $N' > N$. The reward subspaces are $\mathcal{H}^{\pm} = \{\text{Span} |\pm\rangle\} \otimes \mathcal{H}$. Finally, an act U is preferred to an act U' at $|\psi\rangle$ iff

$$\|(|+\rangle\langle+| \otimes 1)U|\psi\rangle\| > \|(|+\rangle\langle+| \otimes 1)U'|\psi\rangle\|. \tag{5.19}$$

I leave readers to satisfy themselves that this system does indeed obey the axioms; the preference order is, of course, the Born Rule.

$r \precsim s$, I will say that a reward t is between r and s iff $s \succeq t \succeq r$; I write $[r, s]$ for the set of rewards between r and s.

If $\mathcal{M}$ consists of some set of orthonormal subspaces (as in the informal proof), then this observation more or less exhausts the usefulness of macrostate indifference. At the other extreme, if $\mathcal{M} = \mathcal{E}$ then macrostate indifference actually entails branching indifference. The distinction between the axioms, then, is a matter of how we mathematically represent the branching structure—which is appropriate, since the motivation for branching indifference itself is that the details of that structure are an unphysical artefact of the mathematics. (See Box 5.2 for details of this.)

Box 5.2. The irrelevance of macrostates

The specification of a particular subset of events as 'macrostates' is actually artificial: in reality, there is no natural finest graining of a decoherent history space, and so no natural choice of macrostates. The branching indifference axiom codifies this, and we should expect that, as such, the choice of macrostates can in some sense be dropped from the formal presentation of the theory. That this is the case will be shown in this section.

To begin, suppose that $\mathcal{P} = \langle \mathcal{H}, \mathcal{E}, \mathcal{M}, \mathcal{U}, \mathcal{R} \rangle$ is a quantum decision problem and that $\succ^{\psi}$ is a state-dependent solution to $\mathcal{P}$. Then for any set $\mathcal{M}'$ with $\mathcal{M} \subset \mathcal{M}' \subset \mathcal{E}$, a state-dependent solution $\succ_*^{\psi}$ of the decision problem $\mathcal{P}' = \langle \mathcal{H}, \mathcal{E}, \mathcal{M}', \mathcal{U}, \mathcal{R} \rangle$ is an extension of $\succ^{\psi}$ if whenever $\succ^{\psi}$ is defined, so is $\succ_*^{\psi}$, and the two agree. An extension is minimal if no restriction of it would count as an extension.

We can now prove the following:

Extension theorem. Suppose that:
  (i) $\mathcal{P} = \langle \mathcal{H}, \mathcal{E}, \mathcal{M}, \mathcal{U}, \mathcal{R} \rangle$ is a quantum decision problem satisfying erasure, branch availability, reward availability and problem continuity;
  (ii) $\succ^{\psi}$ is a state-dependent solution to $\mathcal{P}$ satisfying ordering, diachronic consistency, macrostate indifference, branching indifference, state supervenience, and solution continuity.

Then:
  (a) The quantum decision problem $\mathcal{P}' = \langle \mathcal{H}, \mathcal{E}, \mathcal{E}, \mathcal{U}, \mathcal{R} \rangle$ also satisfies erasure, branch availability, reward availability and problem continuity, and
  (b) there is a unique minimal extension of $\succ^{\psi}$ to $\mathcal{P}' = \langle \mathcal{H}, \mathcal{E}, \mathcal{E}, \mathcal{U}, \mathcal{R} \rangle$ such that it satisfies ordering, diachronic consistency, macrostate indifference, branching indifference, state supervenience, and solution continuity.

(continued)

Box 5.2. Continued

I only sketch the proof, leaving the details to the reader. The first part is fairly trivial, and just follows from the requirement that any event is partitioned by macrostates. For the second part, note firstly that by definition, there will be some set S of states, act-closed with respect to $\mathcal{P}$, on which $\succ^{\psi}$ is defined. There is a unique smallest set S', act-closed with respect to $\mathcal{P}$ and containing S, and we define our extension on this set: if we can show that it exists and is unique on this set, this will suffice to establish that there is a unique minimal extension.

We now construct the extension as follows. If $\psi \in S'$, then by its minimality (and by the composition requirements on acts), there will be a macrostate $M \in \mathcal{M}$, a state $\varphi \in M$, and an act U available at M, such that $U\varphi = \psi$. We can then define $V \succ^{\psi} W$ iff $VU \succ^{\varphi} WU$. This defines a solution to $\mathcal{P}'$; I leave it to the reader to satisfy themselves that the solution obeys the required axioms.

5.7 Formal statement and proof of the representation theorem

Equivalence lemma. Suppose that:

(i) $\mathcal{P}$ is a quantum decision problem satisfying erasure, branch availability and reward availability;

(ii) $\succ^{\psi}$ is a state-dependent solution to $\mathcal{P}$ satisfying ordering, diachronic consistency, macrostate indifference, branching indifference, and state supervenience;

(iii) U and V are available at E, and U' and V' are available at E';

(iv) $\psi \in E$ and $\psi' \in E'$;

(v) $R_{\psi,U} = R_{\psi',U'}$ and $R_{\psi,V} = R_{\psi',V'}$;

(vi) The reward functions of the acts are each non-zero for only finitely many rewards.

Then $U \succ^{\psi} V$ iff $U' \succ^{\psi'} V'$.

Proof. For each reward r for which $R_{\psi,U}(r) \neq 0$, let $\mathcal{M}_r$ and $\mathcal{N}_r$ be partitions of $\mathcal{O}_U \wedge r$ and $\mathcal{O}_{U'} \wedge r$ respectively, and let $\#M_r$ and $\#N_r$ be the number of elements (finite or infinite) in $\mathcal{M}_r$ and $\mathcal{N}_r$ respectively.

Define the sets $\mathcal{P}_r$ (for each r)

$$\mathcal{P}_r = \{\mathcal{W}_{\psi'}(N|U')/\mathcal{W}_\psi(r|U) : N \in \mathcal{V}_r\} \tag{5.20}$$

These are sets of positive real numbers summing to unity, so by branching availability there is an act W available at $\mathcal{O}_U$ such that, for each r and each $M \in \mathcal{M}_r$, $W|_M$ is a $\mathcal{P}_r$-branching of $\Pi_M U \psi$: it splits $\Pi_M U \psi$, which has weight $\mathcal{W}_\psi(M|U)$, into $\#\mathcal{N}_r$ states, one for each $N \in \mathcal{N}_r$, with weights $\mathcal{W}_\psi(M|U) \times \mathcal{W}_{\psi'}(N|U')/\mathcal{W}_\psi(r|U)$. There is therefore[16] a partition $\mathcal{W}$ of $\mathcal{O}_W$ into macrostates, such that:

- For each reward r there are $\#\mathcal{M}_r \times \#\mathcal{N}_r$ elements of $\mathcal{W}$ in r.
- Each such element can be labelled by pairs of elements from $\mathcal{M}_r$ and $\mathcal{N}_r$: let us write it as $K^r_{M,N}$.
- $\mathcal{W}_\psi(K^r_{M,N}|WU) = \mathcal{W}_\psi(M|U) \times \mathcal{W}_{\psi'}(N|U')/\mathcal{W}_\psi(r|U)$.

Furthermore, by branching indifference, $W|_M \sim^{\Pi_M U \psi} 1_M$ for any macrostate M, and hence by diachronic consistency, $WU \sim^\psi U$.

Applying the same procedure with U and U' reversed yields an act W' such that $W'U \sim^\psi U$, and a partition $\mathcal{W}'$ of $\mathcal{O}_{W'}$ by macrostates, such that

- For each reward r there are $\#\mathcal{M}_r \times \#\mathcal{N}_r$ elements of $\mathcal{W}'$ in r.
- Each such element can be labelled by pairs of elements from $\mathcal{M}_r$ and $\mathcal{N}_r$: we write it as $K'^r_{M,N}$.
- $\mathcal{W}_{\psi'}(K'^r_{M,N}|WU) = \mathcal{W}_\psi(M|U) \times \mathcal{W}_{\psi'}(N|U')/\mathcal{W}_{\psi'}(r|U')$.

But since

$$\mathcal{W}_\psi(r|U) \equiv \mathcal{R}_{\psi,U}(r) = \mathcal{R}_{\psi',U'}(r) \equiv \mathcal{W}_{\psi'}(r|U'), \tag{5.21}$$

it follows that $\mathcal{W}_\psi(K^r_{M,N}|WU) = \mathcal{W}_{\psi'}(K'^r_{M,N}|W'U')$.

So we have constructed acts W, W' and partitions $\mathcal{W} = \{W_1, \ldots\}$, $\mathcal{W}' = \{W'_1, \ldots\}$ of $\mathcal{O}_W$, $\mathcal{O}_{W'}$ by macrostates such that:

1. For any i, W_i exists iff W'_i does (i.e., the two partitions have the same number of elements) and there is some reward r such that W_i and W'_i are elements of r.
2. $\mathcal{W}_\psi(W_i|WU) = \mathcal{W}_{\psi'}(W'_i|W'U')$ for all W_i.

---

[16] I appeal here to the irreversibility requirement on decision problems.

Now define

$$\chi_i = \Pi_{W_i} WU\psi / \|\Pi_{W_i} WU\psi\| \qquad (5.22)$$

and

$$\chi_i' = \Pi_{W_i'} W'U'\psi' / \|\Pi_{W_i'} W'U'\psi'\|. \qquad (5.23)$$

By erasure, there exist acts X, X' available at $\mathcal{O}_W$, $\mathcal{O}_{W'}$ such that $(X|_{W_i})\chi_i = (X'|_{W_i'})\chi_i'$. By branching indifference, $X|_{W_i} \sim^{\chi_i} 1_{W_i}$, so by diachronic consistency, $XWU \sim^\psi WU \sim^\psi U$; similarly, $X'W'U' \sim^{\psi'} U'$. Since

$$XWU\psi = \sum_i \mathcal{W}_\psi (W_i | WU)(X|_{W_i})\chi_i, \qquad (5.24)$$

it follows that $XWU\psi = X'W'U'\psi'$.

So: for U and U', we have found acts $Y = XWU$ and $Y' = X'W'U'$ such that $U \sim^\psi Y$, $U' \sim^{\psi'} Y'$, and $Y\psi = Y'\psi'$. Repeating this process for V and V', we can find acts Z, Z' such that $Z \sim^\psi V$, $Z' \sim^{\psi'} V'$, and $Z\psi = Z'\psi'$. The conclusion now follows immediately from state supervenience.

Because of the equivalence lemma, there is a unique total ordering defined on the set of all reward functions, which we once again write as $\succ$ (note that it is state-independent).

Nullity lemma. Suppose that:

(i) $\mathcal{P}$ is a quantum decision problem satisfying erasure, branching availability and reward availability;

(ii) $\succ^\psi$ is a state-dependent solution to $\mathcal{P}$ satisfying ordering, diachronic consistency, macrostate indifference, branching indifference, and state supervenience;

(iii) There exist rewards r, s with $r \succ s$.

Then an event E is null with respect to a state ψ and an act U iff $\langle \psi | U^\dagger \Pi_E U | \psi \rangle = 0$.

Proof. Let $\langle \psi | U^\dagger \Pi_E U | \psi \rangle = \alpha$. An event E is null if and only if, given acts V and W available at $\mathcal{O}_U$ which are identical except on E, $VU \sim^\psi WU$. Given the equivalence lemma, any two such acts are equivalent whenever they have the same weight function, so if E is null for ψ and U, any event E' is null with respect to some U' and ψ' whenever $\langle \psi' | U'^\dagger \Pi_{E'} U' | \psi' \rangle = \alpha$. If $\alpha > 0$, then $\alpha > 1/N$ for some N. By combining branching availability with reward availability, we can construct some act V and state φ with weight function

$$\mathcal{W}_\varphi(E_1|V) = 1/N$$

$$\mathcal{W}_\varphi(E_2|V) = \alpha - 1/N$$

$$\mathcal{W}_\varphi(E_3|V) = 1 - \alpha$$

$E_1 \vee E_2$ is null (wrt φ and V), hence E_1 is, hence any event with weight $1/N$ is. Applying branching availability and reward availability again, we can find φ', W and $F_1, \ldots F_N$ such that $\mathcal{W}_{\varphi'}(F_i|W) = 1/N$. Each F_i is null wrt φ' and W, hence so is $\mathcal{E}$. This contradicts premise (iii), since if all events are null then all rewards are equivalent.

Conversely, suppose that some event has weight zero. Its nullity now follows from state supervenience, since no change to the physical state is enacted by any transformation restricted to that event.

Dominance lemma. Suppose that
   (i) $\mathcal{P}$ is a quantum decision problem satisfying erasure, branching availability and reward availability;
   (ii) $\succ^\psi$ is a state-dependent solution to $\mathcal{P}$ satisfying ordering, diachronic consistency, macrostate indifference, branching indifference, and state supervenience;
   (iii) s, t are rewards with $s \succ t$;
   (iv) $f[\alpha]$ is the reward function defined by $f[\alpha](s) = \alpha$, $f[\alpha](t) = 1 - \alpha, f[\alpha](r) = 0$ for all other r.
Then $f[\alpha] \succ f[\beta]$ iff $\alpha > \beta$.

Proof. This is an easy corollary of the nullity lemma. Suppose $\alpha > \beta$, then by branching availability and reward availability, there will be some act A and state φ with weight function

$$\mathcal{W}_\varphi(E_1|A) = \beta$$

$$\mathcal{W}_\varphi(E_2|A) = \alpha - \beta$$

$$\mathcal{W}_\varphi(E_3|A) = 1 - \alpha$$

By reward availability there exist sets of compatible acts $\{U_1, U_2, U_3\}$ and $\{V_1, V_2, V_3\}$ such that U_i and V_i are available at E_i, and such that U_1, V_1 and U_2 have outcomes all lying in s and V_2, U_3 and V_3 have outcomes all lying in t. By macrostate indifference and branching indifference $U_i \simeq^\chi V_i$ for any $\chi \in E_i$ and in particular $U_2 \succ^\chi V_2$ for any $\chi \in E_2$.

If we define

$$W_\alpha = U_1 \Pi_{E_1} + U_2 \Pi_{E_2} + U_3 \Pi_{E_3} \tag{5.25}$$

and

$$W_\beta = V_1 \Pi_{E_1} + V_2 \Pi_{E_2} + V_3 \Pi_{E_3} \tag{5.26}$$

then by diachronic consistency, since E_2 is not null then $W_\alpha \cdot A \succ^\psi W_\beta \cdot A$. But the reward functions of $W_\alpha \cdot A$ and $W_\beta \cdot A$ are $f[\alpha]$ and $f[\beta]$ respectively, and the conclusion follows.

Utility lemma. Suppose that:

(i) $\mathcal{P}$ is a quantum decision problem satisfying erasure, branching availability, and reward availability;

(ii) $\succ^\psi$ is a state-dependent solution to $\mathcal{P}$ satisfying ordering, diachronic consistency, macrostate indifference, branching indifference, and state supervenience;

(iii) s, t are rewards with $s \succ t$;

(iv) u_s, u_t are real numbers with $u_s > u_t$.

Then there is a unique real function u on the set $\mathcal{S} = [t, s]$ of rewards between t and s such that for any macrostate E, any state $\psi \in E$, and any two acts U, V available at E whose rewards lie in a finite subset of $\mathcal{S}$,

$$U \succ^\psi V \text{ whenever } EU_\psi(U) > EU_\psi(V) \tag{5.27}$$

(where the expected utilities are defined with respect to u, of course) and such that $u(s) = u_s$ and $u(t) = u_t$.

Proof. For simplicity we assume $u_s = 1$ and $u_t = 0$ (other values lead to a simple positive affine transformation of the utility function). I define the following reward functions: $f[\alpha]$ is defined as in the dominance lemma, and $g[r]$ is defined by $g[r](r') = \delta_{r,r'}$.

I now define $u(r)$ by

$$u(r) = \text{lub}\{\alpha : g[r] \succ f[\alpha]\}. \tag{5.28}$$

Let $\{u_n(r)\}$ be a sequence of functions such that $u_m(r) \leq u(r)$ and $\lim_{n\to\infty} u_n(r) = u(r)$, and let U be any act available at E whose rewards lie in $\mathcal{S}$. We write E_r for $\mathcal{O}_U \wedge r$ and χ_r for the normalized projection of ψ onto E_r.

From branching availability and reward availability, for each n we can find a compatible set of acts $\{A_n(r) : R_{\psi,U}(r) \neq 0\}$ such that $A_n(r)$ is available at E_r and A_n has reward function $f[u_n(r)]$; we define $\mathcal{A}_n = \sum_{r \in \mathcal{S}} A_n(r) \Pi_{E_r}$. By construction, $1_{E_r} \succeq^{\chi_r} A_n(r)$ for all r and n, so by diachronic consistency $U \succeq^{\psi} \mathcal{A}_n \cdot U$.

By definition, the reward function of $\mathcal{A}_n \cdot U$ (with respect to ψ) is $f[\lambda_n]$, where

$$\lambda_n = \sum_{r \in \mathcal{S}} \mathcal{W}_\psi(r|U) u_n(r). \tag{5.29}$$

So if $f[U]$ is the reward function of U (with respect to ψ), we have established that $f[U] \succeq f[\lambda_n]$, and hence by the dominance lemma, $f[U] \succeq f[\lambda]$ whenever $\lambda < \lambda_n$ for some n. Since $u_n(r) \to u(r)$ as $n \to \infty$ for each r, $\lambda_n \to \mathrm{EU}_\psi(U)$ as $n \to \infty$, and hence $f[U] \succ f[\lambda]$ whenever $\lambda < \mathrm{EU}_\psi(U)$. Applying the same argument with a decreasing sequence, $f[U] \prec f[\lambda]$ whenever $\lambda > \mathrm{EU}_\psi(U)$.

Now suppose that U and V are two such acts with $\mathrm{EU}_\psi(U) > \mathrm{EU}_\psi(V)$. Then for any α lying between the two expected utilities, there will exist an act W with reward function (wrt ψ) $f[\alpha]$. We have proved that $U \succ^{\psi} W$, and $W \succ^{\psi} V$, so it follows that $U \succ^{\psi} V$.

To see that this utility function is unique, note that if there were another utility function u' we could construct acts whose utilities were the same as calculated by this second utility, but not as calculated by the first; this contradicts the requirements on u'.

Born Rule theorem. Suppose that:
    (i) $\mathcal{P}$ is a quantum decision problem satisfying erasure, branching availability, reward availability and problem continuity;
    (ii) $\succ^{\psi}$ is a state-dependent solution to $\mathcal{P}$ satisfying ordering, diachronic consistency, macrostate indifference, branching indifference, state supervenience, and solution continuity.

Then there is a function u on the rewards of $\mathcal{P}$, unique up to positive affine transformations, such that if EU denotes the expected utility with respect to this function,

$$U \succ^{\psi} V \text{ iff } \mathrm{EU}_\psi(U) > \mathrm{EU}_\psi(V). \tag{5.30}$$

Proof. Note that problem continuity and solution continuity jointly entail that if $U \succ^{\psi} U'$, there are neighbourhoods $\mathcal{N}, \mathcal{N}'$ of U, and U' respectively such that all acts in $\mathcal{N}$ and $\mathcal{N}'$ are available and all acts in $\mathcal{N}$

are preferred (given ψ) to all acts in $\mathcal{N}'$. For simplicity I shall refer to this simply as continuity.

We begin by proving that if $s \succ r_1 \succeq r_2 \succ t$, then if the utilities determined by the utility lemma (via this choice of s and t) for r_1 and r_2 coincide, then $r_1 \sim r_2$. Let this utility function be u and again, for convenience take $u(s) = 1$ and $u(t)=0$. Fix E and $\psi \in E$, and let U_1 and U_2 be acts available at E whose ranges lie in r_1 and r_2 respectively (by reward availability, some such acts exist). If $r_1 \succ r_2$, then $U_1 \succ^\psi U_2$. By continuity, there must exist neighbourhoods $\mathcal{N}_1$, $\mathcal{N}_2$ of U_1 and U_2 such that any available act in $\mathcal{N}_1$ is preferred (given ψ) to any available act in $\mathcal{N}_2$.

Now let $f_1[\alpha]$ and $f_2[\alpha]$ be reward functions with $f_1[\alpha](r_1) = 1 - \alpha$, $f_1[\alpha](t) = \alpha$ and $f_2[\alpha](r_2) = 1 - \alpha, f_2[\alpha](s) = \alpha$. By branching availability and reward availability, there must exist some α, and some acts $U_{i,\alpha}$, such that $U_{i,\alpha} \in \mathcal{N}_i$ and the reward function of $U_{i,\alpha}$ (with respect to ψ) is $f_i[\alpha]$.

So we have that $U_{1,\alpha} \succ U_{2,\alpha}$. But $EU_\psi(U_{1,\alpha}) < EU(U_1) \equiv u(r_1)$, and $EU_\psi(U_{2,\alpha}) > EU(U_2) \equiv u(r_2)$. So by the utility lemma we must have that $u(r_1) > u(r_2)$.

We can now define a utility function for the whole of $\mathcal{R}$. For any rewards r_1, r_2 with $r_1 \succ r_2$, and any real numbers x_1, x_2 with $x_1 > x_2$, I will write $u[r_1, r_2, x_1, x_2]$ for the unique utility function determined on $[r_2, r_1]$ by setting the utility of r_i to x_i.

Now, let s, t be any two rewards with $s \succ t$ (if there are no such rewards, the theorem is true trivially). I define the utility of any reward r by:

- If $s \succeq r \succeq t$, $u(r) = u[s, t, 1, 0](r)$.
- If $r \succ s$, $u(r)$ is the unique value fixed by requiring that $u[r, t, u(r), 0](s) = 1$.
- If $t \succ r$, $u(r)$ is the unique value fixed by requiring that $u[s, r, 1, u(r)](s) = 0$.

(Notice that this definition relies on the assumption that the utilities of s and t are guaranteed to be distinct.)

I now prove that for acts with finitely many rewards, if $U_1 \succ^\psi U_2$ then $EU_\psi(U_1) > EU_\psi(U_2)$. For suppose that $U_1 \succ^\psi U_2$. By continuity, if f is the reward function of U_1 (with respect to ψ) then it will be possible to find some act V with reward function g such that, for some rewards r_1 and r_2 with $r_1 \succ r_2$:

- $V \succ^{\psi} U$;
- If $r \neq r_1$ and $r \neq r_2$, $g(r) = f(r)$;
- $g(r_1) < f(r_1)$; $g(r_2) > f(r_2)$.

This means that we must have $\mathrm{EU}_{\psi}(V) \geq \mathrm{EU}_{\psi}(U_2)$; since $\mathrm{EU}_{\psi}(V) < \mathrm{EU}_{\psi}(U_1)$, it follows that $EU_{\psi}(U_1) > EU_{\psi}(U_2)$.

This suffices to prove the Born Rule theorem under the assumption that any act has only finitely many non-null rewards. To extend to the infinite case, let U_1 and U_2 be arbitrary acts, and suppose for some ψ that $U_1 \succ^{\psi} U_2$. By continuity, if f_1 and f_2 are the reward functions (given ψ) of U_1 and U_2, it will be possible to find a finite subset $\mathcal{R}_0$ of $\mathcal{R}$, and acts V_1, V_2 with reward functions g_1, g_2, such that:

- $V_1 \succ^{\psi} V_2$;
- $g_i(r) = f_i(r)$ for $r \in \mathcal{R}_0$;
- If $r \notin \mathcal{R}_0$, then $g_1(r) = s$, and $g_2(r) = t$, where $s \succ t$.

Since V_1 and V_2 have only finitely many non-null rewards, $\mathrm{EU}_{\psi}(V_1) > \mathrm{EU}_{\psi}(V_2)$. But by construction $\mathrm{EU}_{\psi}(U_1) > \mathrm{EU}_{\psi}(V_1)$ and $\mathrm{EU}_{\psi}(U_2) < \mathrm{EU}_{\psi}(V_2)$, so $\mathrm{EU}_{\psi}(U_1) > \mathrm{EU}_{\psi}(U_2)$.

5.8 Other proposed strategies for action

In the nine years since Deutsch's original paper on decision-theoretic probability, a bewildering variety of alternative strategies for rational action have been proposed in the literature and in discussion. Some of these strategies have independent motivations; some are purely meant as counterexamples; all contradict the Born Rule, and so all violate the decision-theoretic axioms of this chapter.

This being the case, perhaps there is little need to discuss the alternative strategies: a proof is a proof. On the other hand, it may be instructive to show exactly how some of these alternative proposals violate my axiom scheme: apart from casting light on the motivation for the axioms, this may show how what appear to be coherent and even plausible strategies come apart on close inspection.

The proposed counterexamples, as will become apparent, break into four categories. There are the 'wrong-probability' rules, which also require an agent to maximize expected utility but with respect to some probability measure other than the Born Rule. There are the 'no-probability' rules, which (purportedly) cannot be represented in terms of expected utilities at all. There are what might be called the 'I-don't-want-to-play' rules, which are not so much positive strategies as arguments against the existence of any strategy. And one special group, the contextual strategies, deserve a category of their own.

5.8.1 Branch counting

Description. each branch is given an equal probability, so that if there are N branches following a particular experiment, each branch is given probability $1/N$. Utility is then maximised with respect to this probability.

Origin. Has been reinvented innumerable times, but the first proponent may have been Graham (1973).

Rationale. Each branch contains a copy of me; none of them can detect, nor care about, their quantum-mechanical weight; so I should not care about that weight either, and so I have no reason to prefer one over another.

Why it is irrational. As I discussed in section 4.3, branch number cannot actually be defined; even if it could, the branch counting rule leads to inconsistency over time. Within my decision theory, this shows up in the fact that branch counting violates the combination of branching indifference and diachronic consistency. For consider two acts $A1$ and $A2$: $A1$ consists of a two-outcome measurement (a spin measurement, say) followed by a reward of utility r in the spin-up branch. $A2$ consists of $A1$, followed by another two-outcome measurement in the spin-up branch. By branching indifference, the agent who gets the reward is indifferent about whether or not he makes a further measurement; by diachronic consistency, then, the original agent is indifferent between $A1$ and $A2$. But the utility of $A1$ (in which there are two branches, one of which provides a reward) is $r/2$; the utility of $A2$ is $2r/3$.

5.8.2 The fatness rule

Description. each branch is given a probability proportional to its quantum-mechanical weight multiplied by the mass of the agent in kilograms (such that the total probability is equal to one). Utility is maximized with respect to this probability.

Origin. David Albert (2010).

Rationale. Albert says, tongue in cheek, that an agent should care about branches where he is fatter because 'there is more of him' on that branch. He isn't serious, though: the rule is purely presented as a counterexample.

Why it is irrational. It violates diachronic consistency. Albert's agent is (*ex hypothesi*) indifferent to dieting. But he is not indifferent to whether his future selves diet: he wants the ones on branches with good outcomes to gain weight, and the ones on branches with bad outcomes to lose weight.

This is perhaps a good point to recall the rationale for diachronic consistency: rational action takes place over time and is incompatible with widespread conflict between stages of an agent's life. In the case of the fatness rule, agents have motivation to coerce their future selves—by hiring 'minders', say—into dietary programs that they will resist. Multiply this conflict indefinitely many times (for branching is ubiquitous) and rational action becomes impossible.

(To object 'maybe rational action is impossible in the Everett interpretation' is acceptable only if some reason can be given for the Born Rule being irrational; see below under the 'curl-up-and-die' rule for discussion of this possibility.)

5.8.3 The fake-state rule

Description. The agent maximizes expected utilities as for the Born Rule, but using a quantum state other than the physically real one.

Origin. Suggested many times in conversation.

Rationale. None in particular, though it is often intended to undermine the connection between the 'real' state and the physics.

Why it is irrational. It violates state supervenience. There will be cases where two acts produce the same physical state but where one produces a different fake state than the other. (This is inevitable: any two distinct quantum states are invariant under different sets of transformations.) The fake-state rule will then give the acts different utilities; state supervenience rules this out. Or, put another way: the fake-state rule assigns different values to the same physical state under two different descriptions.

Note that it is crucial here—as elsewhere in decision theory—that the agent has a choice between different actions, and therefore between different sets of histories and weights. Of course, in a deterministic Universe it is fixed which action will actually occur, but this does not remove the necessity of defining preferences, and hence indirectly probabilities, over a wide range of actions.

5.8.4 The distributive-justice rule

Description. The agent does not maximize expected utilities at all. He treats his various successors in rather the way that a just ruler would treat his various subjects: in particular, he will not allow the suffering of one even if it brings great advantage to others.

Origin. Huw Price (2010).

Rationale. Any action we choose generates a multitude of individuals; we have a duty to treat them all ethically, and in particular we would not be morally justified in letting one suffer unduly for the others' benefit.

Why it is irrational. The rule is very underspecified, so it isn't easy to answer this, but on natural precisifications it either violates continuity or is not actually a counterexample to the Born Rule.

To expand: a large part of what Price wants can be achieved by an appropriate utility function. An agent moved by Price's concerns can drastically increase the disutility of bad consequences and scale down the utility of good consequences, with the effect that trade-offs of the sort he considers get a much lower utility and so will tend to be rejected in favour of more equitable options. There is nothing in Everettian decision theory that prevents an agent from making such modifications to their utility function on recognizing the ethical consequences of the Everett interpretation.[17]

If Price wants to hold that *no* amount of suffering, however low-weight the branch on which it occurs, is acceptable, then this strategy will not work, but there is a clash with continuity. Suppose there are three rewards r_1 and r_2 with $r_1 \succ r_2$, and a (dire) punishment p. Price will prefer r_1 to r_2 but will prefer r_2 to $(1 - w)r_1 + wp$, whatever the value of w; clearly this violates continuity.

---

[17] Personally, though, I don't feel inclined to. Call me callous.

Now, I think the physical arguments for continuity are pretty unassailable, but it is worth noting that the principle is only really used in my proof precisely to rule out infinite or infinitesimal utilities. (The only other use is for the mathematically convenient but physically tangential purpose of extending the Born Rule to the case of infinitely many rewards.) If such utilities are allowed, there is no problem with extending the Born Rule to cover even Price's case (though the utility function will have to be modelled in non-standard analysis and the maths will start getting fiddly). And in fact, precisely the same situation has arisen in *classical* decision theory, and the structure axioms of classical decision theory are selected precisely to rule out the case of infinite (dis)utility. (See Savage 1972: 81–2.)

5.8.5 The variety rule

Description. An agent prefers A to B, but prefers receiving A in half the branches and B in the other half to either A or B.

Author. Suggested in a seminar by Adam Elga in 2004; has not appeared in print as far as I am aware.

Rationale. An agent may regret having to make one choice or another, and may rather like the idea that one version of himself makes one choice, one another. (In Elga's example, a student prefers physics to history but likes both; that student might prefer to do history in one branch, physics in the other.)

Why it is irrational. It either violates diachronic consistency, or it isn't a counterexample to the Born Rule.

To expand: suppose you are the agent who chose history. What prevents you changing your mind and switching to physics? It doesn't, after all, hurt your counterpart in the physics branch. This would clearly violate diachronic consistency.

But perhaps you wouldn't choose to switch back. That's to say that although you prefer doing physics to doing history, you prefer doing history *as a result of a situation in which a certain process chose history for you* rather than doing physics *against the result of that process*. In that case, the utility you are assigning to (history-after-process) is higher than the utility you assign to (physics-against-process), and indeed higher than (physics-without-process). The different situations in which you end up doing history count as different rewards.

Exactly analogous situations can arise in classical decision theory. A student might decide that on balance he'd rather do physics than history, but nonetheless resolves to decide by the toss of a coin (because, say, he finds it comforting to have the decision taken from his hands; the reader can probably supply other motivations). That student, again, will place a higher utility on (history after coin toss) than on (physics).

Of course, if every outcome's utility depended sensitively on the circumstances in which that reward arose, decision theory couldn't get off the ground: there would be no way to define probability without being able to have the same reward available in different acts. But again, this is not specific to quantum decision theory.

5.8.6 The anything-goes rule

Description. Not so much a 'rule' as a rejection of the need to have one: according to this position, any transitive preference ordering over acts is rationally acceptable.

Origin. Suggested by Tim Maudlin in seminars on multiple occasions; frequently suggested in conversations.

Rationale. Everettian quantum theory is deterministic, and we already have a perfectly acceptable deterministic decision theory: its only axiom is ordering. So any transitive ordering should be fine.

Why it is irrational. Even in deterministic decision theory, transitivity is not the only constraint. Rational agency is not possible without diachronic consistency; in addition, preference orders have to be defined on actual physical acts, so mathematical modelling of those orders should require an agent to be indifferent between the same state of affairs differently defined. Furthermore, the only interesting decision-theoretic strategies are those which are physically performable in at least an idealized sense. All of the rationality axioms of this chapter fit into one of these categories; even in deterministic decision theory, then, they are rationally required.

5.8.7 The curl-up-and-die rule

Description. The converse of the anything-goes rule, this is not so much a 'rule' for rational action as the claim that *no* rational strategy is possible in Everettian quantum theory.

Origin. Frequently suggested in conversation.

Rationale. Various; see below.

Why it is irrational. Unless there is something concretely wrong with the Born Rule, there is no case to be made that no rational strategy is available: the Born Rule is available.

I am aware of two general objections to the rationality of the Born Rule, though. The first is that it is rationally compulsory for an agent to weight each branch equally; since the Born Rule violates this requirement, it cannot be rational (and if only the Born Rule is rational, rationality is impossible in an Everettian universe). Arguments are seldom given for the suggestion that this is a rational requirement (I can see that at best it might be a rational *desideratum*, but it's not at all clear to me why, in a universe where it isn't physically possible to obey the requirement, we should be unable to settle for some second-best option). In any case, though (at the risk of repetitiveness), there is no coherent notion of branch count available in quantum mechanics, so it's not even meaningful to talk of 'weighting each branch equally'.

The other objection (frequently made in discussions, and made in print by Hemmo and Pitowsky 2007) is that no strategy can be rational if it can be known in advance by those adopting it that some of them (or some of their successors) will make wrong decisions. So in particular, it is a corollary of the Born Rule that an agent measuring a long succession of identical quantum systems should regard the observed frequencies as a guide to what state each system is in; but since all sequences of results occur somewhere, some of the agent's successors will get the wrong outcome.

Now, it is true that some agents will indeed be misled in this way. But there is nothing particularly quantum-mechanical about this. If the universe is spatially infinite (as current observations support), we can guarantee that somewhere in the universe are people as similar to us as you like but whose observed statistics have systematically misled them. Even on Earth, one can fairly easily come up with similar examples. Suppose that the British government declared that it puts some people under (non-covert) surveillance at random, but that there are very few such people: only one in ten million. And suppose it is claimed that the government is lying, and actually puts many more people than that (tens of thousands, say) under surveillance. Then each person in Britain is rational to adopt the strategy: if I am under surveillance, the government is (almost certainly) lying—even though they know that if the government is not lying, five or six people in Britain will be misled into thinking it was.

Ultimately, some people get unlucky. There is no contradiction between this and the rationality of a decision-theoretic strategy, provided that strategy tells us not to care about the unlucky cases. The Born Rule tells us exactly that.

5.8.8 Contextual rules

Description. An agent's preferences conform to a probability rule that violates the principle of non-contextuality: that is, it assigns different probabilities to the outcomes of a measurement of operator X according to whether or not a compatible operator Y is measured at the same time.

Origin. Various, but a particularly forceful advocacy can be found in Hemmo and Pitowsky (2007).

Rationale. As is well known, any non-contextual quantum probability rule (and hence, any strategy for rational action expressible in terms of such a rule) can be proved to be the Born Rule applied to some (possibly mixed) state.[18] The suspicion, then, is that the decision-theoretic arguments are just a combination of Gleason's theorem (or a relative of it) with an unjustified assumption of non-contextuality.

Why it is irrational. Probably the easiest way to explain what is wrong with contextual rules is that they violate state supervenience. If we regard measurements as physical processes rather than as primitive, which operator(s) are being measured in a given process is dependent on the interests of the experimenter, and cannot simply be read off from the physics. (Consider the Stern–Gerlach experiment, for instance: is it a measurement of spin, or of position?) For a decision rule to be contextual, then, is for a rational agent to prefer a given act to the

---

[18] This is usually explained in terms of Gleason's Theorem, but this is a rather outdated approach now that POVMs, not PVMs, are widely—and in my view correctly—seen as the best way to represent measurements in quantum theory. Most of the mathematical complexity of Gleason's theorem can be dispensed with if we require our probability function to be defined on POVMs and not just PVMs. See Caves et al. (2004) for further discussion.

same act (knowably the same act, in fact) under a different description, which obviously violates state supervenience (and, I hope, is obviously irrational).

It is fair to note, though, that just as a non-primitive approach to measurement allows one and the same physical process to count as multiple abstractly construed measurements, it also allows one and the same abstractly construed measurement to be performed by multiple physical processes. It is then a nontrivial fact, and in a sense a physical analogue of noncontextuality, that rational agents are indifferent to which particular process realizes a given measurement.

In earlier work (Wallace 2002a; 2003b; 2007) I called this fact *measurement neutrality*. It is indeed a tacit premise in Deutsch's original (1999) proof of the Born Rule, as I argued in Wallace (2003b); it was an explicit premise in early versions of my proof (Wallace 2002a; 2003c). Here, though, it is a theorem (a trivial corollary of the Born Rule theorem, in fact) that measurement neutrality is rationally required. The short answer as to why is that two acts which correspond to the same abstractly construed measurement can be transformed into the same act via processes to which rational agents are indifferent. To see the long answer, reread sections 5.3–5.7.

Incidentally, Gleason's theorem (or, more accurately, its POVM generalization) is much more directly needed if we wish to generalize the results of this chapter to situations where the quantum state is unknown to the agent, as we will see in Chapter 6.

CHAPTER 5.   A rational agent, believing that the Everett interpretation is true and that the quantum state of a given system is $|\psi\rangle$, knows that measurements on that state will generally split his part of the multiverse into multiple branches, with different measurement outcomes, and different versions of the agent, on different branches; he also knows that the relative weights of these branches are given by the Born Rule, applied to the post-measurement state of the system and measurement device. Rationality considerations not different in kind to those which apply in single-universe decision making then compel the agent to act as if a set of branches of relative weight w has probability w. In other words, he is rationally required to act as if the Born Rule is true.

CHAPTER 6. Most of the relevance of probability in science is not via the decision-theoretic problem of what to predict given known probabilities, but via its inverse, the inferential problem of what to infer, about the probabilities and the physics in general, given the observed data. How is this problem addressed in Everettian quantum mechanics?

6

Everettian Statistical Inference

"But do you really mean, sir," said Peter, "that there could be other
worlds—all over the place, just round the corner—like that?"

"Nothing is more probable", said the Professor, taking off his
spectacles and beginning to polish them, while he muttered to himself, "I wonder what they *do* teach them in these schools."

C. S. Lewis, *The Lion, the Witch, and the Wardrobe*[1]

6.1 The problem of statistical inference

The Born Rule theorem which was the central result of Chapter 5 establishes that a rational agent, certain that unitary quantum mechanics is true
and that the quantum state of his branch is $|\psi\rangle$, will act as if branches in his
future had probabilities and those probabilities were equal to the (relative)
branch weights. This already suffices to give us a greater understanding
of quantum probability than we ever had of classical probability: in the
latter case, the connection between (objective) probability and action was
always simply postulated.

Nevertheless, the rather stylized betting framework used in Chapter 5
does seem to have rather little in common with the sorts of situations
in which probability is actually used in scientific practice. My goal in
this chapter is to bridge this gap, by showing how Everettian probability
understood in the sense of Chapter 5 is in fact sufficient to handle more
'realistic' uses of probability. In doing so, I will show that both the practical
and epistemic problems can be fully solved within Everettian quantum

---

[1] Lewis (1987: 49).

mechanics. Since the practical problem is largely (arguably, completely) solved by the Born Rule theorem, and since the epistemic problem seems closer to actual scientific practice, the latter will be my main focus.

It will be useful, in fact, to consider three different statistical problems from an Everettian perspective. The first might be called the *unknown state problem*: supposing that we do not in fact know the quantum state of a system, how should we use statistical data to infer it? The second is the *unknown dynamics problem*: we know the state but not some feature of the dynamics, such as the magnetic moment of the electron. The third is the *unknown theory problem*: we do not know whether quantum mechanics is correct, and we wish to see whether it is supported by experimental data. For my purposes, the unknown dynamics problem generally reduces to the unknown state problem, since experiments to measure (say) the magnetic moment of the electron generally involve starting with some known state and seeing what state it evolves into; as such, I mostly restrict the discussion to the other two problems.

Statistical inference itself is not uncontroversial, even when the Everett interpretation is set aside. For this reason (and to make the point that Everettian quantum mechanics is not hostage to a particular theory of inference), in this chapter I consider three different perspectives on it. First, and most straightforwardly (sections 6.1–6.3) I consider a fairly direct approach to the problem. Then (sections 6.4 and 6.5) I introduce the mechanism of Bayesian inference to show how an agent can be represented as assigning prior personal probabilities to various possible theories, and updating those personal probabilities upon collecting evidence. Thirdly (sections 6.6–6.9) I consider the whole problem formally, within a decision-theoretic framework which is an extension of the previous chapter's: in this framework, I prove a general representation theorem which shows that Everettian branch weight plays exactly the role that 'objective probability' is supposed to play (and which has the Born Rule theorem as a special case). The framework will also provide some insight into the status of noncontextuality within the Everett interpretation, and into what about Everettian quantum mechanics, in particular (and as opposed, say, to hidden-variable theories, operationalist readings of quantum theory, or classical mechanics), allows the Born Rule theorem and its generalizations to be proved. In the final section (6.10), I place this chapter's results in the broader context of other work on the Everettian probability problem (including my own earlier work) and of classical decision theory.

6.2 The unknown state problem: a direct approach

As a simple example of the unknown state problem, suppose that we have a large number of identical qubits (i.e. two-state systems) which we know to have all been prepared in the same pure state (which we can write without loss of generality as $\alpha|0\rangle + \beta|1\rangle$), but that we do not know what the state is. How should we go about determining that state?

If we set aside conceptual questions, and any Everett-specific worries, the answer is of course very straightforward. We should pick a basis for the system $(\{|0\rangle, |1\rangle\}$, say) and measure many copies of it with respect to that basis. If we make N measurements in total, where N is large, and we get $|0\rangle$ N_0 times, we assume that $|\alpha|^2 \simeq N_0/N$ and $|\beta|^2 \simeq (N - N_0)/N$. We then apply the same process with respect to a different basis to fix the phase.

We can describe essentially the same process from a different perspective: suppose that we tentatively believe that $|\alpha|^2 = \lambda$. To test this hypothesis, we make many measurements in the $\{|0\rangle, |1\rangle\}$ basis, and reject the hypothesis if the fraction of measurements which give result $|0\rangle$ is not $\simeq \lambda$. (Exactly how close it must be will depend on how many times we make the measurement and how confident we wish to be in our answer.) This approach to inference is sometimes called *Fisherian* after Fisher (1925).

Both of these processes define *strategies* which the experimenter has decided to adopt, with the details of the strategy dependent on what the result is. As such, both can be defined perfectly well in the Everett interpretation. Indeed, it will be possible to write down the quantum state $|\psi\rangle$ of the experimenter's branch after the strategy has been carried out (as a function of the number N of repetitions of the measurement, and of the unknown parameters α and β).

In the first example, this gives

$$|\psi\rangle = \sum_{M=0}^{N} \sqrt{\frac{N!}{M!(N-M)!}} \alpha^M \beta^{(N-M)} |\text{`estimate is } \simeq M/N\text{'}\rangle \quad (6.1)$$

where $|\text{`estimate is } \simeq M/N\text{'}\rangle$ is a post-measurement joint state of system and experimenter in which the experimenter's estimate of $|\alpha|^2$ is that it is approximately M/N. Applying the normal approximation to the binomial distribution (and exploiting the phase ambiguity in the definition of $|\text{`estimate is } \simeq M/N\text{'}\rangle$) simplifies this to

$$|\psi\rangle \simeq \mathcal{N} \sum_{M=0}^{N} \sqrt{\exp\left(-\frac{1}{2|\sigma|^2}(M/N - |\alpha|^2)^2\right)} |\text{'estimate is} \simeq M/N\text{'}\rangle$$

(6.2)

where $\sigma^2 = |\alpha|^2|\beta|^2/N$ and $\mathcal{N}$ is a normalization constant.

It is easy to see from this that, as N becomes large, branches in which the experimenter's estimate is not extremely accurate become very low weight; as such—given the Born Rule theorem—the experimenter will rationally neglect those branches, and regard the strategy as almost as good as a strategy which just delivers the correct answer with certainty.

6.3 A direct approach to theory confirmation

The methods of the last section can be applied to the unknown theory problem too. A straightforward model for the testing of scientific theories, advocated most famously by Karl Popper,[2] is *falsificationism*, sometimes called the *hypothetico-deductive method*: a theory is tested experimentally by attempting to confirm as many as possible of its predictions—in particular, those predictions which we would not otherwise have been inclined to accept. If any prediction is disconfirmed, the theory has been falsified, and should be rejected; if it passes a given test, this is generally taken as evidence in its favour.[3]

The falsification model clearly runs into problems when we try to test theories which make only probabilistic predictions. If theory T says that something will *probably* occur, and it doesn't, this is still compatible with T being true; conversely, even if the something does occur, it could just have been a coincidence. In practice, though, we usually get around this by examining situations where the probabilities are very close to zero or one, usually by repeating experiments many times.

Suppose, for instance, that theory T says that a given two-outcome experiment actually has a 60% chance of giving outcome one rather than outcome two; and suppose that conditional on T being false, we expect

---

[2] In e.g. Popper (1963).

[3] Popper infamously claimed that successful tests of a theory provided no reason at all to believe this theory, although prima facie this is very hard to square with scientific practice. See Deutsch (1997: 156–7) and Newton-Smith (1981: 44–76) for further discussion of this aspect of Popper's philosophy.

the two results to be equiprobable.[4] Applying the normal approximation again, we can estimate that if T is true, the probability, after (say) 10,000 runs of the experiment, of finding that outcome one occurs $6,000 \pm 500$ times is equal[5] to $1 - \epsilon$, where $\epsilon \sim e^{-25}$. If the 'null hypothesis' that the two outcomes are equiprobable is true, the probability of this result is $\epsilon' \sim e^{-25}$. So in practice, we take a result of $6,000 \pm 500$ to confirm theory T and any other result to disconfirm it, and tolerate the negligibly small probability that we draw the wrong conclusion from the data.

This strategy, too, can be considered in Everettian terms, taking the theory T to be EQM, Everettian quantum mechanics. Suppose first that EQM is false. Then an experimenter following the strategy will be almost certain to correctly conclude that EQM is false, so the strategy looks good so far.

Now suppose that EQM is true. Then the strategy causes branching, but in branches whose aggregate weight is very close to 1, the agent concludes correctly that EQM is true. Of course, in some branches, of total weight $\sim e^{-25}$, he instead draws the wrong conclusion—but this is not a problem for the strategy, because the Born Rule theorem tells us that the agent is rationally justified in neglecting those branches given their very low aggregate weight.

I conclude that, insofar as the methods of this section and the last are appropriate to non-Everettian statistical inference, they are equally appropriate to inferences made on the assumption that Everettian quantum mechanics is true, and to inferences about whether or not Everettian quantum mechanics *is* true. However, the methods described so far, although still widely used in experimental science, have somewhat fallen from favour amongst statisticians and philosophers of science. An alternative, and widely[6] popular, alternative approach uses the framework of personal probability sketched in section 4.8, and is normally called a *Bayesian* approach. This approach will be the topic of the next two sections.

---

[4] A more realistic situation would be complicated somewhat by the fact that presumably there are hypotheses other than T which give the same predictions as T in this case.

[5] These are crude estimates, but you get the idea.

[6] But not universally: see e.g. Glymour (1981), (a widely discussed dissent) or Norton (2011), (for more recent concerns).

6.4 The Bayesian approach to inference

The Bayesian approach relies on the idea, introduced in Chapter 4, that an agent's approach to decision-making in the face of uncertainty can be modelled by a *personal probability* over possible outcomes, such that the agent adopts that course of action which maximizes utility with respect to that probability. To apply this to scientific inference, we need to make allowance for situations where an agent (or, in this case, an experimenter) needs to update his or her personal probabilities upon receiving new evidence. For this reason, I will represent those personal probabilities not by a function $\mathrm{Cr}(\cdot)$ of one argument, but by a function $\mathrm{Cr}(\cdot|\cdot)$ which, for any propositions A and B gives their personal probability $\mathrm{Cr}(A|B)$ in A being true given that B is true. (I use 'Cr' (for *credence*, Lewis's name for personal probability) to make clear that this is one of two notions of probability in use.)

Operationally, we can as usual make sense of $\mathrm{Cr}(A|B)$ as giving the odds that the experimenter would place on A conditional on B being true. We assume that for any B, $\mathrm{Cr}(\cdot|B)$ satisfies the probability calculus; more crucially for our purposes, we assume that $\mathrm{Cr}(\cdot|\cdot)$ satisfies

$$\mathrm{Cr}(A|BC) = \frac{\mathrm{Cr}(AB|C)}{\mathrm{Cr}(B|C)} \tag{6.3}$$

for any propositions A, B, C for which $\mathrm{Cr}(B|C) > 0$, a principle known as *Bayesian updating*.

It is a direct consequence of Bayesian updating that

$$\mathrm{Cr}(A|BC) = \frac{\mathrm{Cr}(B|AC)\mathrm{Cr}(A|C)}{\mathrm{Cr}(B|C)}. \tag{6.4}$$

So if H is some scientific hypothesis (that dinosaurs existed, that the Higgs boson has mass > 120 GeV, that the quantum state of a system lies in such-and-such region of state space, etc) and E is some particular set of evidence (fossils, cloud-chamber tracks, a certain frequency of results in a set of quantum measurements, etc.), we have

$$\mathrm{Cr}(H|EC) = \left(\frac{\mathrm{Cr}(E|HC)}{\mathrm{Cr}(E|C)} \right) \mathrm{Cr}(H|C). \tag{6.5}$$

That is, the parenthetical term on the right hand side is the factor by which we should increase our personal probability in H upon learning E. In scientific situations, often $\mathrm{Cr}(E|HC)$ is determined by H itself, so this formula suffices to tell us, given $Cr(\cdot|C)$ (i.e. given the experimenter's pre-experiment personal probabilities) how those personal probabilities should be updated.

For instance, let H be the hypothesis that the dinosaurs existed and let E be the proposition that dinosaur-shaped fossils have been found. Given our auxiliary evidence (about fossil formation and the like), the probability $Cr(E|HC)$ of fossils forming given that there were dinosaurs is close to one; on the other hand, $Cr(E|\neg HC)$ would normally be close to zero (rival formation methods for fossils include widespread hoaxing, a deceiver God, hallucinations, or blind chance, none of which is assigned a high prior probability by most scientists).

Now $Cr(E|C) = Cr(EH|C) + Cr(E\neg H|C)$, and $Cr(EA|C) = Cr(E|AC)/Cr(A|C)$, so $Cr(E|C) \simeq Cr(E|HC)Cr(H|C) \simeq Cr(H|C)$. Equation 6.5 then yields

$$Cr(H|EC) \simeq \left(\frac{1}{Cr(H|C)} \right) Cr(H|C) = 1 : \tag{6.6}$$

in other words, the existence of fossils makes it almost certain that the dinosaurs existed, except for experimenters with very strange initial personal probability functions.

The Bayesian approach to *objective* probability (which I called Lewisian dualism in section 4.9, since Lewis 1980 is its clearest and most widely known statement) rests on three pillars:

1. *Synchronic personal probability.* At any given time, at which their information is represented by A, a rational agent acts in accordance with a maximum-expected-utility function, with respect to some personal probability function $Cr(\cdot|A)$ which satisfies the axioms of probability theory.

2. *Bayesian updating.* Rational agents update their probability function, upon receiving new evidence, in accordance with the Bayesian updating principle stated above.

3. *The Principal Principle.* For any proposition A and any $x \in [0, 1]$ if X is the proposition that the objective probability of A is x and E is any other (admissible[7]) information, then

$$Cr(A|XE) = x. \tag{6.7}$$

---

[7] See section 4.12's discussion of the admissibility condition.

These three principles appear to capture 'all we know about [objective probability]' (Lewis 1980: 86), or at least everything that we use objective probability for in science.[8] In particular (as Lewis shows) they capture the principle that objective probabilities can be measured, and so probabilistic theories tested, by observations of long-run, but finite, relative frequencies. For if theory T predicts that the probability of some outcome of an experiment is p, then standard arguments (of the form reviewed in section 6.2) tell us that after a large number of repetitions of that experiment, the probability of a relative freqency close to p is ~ 1. So if X is the proposition that the relative frequency is indeed close to p, and E is the experimenter's prior information, then the Principal Principle tells us that $Cr(X|TE) \sim 1$. Bayesian updating now gives

$$Cr(T|XE) = Cr(X|TE)Cr(T|E)/Cr(X|E), \qquad (6.8)$$

or

$$Cr(T|XE)/Cr(T|E) \sim 1/Cr(X|E), \qquad (6.9)$$

so in general (that is, unless the experimenter was already very confident that the relative frequency was $\sim p$), recording a relative frequency of p significantly increases the experimenter's confidence in theory T.

6.5 Bayesian inference and Everettian quantum mechanics

Greaves (2007) has worked out how the Bayesian framework should be applied to inferences involving the Everett interpretation (including inferences about the truth of Everettian quantum mechanics itself). The basic requirement is that Cr be reinterpreted as what Greaves calls a *quasi-credence*: a probability measure defined over sets of classical histories, whether those histories are non-branching (as in the classical case) or branching (as in the quantum case). In some cases (such as an agent's quasi-credence that Everettian quantum mechanics is true or that the universal quantum state is $|\psi\rangle$), quasi-credences are clearly just personal

---

[8] I should concede that I quote Lewis slightly out of context: he actually claimed that the Principal Principle alone captured 'all we know about chance', but this is to be understood against a tacit background of assumptions about subjective probability which I take to be captured by items 1 and 2 listed above.

probability; in others (those where the quasi-credence is for some particular set of histories), it is far from obvious that quasi-credence is just personal probability.[9] In any case, though, the operational significance of quasi-credence is intended to be the same as personal probability: it synchronically governs an agent's actions via the maximum-expected-utility principle, it is updated via Bayesian updating, and it is connected to quantum-mechanical probability (and any other kinds of probability used in other theories which the agent is considering) via the Principal Principle.

Why accept these principles for quasi-credence; indeed, why assume quasi-credence exists in the first place? As I sketched in Chapter 4, the answer is largely that the justifications used for ordinary personal probability carry straightforwardly across to quasi-credence, a point made in more detail by Greaves and in Greaves and Myrvold (2010). Beginning with the existence of quasi-credence and the maximum-expected-utility principle, this is normally justified in the classical case by either Dutch-book arguments (reviewed in section 4.9) or (more satisfactorily) by decision-theoretic representation theorems (reviewed in Appendix B). Both of these justifications carry over *mutatis mutandis* to the quasi-credence case (as is discussed in some detail, in the case of representation theorems, in Greaves and Myrvold (2010)).

A rather wider variety of justifications have been given for Bayesian updating. The most plausible, in my view, are the diachronic versions of the justifications given for synchronic personal probability: diachronic versions of the Dutch book (Lewis 1997) and appeal to diachronic principles of rationality (e.g. Greaves and Wallace (2006)). Again, both apply just as well to the Everettian case: for detailed arguments to this effect, see Greaves (2007); for a formal representation theorem defended in the Everettian context, see Greaves and Myrvold (2010).

As for the Principal Principle, I was at pains to stress in Chapter 4 that no classical justification is actually available: within non-quantum physics, the Principal Principle is a bare posit. So advocates of Everettian quantum mechanics need feel no qualms about taking it in the same sense. Nonetheless, the point of Chapter 5 is that we can do better.

---

[9] In my view (cf. Ch. 7), quasi-credence is indeed just personal probability; in this chapter, though, I follow Greaves in not assuming this at the outset.

It might, in fact, seem that we have already done better: isn't the conclusion of the Born Rule theorem just that the Principal Principle is true? Not quite, for slightly subtle reasons. Let's consider an agent who accepts the following two principles:

1. Given a choice between various actions each of which leads to different possible rewards (from some set $\mathcal{R}$) with different chances, then conditional on the objective probability of reward r given that action U is taken being $\Pr(r|U)$, make the choice by maximizing expected utility with respect to these probabilities and to some utility function $\mathcal{V}$ over $\mathcal{R}$. In the particular case of quantum mechanics, if $\mathcal{W}_\psi(r|U)$ is the relative weight of reward r given that the quantum state is ψ and that act U is performed, then this means that conditional on the quantum state being ψ, maximise the quantity

$$\mathrm{EV}(U|\psi) = \sum_{r\in\mathcal{R}} \mathcal{W}_\psi(r|U)\mathcal{V}(r). \qquad (6.10)$$

   I call this principle the *Minimal decision-theoretic link* (link, that is, between probability and action).

2. Given a choice between actions where the outcomes are dependent on unknown facts about the Universe as a whole (or at least: about the entire part of the Universe in the agent's causal future), make the choice by maximizing the expected utility with respect to personal probability and a utility function. In particular, in evaluating acts where the quantum state is unknown, maximize the quantity

$$\mathrm{EU}(U|h) = \int \mathrm{d}\psi\, \mathrm{Cr}(\psi|h)\mathcal{U}(U|\psi) \qquad (6.11)$$

   where h is a history representing the agent's current information, $\mathrm{Cr}(\psi|h)$ is a personal probability function, and $\mathcal{U}(U|\psi)$ is the utility of act U given that the state is ψ.

The first of these principles (the minimal decision-theoretic link) has no known justification in non-Everettian physics; in Everettian quantum mechanics, it just states the conclusion of the Born Rule theorem. The second principle is a standard assumption of classical decision theory (note that the Cr function is an ordinary personal probability, not a quasi-credence). Neither principle, though, is the Principal Principle: the latter makes no mention of objective probability, and the former is silent on what to do when the quantum state is not known, or indeed when the agent is uncertain about whether quantum mechanics is true.

Very well: can they be combined? Well, suppose that we could assume that the utility $\mathcal{U}(U|\psi)$ of action U given that the quantum state is ψ (the quantity plugged into the right hand side of equation 6.11) is equal to $EV(U|\psi)$ (the left hand side of equation 6.10). Then we would have

$$EU(U|h) = \sum_{r \in \mathcal{R}} \int d\psi \, Cr(\psi|h) \mathcal{W}_\psi(r|U) \mathcal{V}(r). \qquad (6.12)$$

If we define

$$QCr(Ur|h) = \int d\psi \, Cr(\psi|h) \mathcal{W}_\psi(r|U) \qquad (6.13)$$

then (6.12) becomes

$$EU(U|h) = \sum_{r \in \mathcal{R}} QCr(Ur|h) \mathcal{V}(r). \qquad (6.14)$$

So if we can justify our assumption, we could deduce that an agent acting according to this rule is acting just as if he assigns quasi-credence $QCr(U|h)$ to obtaining reward r given that action U is performed; that is, his actions conform to the Principal Principle (and as a bonus, we would have derived rather than assumed that he acts by maximizing expected utility with respect to a quasi-credence function, though we would still need to justify the fact that he updates that quasi-credence function via Bayesian updating).

So, *can* we justify this assumption? Well, the utility functions $\mathcal{U}$ and $\mathcal{V}$ have been defined operationally in rather different contexts ($\mathcal{V}$ via the Born Rule Theorem, $\mathcal{U}$ via Dutch Book arguments, the representation theorem of appendix B, or whatever your preferred derivation of personal probability might be). So it does not seem prima facie obvious that the utility of U (given that the state is ψ) coincides with $EV(U|\psi)$. We could argue for their being the same function via the philosophical strategy of explaining why quasi-credence is just another form of personal probability (Chapter 7 pursues this strategy; see also Saunders 1998, Saunders and Wallace 2008, and Saunders 2010). But the problem can be approached more directly: by considering a thought experiment, it is possible to prove that the two functions do coincide (up to a harmless positive affine transformation) in just the required way, a result I call the *utility equivalence lemma* (though strictly speaking, it is an informal argument combined with a formal lemma). The details, however, are mathematically rather messy, and I relegate them to Appendix D.

Given the utility equivalence lemma and the minimal decision-theoretic link, any agent who acts according to the rules of classical decision theory (i.e. who assigns personal probabilities to distinct possibilities and acts in accordance with a maximum-expected-utility rule) will conform to the Principal Principle. Given this—and given that Bayesian updating can be justified in branching situations just as well as in non-branching ones—the Everett interpretation poses no particular problems for a Bayesian picture of inference and theory testing. Indeed, the theory is better off than in the parallel case of a non-branching probabilistic theory being tested, for the minimal decision-theoretic link is a bare posit in that case but is provable via the Born rule theorem in the Everettian case.

6.6 A unified approach: the unknown state problem

The previous section provided a synthesis of the Born Rule theorem and Greaves and Myrvold's arguments about theory confirmation in the Everettian context; this synthesis establishes that Everettian quantum mechanics, and specific predictions within Everettian quantum mechanics, can be confirmed or refuted in the same way as for non-Everettian theories. The overall account is quite complicated, though, and much of the complication arises from the fact that the concepts of probability and utility are in effect derived twice independently from considerations of rationality—once in the Born Rule theorem, once in deriving quasi-credence and its update rule (informally in Greaves 2007, formally in Greaves and Myrvold 2010)—and then work is needed to link them together. It would be significantly more elegant to find a unified approach: this is the goal of the next three sections.

My account necessarily involves a significant amount of classical decision theory (in particular, the classical diachronic decision theory which I develop in Appendix B); as such, although the account should be understandable in a self-contained way, I relegate some of the formal proofs to an appendix (Appendix C). Readers of Appendix B will note that many concepts defined there in a general context are defined again here in a quantum-mechanical context; the definitions are compatible in all cases.

My framework is the quantum decision problem introduced informally in sections 5.3 and 5.4 and formally in section 5.6. Recall that I introduced the concept of a *state-dependent solution* to a quantum decision problem: a solution, recall, assigns to each macrostate M and to each $\psi \in M$ a total ordering $\succ^\psi$ of the acts available at M. I showed that, provided that the solution obeyed certain assumptions about rationality and provided that the problem itself was sufficiently rich, $\succ^\psi$ was uniquely specified by $U \succ^\psi V$ iff $\mathrm{EU}_\psi(U) > \mathrm{EU}_\psi(V)$, where

$$\mathrm{EU}_\psi(U) = \sum_{r \in \mathcal{R}} \mathcal{W}_\psi(r|U)\mathcal{U}(r), \qquad (6.15)$$

and where $\mathcal{R}$ is the set of rewards and $\mathcal{U}$ is a real function on $\mathcal{R}$ specified quasi-uniquely (i.e. uniquely up to positive affine transformations) by $\succ^\psi$ itself.

However, the idea of a state-dependent solution does rather require us to know that the state is ψ (or, more precisely, it tells us only what is rationally required conditional on the state being ψ). What is needed here, on the other hand, is a type of solution which does not depend on ψ. The naive choice for such a function would be simply some ordering, for each macrostate, of the acts available at that macrostate. But this fails to allow for the fact that rational agents may condition their preferences on previous information obtained (in the state-dependent case, this was not an issue: knowing the current state superseded any such information); it also fails to allow for the possibility that the agent may not *know* the current macrostate precisely. An agent's preferences, then, ought to be expressed by some function which determines, from the information an agent has, the agent's preferences between future actions.

To represent the agent's information, it is useful to define a *history* as a sequence

$$h = \langle U_1, E_1, U_2, E_2, \ldots U_n, E_n \rangle \qquad (6.16)$$

where for each i, U_{i+1} is an act available at E_i and E_i is an subevent of the outcome event of U_i, and to define the *history operator* C_h by

$$C_h = \Pi_{E_n} U_n \cdots \Pi_{E_1} U_1 \Pi_{E_0} \qquad (6.17)$$

where E_0 is the domain of U_1.

212 PROBABILITY IN A BRANCHING UNIVERSE

Box 6.1. Two definitions of 'history'

In Chapter 3 I defined a history as a string of projectors $h = \langle \Pi_{E_1}, \ldots \Pi_{E_n} \rangle$ (actually, technically as a string of indices of projectors) and defined the history operator C_h by

$$C_h = \Pi_{E_n} \cdots \Pi_{E_1}; \qquad (6.18)$$

I allowed any old string of projectors to constitute a history, but confined my interest only to sets of histories that were consistent. In this chapter I have defined a history as a string of alternating subspaces and unitary maps $h = \langle E_0, U_1, E_1, \ldots, U_n, E_n \rangle$, with the subspaces and maps each taken from a fixed set satisfying certain constraints, and I defined the history operator by

$$C_h = \Pi_{E_n} U_n \cdots \Pi_{E_1} U_1 \Pi_{E_0}. \qquad (6.19)$$

To see the link between these definitions, start with the second, and suppose that we fix in advance an agent's strategy: which acts he will choose given the data he receives. This amounts to selecting a sequence of acts $V_1, \ldots V_n$, with V_{i+1} available at $\mathcal{O}_{V_i}$. With the sequence fixed, a history is determined entirely by the sequence of events: the sequence $E_0, \ldots E_n$ determines the history $\langle E_0, V_1, E_1, V_2|_{E_1}, \ldots, V_n|_{E_{n-1}}, E_n \rangle$. The relation between the two definitions is now just the relation between the Schrödinger and Heisenberg pictures, and the constraints on acts and events serve to guarantee that the set of histories has a branching structure and therefore is consistent.

In other words: an agent's strategy determines a consistent set of histories, and an agent's choice of strategy is the choice of which set to bring about. (Of course, ultimately the laws of physics fix which choice will be made; this is the old problem of free will (cf. Box 4.1) but it does not prevent us considering an agent's rational preferences between those choices.)

(This definition of history is related to the one in Chapter 3, but differs by being presented in the Schrödinger picture and, more significantly, by taking the dynamics as not necessarily fixed; see Box 6.1.) A history is a very general way of encoding the information available to an agent prior to choosing between acts. I write end(h) for the last event in h (for E_n in the above example), and start(h) for the domain of the first act in h; given two histories h_1, h_2 with end(h_1) = start(h_2), I write $h_2 \cdot h_1$ for their concatenation.

I now define a *solution* to a quantum decision problem as specified by:

1. A preferred event E, the *starting point*, which I require to be a macrostate.
2. For each history h with start(h) = E, a two-place relation $\succ_h$ on the acts available at end(h).

For a given solution, I call a history h *available* if start(h) = E: thus by definition $\succ_h$ is defined iff h is available.

One particular class of solutions can be quickly constructed. Suppose that $\succ^\psi$ is a state-dependent solution and that $\varphi \in M$ is some particular state in a macrostate M: then we can take M as a starting point and define

$$U \succ_h V \text{ iff } U \succ^{C_h\varphi} V, \tag{6.20}$$

where C_h is the history operator defined above. This is the solution appropriate to an agent who initially knows that the state is φ (I rely on the extension theorem discussed in Box 5.2 to ensure that $\succ^{C_h\varphi}$ is well-defined).

A more general set of solutions can be stated using the density operator formalism. Let M be any macrostate, suppose that ρ is an arbitrary density operator for that macrostate, and define

$$\rho_h = \frac{C_h\rho C_h^\dagger}{\text{Tr}(C_h\rho C_h^\dagger)}. \tag{6.21}$$

Then a solution is defined by

$$U \succ_h V \text{ iff } \text{EU}_{\rho_h}(U) > \text{EU}_{\rho_h}(V), \tag{6.22}$$

where $\text{EU}_\rho(U)$ is defined by

$$\text{EU}_\rho(U) = \sum_{r \in \mathcal{R}} \text{Tr}(\rho U^\dagger \Pi_r U)\mathcal{U}(r), \tag{6.23}$$

and $\mathcal{U}$ is some real function on $\mathcal{R}$. I say that a solution with starting point M is *represented* by a density operator ρ on M if the above holds for some $\mathcal{U}$ which is unique up to positive affine transformations.

Clearly, Bayesian methods of inference in quantum mechanics are justified insofar as all rationally acceptable solutions to quantum decision problems are represented by some density operator: the question, then, is under what circumstances that is true. As with the state-dependent solutions of Chapter 5, specifying these circumstances in full mathematical rigour gets unavoidably technical, and so I follow a similar strategy: in this chapter I give only a summary of the main results, and leave a fully rigorous statement and proof to the appendices.

To specify our conditions, first note that for any act U available at E, and any partition $\mathcal{F}$ of the event space $\mathcal{E}$, the set

$$\{U^{\dagger} \Pi_F U : F \in \mathcal{F}\} \tag{6.24}$$

is a positive operator valued measure (POVM) on E, which I call the $\mathcal{F}$-POVM of U. I say that a POVM is available at E iff for some partition $\mathcal{F}$, some act available at E has that POVM as its $\mathcal{F}$-POVM.

In particular, given the set of rewards $\mathcal{R}$, each act specifies a unique POVM labelled by $\mathcal{R}$. I say that a solution to a quantum decision problem is *noncontextual* iff, for any available history h and any U, V available at end(h), if U and V have the same $\mathcal{R}$-POVM then $U \sim_h V$. This is not exactly the principle of noncontextuality used in Gleason's theorem, but it embodies essentially the same ideas.

The POVM formalism also allows me to state one further constraint on the decision problem itself: I say that a decision problem is *richly structured* if whenever some set of positive operators whose sum is less than or equal to 1_E (i.e. such that 1_E minus their sum is a positive operator or zero) are all separately available at E, then there is a POVM available at E which contains the whole set. (As usual, this is just an idealization on the capabilities of experimenters: it would be cleaner, but not mathematically required, just to stipulate that all POVMs are available at E.)

In Appendix C.4, I prove what I call the *non-contextual inference theorem*. This theorem states that any solution to a quantum decision problem, provided that the problem is richly structured and satisfies the assumptions of Chapter 5 and that the solution satisfies certain rationality constraints similar to those discussed in Chapter 5, is represented by a density operator iff it is noncontextual.

This leaves the question of how noncontextuality itself is to be justified, and providing a justification will require us to make some connection between the notion of solution presented here and the state-dependent solutions already discussed. Such a connection was already appealed to in section 6.5 when I argued that the utility of an action was the average over all quantum states of its utility given a quantum state ψ, weighted by the personal probability of each ψ. However, such a rule already assumes a great deal of the apparatus of statistical inference, and it would be nice to find a more qualitative constraint.

As a first step, consider the following: suppose that an agent's preferences are represented both by a solution $\succ_h$ and by a state-dependent solution $\succ^\psi$; and suppose that he is choosing between acts U and V such that *for every ψ* (consistent with h), $U \succeq^\psi V$. In this situation at least, it seems clear that we must have $U \succeq_h V$: if the agent does not know the physical state, but prefers U to V irrespective of what the state is, then it seems rationally required that the agent prefers U to V, period. I call this condition *compatibility* between a solution and a state-dependent solution.

In fact, this 'first step' is the only step that we will require. It is possible to prove (given, as usual, certain constraints on problem and solution) that any solution to a quantum decision problem which is compatible with a state dependent solution must be noncontextual. In fact, the proof is fairly straightforward: it suffices to observe that if U and V generate the same $\mathcal{R}$−POVM, then (given that we know that $\succ^\psi$ is represented by the Born Rule) $U \sim^\psi V$ for every ψ.

This result, taken together with the noncontextual inference theorem, suffices[10] to prove the

Everettian Inference theorem. Suppose that:

- $\mathcal{P}$ is a quantum decision problem satisfying reward availability, problem continuity, branching availability and rich structuring;
- $\succ_h$ is a solution to $\mathcal{P}$ satisfying ordering, diachronic consistency and solution continuity;[11]

---

[10] Again, see Appendix C for details.

[11] Strictly speaking, I have defined ordering, diachronic consistency and solution continuity only for state-dependent solutions. Their definitions for solutions are given in Appendix C.4, but are essentially the same as for state-dependent solutions.

- $\succ^\psi$ is a state-dependent solution to $\mathcal{P}$, satisfying ordering, diachronic consistency, solution continuity, branching indifference, macrostate indifference, and state supervenience;
- $\succ^\psi$ and $\succ_h$ are compatible.

Then $\succ_h$ is represented by a density operator.

The Everettian Inference theorem is not *quite* enough to solve the epistemic problem, though. It is well-suited to handle situations in which an agent wishes to determine the quantum state (the 'unknown state problem' mentioned in section 6.1), but needs generalization to model situations where the agent wants to gain information on the dynamical laws (the 'unknown dynamics problem') or on whether quantum mechanics itself is supported by the evidence (the 'unknown theory problem'). In the next two sections, I generalize it to handle these cases too.

6.7 A unified approach: the unknown dynamics problem

Handling the unknown dynamics problem in the formal decision-theoretic framework requires only minor additions to that framework. Suppose we have a decision problem; then a partition $\mathcal{K}$ of $\mathcal{H}$ into mutually orthogonal subspaces is *event-compatible* iff for any event E,

$$\vee\{K \wedge E : K \in \mathcal{K}\} = E$$

(this will be true, in particular, if $\mathcal{K} \subset \mathcal{E}$, i.e. if all elements of $\mathcal{K}$ are events). I call an event-compatible partition $\mathcal{K}$ a set of *permanent hypotheses* if for any act U and any $K \in \mathcal{K}$, if $\psi \in K$ and ψ is in the range of U then $U\psi \in K$.

Why is this framework relevant to the unknown dynamics problem? Because we can model the experimenter's problem as follows: he knows that the true Hilbert space is some $K \in \mathcal{K}$, but not which one. Every event is therefore a direct sum of subspaces of the elements of $\mathcal{K}$, and every act is a sum of transformations of those subspaces which remain within K. The experimenter's problem, then, is to find which permanent hypothesis is correct—i.e. to find K.

Now, the Everettian Inference theorem tells us that there exists a density operator ρ which represents any rational solution to the decision problem. This may not be obvious: after all, under this interpretation of the decision problem the agent knows that ψ is definitely inside one element or another of $\mathcal{K}$, so there is no need for him to have a state-dependent solution defined for states which are superpositions of elements of different Ks. However, I am not requiring that $\succ^\psi$ *is* defined for all states. It suffices for the Born Rule theorem that $\succ^\psi$ is defined for a non-empty act-closed set of states (that is, a set of states closed under the actions of any act). In the unknown state problem, the appropriate act-closed set of states is the smallest such set containing all states in the starting point of the solution. For the unknown dynamics problem, it is instead appropriate to consider the smallest act-closed set containing all states in the starting problem that are also in some $K \in \mathcal{K}$. In either case, the Everettian Inference theorem still applies.

Because every act preserves each element of $\mathcal{K}$, we can always choose ρ to have the form

$$\rho = \sum_{K \in \mathcal{K}} p_K \Pi_K \rho_K \Pi_K \tag{6.25}$$

where ρ_K is a density operator on K (this is just the observation that each K is a superselection sector) and $\{p_K : K \in \mathcal{K}\}$ is a set of positive numbers summing to unity. Furthermore, if $K \in \mathcal{K}$ then Π_K commutes with the history operator C_h for any history h, and so

$$\rho_h = \sum_{K \in \mathcal{K}} p_{K,h} \Pi_K \rho_{K,h} \Pi_K \tag{6.26}$$

where

$$\rho_{K,h} = \frac{C_h \rho_K C_h^\dagger}{\mathsf{Tr}(C_h \rho_K C_h^\dagger)} \tag{6.27}$$

(that is, $\rho_{K,h}$ is obtained from ρ_K by the usual state-update rules) and $p_{K,h} = \mathsf{Tr}(\Pi_K \rho_h \Pi_K)$.

Now: let h be an available history, U an act available at end(h), and E a subevent of $\mathcal{O}_U$. Then (recalling that $\langle E, U \rangle \cdot h$ is the history obtained from h by concatenating to it the one-act history $\langle E, U \rangle$) we have

$$
\begin{aligned}
p_{K, \langle E, U \rangle \cdot h} &= \mathsf{Tr}(\Pi_K \rho_{\langle E, U \rangle \cdot h} \Pi_K) \\
&= \frac{\mathsf{Tr}(\Pi_K \Pi_E U \rho_h U^\dagger \Pi_E \Pi_K)}{\mathsf{Tr}(\Pi_E U \rho_h U^\dagger \Pi_E)} \\
&= \frac{\mathsf{Tr}(\Pi_E U \Pi_K \rho_h \Pi_K U^\dagger \Pi_E)}{\mathsf{Tr}(\Pi_E U \rho_h U^\dagger \Pi_E)} \\
&= \frac{p_{K,h} \mathsf{Tr}(\Pi_E U \Pi_K \rho_{K,h} \Pi_K U^\dagger \Pi_E)}{\mathsf{Tr}(\Pi_E U \rho_h U^\dagger \Pi_E)}.
\end{aligned}
\tag{6.30}
$$

If we write $\Pr(E|U, h, K)$ for the probability of event E (given that act U is to be performed and given information h) which is generated by the solution represented by ρ_K—i.e. if we define it by

$$
\Pr(E|U, h, K) = \mathsf{Tr}(\rho_{K,h} U^\dagger \Pi_E U)
\tag{6.31}
$$

—and write $\Pr(K|h)$ for $p_{K,h}$, we have

$$
\Pr(E|U, h) = \sum_{K \in \mathcal{K}} \Pr(E|U, h, K)\Pr(K|h).
\tag{6.32}
$$

and

$$
\Pr(K|\langle E, U \rangle \cdot h) = \frac{\Pr(E|U, h, K)\Pr(K|h)}{\Pr(E|U, h)}.
\tag{6.33}
$$

This is just the Bayesian updating rule, cast into our formal framework. So (6.32) and (6.33), together, suffice to tell us that in both decision-theoretic and inferential contexts $\Pr(K|h)$ functions operationally as the probability of the hypothesis represented by K given the information represented by h, whether or not it is operationally possible to bet on the truth of that hypothesis.

6.8 A unified approach: the unknown theory problem

The formal decision-theoretic framework developed above cannot be applied straightforwardly to the unknown theory problem, for the simple reason that that framework builds in explicitly quantum-theoretic assumptions. To provide a formal solution of this problem we will need a generalized concept of decision problem, one which does not automatically presume that some form of Everettian quantum mechanics is correct.

The fact that we cannot presume quantum mechanics creates significant *technical* problems for the definition of a decision problem. As I have argued throughout part II, various probabilistic concepts actually work more naturally in an Everettian than a classical context, and so our generalized notion of a decision problem will need to build in significant amounts of classical decision theory which were not needed in a framework which presumed quantum mechanics. The central concept is that of a *diachronic decision problem*, an abstract version of the quantum decision problem which keeps the purely decision-theoretic parts of the problem but strips away any connection to Hilbert space.[12]

To give a properly rigorous definition of a diachronic decision problem would require too lengthy an excursion into classical decision theory, so I relegate such a definition to Appendix B. But in outline, it is specified by:

- a complete Boolean algebra $\mathcal{E}$, the *event space*;
- for each $E \in \mathcal{E}$, a set $\mathcal{A}_E$ of *acts available at* E;
- for each act $A \in \mathcal{A}_E$, an event $\mathcal{O}_A$, the *outcome event* for A;
- a partition of $\mathcal{E}$ into *rewards*;
- composition and restriction rules for acts, so that $A \cdot B$ and $A|_E$ are well-defined.

There is no requirement for events to be sets of states, nor for acts to be maps from states to states. Nonetheless, this formal structure makes it perfectly possible to define histories and solutions in the same way as for the quantum (or the classical) decision problem; most of the axioms for problems and solutions (in particular: macrostate indifference, reward availability, ordering, and diachronic consistency) also make sense in this more abstract framework.

---

[12] Why 'diachronic'? Because—unlike the settings for most representation theorems—this decision problem involves decision-making over a period of time, and requires an agent to adopt a multi-time strategy and not just express instantaneous preferences. The contrast with synchronic decision problems should be clear to readers of Appendix B, but perhaps not to others, since the quantum decision problems we have considered have all been diachronic.

Fairly obviously, a quantum (or classical) decision problem is also a diachronic decision problem. It will be helpful to make this more explicit: define a *quantum structure* for a diachronic decision problem to be specified by:

- a Hilbert space $\mathcal{H}$;
- an isomorphism $E \to E_Q$ of $\mathcal{E}$ onto a complete Boolean algebra of subspaces of $\mathcal{H}$;
- for each $E \in \mathcal{E}$, a map $A \to U_A$ from $\mathcal{A}_E$ into the set of unitary maps from E_Q into $\mathcal{H}$, such that $U_{B \cdot A} = U_B U_A$, $U_{A|_F} = (U_A)|_{F_Q}$, and $(\mathcal{O}_A)|_Q$ is the smallest event containing the range of U_A.

A quantum structure makes a diachronic decision problem into a quantum decision problem whose set of macrostates is equal to its event space (we could allow explicitly for the macrostates, but as we have seen, they are ultimately dispensable). Equally, we can think of a quantum decision problem as an ordered pair of a diachronic decision problem and a quantum structure for that problem.

We can also define a concept of *subproblems* of a diachronic decision problem: a subproblem is just an event H such that if A is an act available at an event E, then

$$\mathcal{O}_{A|_{H \wedge E}} = \mathcal{O}_A \wedge H. \tag{6.34}$$

Subproblems can be thought of as representing potentially unknown and unchangeable information such as the laws of physics; they are the analogues of 'permanent hypotheses' of quantum decision problems introduced in section 6.7 (and any subproblem of a quantum decision problem is a permanent hypothesis), although given the more impoverished and abstract framework of diachronic decision problems, I cannot allow here as I did there for the possibility that a subproblem is not also an event. Operationally speaking, ruling out this possibility commits us in principle to the strong idealization that the agent can bet on what the subproblem is (since, by construction, all events can be bet upon[13]). Clearly, if H is a subproblem of a decision problem $\mathcal{P}$, there is a natural decision problem $\mathcal{P}_H$ defined by considering only those events which are subevents of H.

---

[13] Once again, this is no more than must be assumed to give an operational definition of the probability of a theory in non-quantum decision theory.

If $\succ_h$ is a solution to $\mathcal{P}$ with starting point E, and if $E \wedge H \neq \emptyset$, we can define a solution on $\mathcal{P}_H$ with starting point $E \wedge H$. We do so as follows:

$$U \succ_h V|_H \text{ iff } U \succ_{h^*} V, \tag{6.35}$$

where $h^* = h \cdot \langle H, 1_E \rangle$. I call this the *restriction* of $\succ_h$ to H. (It is easy to see that if $\succ_h$ is diachronically consistent, and if for any history $h = \langle A_1, E_1, \ldots, A_n, E_n \rangle$ with $\text{start}(h) \wedge H \neq \emptyset$ we define $h|_H$ by

$$h|_H = \langle A_1|_{H \wedge \text{start}(h)}, E_1 \wedge H, \ldots, A_n|_{H \wedge E_{n-1}}, E_n \wedge H \rangle, \tag{6.36}$$

then

$$U|_H \succ_{h|_H} V|_H \text{ iff } U \succ_h V. \tag{6.37}$$

Hence $\succ_{h|_H}$ really does seem to capture the notion of an agent's preferences conditional on H being true.)

All this now allows us to give the framework we need for the epistemic problem: I define a *partially quantum decision problem* to be a triple $\langle \mathcal{P}, H, \mathcal{X} \rangle$:

- a diachronic decision problem $\mathcal{P}$;
- a subproblem H of $\mathcal{P}$;
- a quantum structure $\mathcal{X}$ for H.

Most of the assumptions we will require have already been stated. Three, however, are taken from classical decision theory and are required because of the non-quantum part of the decision problem. Their formal statements may be found in Appendix B. Informally, though, they are:

Probabilism. If an agent prefers a act whereby act A is performed and a reward is given if the result is E to an act whereby A' is performed and the same reward is given if the result is E', this preference does not change if the reward itself changes.[14]

Partitioning. At any event E and for any N, there is an act A available at E and a partition $F_1, \ldots F_N$ of $\mathcal{O}_A$ such that the agent judges each F_i to be equally likely.

---

[14] The name (which I take from Savage 1972) is natural because this assumption guarantees that an agent's preferences define a well-defined likelihood order over outcomes.

Non-infinitesimality. The agent does not judge any bet as infinitesimally or infinitely better than another bet (this rules out the possibility that the real numbers are not fine enough to represent an agent's preferences).

All three assumptions are derivable in a purely quantum decision problem, but the more abstract framework of the diachronic decision problem requires them to be stated explicitly; they (or similar axioms) are standard parts of any decision theory.

I can now state my final theorem (the formal statement, and proof, can be found in Appendix C):

Everettian Epistemic theorem. Suppose that:

- $\langle \mathcal{P}, Q, \mathcal{X} \rangle$ is a partially quantum decision problem satisfying reward availability;
- $\succ_h$ is a solution to $\langle \mathcal{P}, Q, \mathcal{X} \rangle$ satisfying ordering, diachronic consistency, macrostate indifference, probabilism, partitioning and non-infinitesimality;
- the quantum decision problem $\langle Q, \mathcal{X} \rangle$ satisfies branching availability and rich structuring;
- $\succ^\psi$ is a state-dependent solution to $\langle Q, \mathcal{X} \rangle$ satisfying ordering, diachronic consistency, macrostate indifference, and state supervenience;
- the restriction of $\succ_h$ to $\langle Q, \mathcal{X} \rangle$ is compatible with $\succ^\psi$.

Then:

(i) There exists a real function $\mathcal{U}$ on $\mathcal{R}$ unique up to positive affine transformations, and for each h and $A \in \mathcal{A}_{\text{end}(h)}$ a unique probability measure $\Pr(\cdot|A, h)$ on $\mathcal{O}_A$, such that if we define

$$\text{EU}_h(A) = \sum_{r \in \mathcal{R}} \Pr(r|A, h)\mathcal{U}(r) \qquad (6.38)$$

then $A \succ_h B$ iff $\text{EU}_h(A) > \text{EU}_h(B)$.

(ii) Pr satisfies the Bayesian update principle:

$$\Pr(F|B|_E, \langle E, A \rangle \cdot h) = \frac{\Pr(F|BA, h)}{\Pr(E|A, h)}. \qquad (6.39)$$

(iii) If $E \wedge Q \neq \emptyset$, there is a density operator ρ_h on $E \wedge Q$ such that if we write $E_Q \equiv E \wedge Q$ and $E_C \equiv E \wedge \neg Q$), we have

$$\Pr(E|A, h) = \Pr(E_C|A|_{E_C}, \langle C, \widehat{1}_{\text{end}(h)} \rangle \cdot h)\Pr(\neg Q|\widehat{1}_{\text{end}(h)}, h)$$

$$= \text{Tr}(\rho_h U^{\dagger}_{A|_{E_Q}} \Pi_{E_Q} U_{A|_{E_Q}})\Pr(Q|\widehat{1}_{\text{end}(h)}, h). \quad (6.40)$$

(iv) ρ_h obeys the update rule

$$\rho_{h_E} = \frac{\Pi_{E_Q} U_{A|_{\text{end}(h)_Q}} \rho_h U^{\dagger}_{A|_{\text{end}(h)_Q}} \Pi_{E_Q}}{\text{Tr}(\Pi_{E_Q} U_{A|_{\text{end}(h)_Q}} \rho_h U^{\dagger}_{A|_{\text{end}(h)_Q}})} \quad (6.41)$$

For any subproblem H, and in particular for $H = Q$, (6.39) entails that $\Pr(H|U, h) = \Pr(H|V, h)$, so we may as well just define $\Pr(H|h) = \Pr(1_{\text{end}(h)}, h)$. (6.39) then entails that

$$\Pr(Q|\langle E, A \rangle \cdot h) = \frac{\Pr(Q \wedge E|A, h)}{\Pr(E|A, h)}. \quad (6.42)$$

The Everettian Epistemic theorem, within the setting of a fully formal decision theory, gives Everettians everything they need. A rational agent unsure of the truth of (Everettian) quantum mechanics will have their preferences represented by a utility and probability function, and conditional on Everettian quantum mechanics being true, that probability function is represented in turn by a density operator. The agent's probability of Everettian quantum mechanics being true is updated according to standard Bayesian inference, using the density operator to calculate probabilities conditional on it being true.

The Inference and Epistemic theorems are in a sense the central result of this part of the book. The Inference theorem shows, in a fully formalized way, that Everettian quantum mechanics provides a framework for inference, probability, and decision-making which improves on the non-quantum situation both through the increased elegance and simplicity of its decision-theoretic axioms and, more importantly, through the status of objective probability as an entirely derived concept. The Epistemic theorem shows, with equal formal rigour, that there is no circularity in understanding how the evidence for quantum mechanics indeed supports Everettian quantum mechanics: all that is needed are basic assumptions of classical decision theory together with the formal quantum structure.

In section 6.10, I link these results to the wider set of arguments (including my own earlier arguments) that have been presented to make sense of Everettian probability in the recent literature. First, though, the decision-theoretic framework provides an interesting opportunity to explore just what about Everettian quantum mechanics in particular allows such a thorough explication of probability, and to expand further on the significance of noncontextuality.

6.9 Digression: why Everett in particular?

In this section, I will consider three other kinds of theories: the operational versions of quantum mechanics in which the quantum state is to be understood simply as summarizing an agent's information; classical mechanics; and the class of deterministic hidden-variable theories which includes the de Broglie–Bohm pilot-wave theory. (Another class of theories—the stochastic theories—of course introduce the probability rule explicitly.) The formal decision theory developed in Chapter 5 and section 6.6 will allow us to see in just what ways Everettian quantum mechanics actually improves on those theories in its account of probability.

In an operational interpretation of quantum mechanics, the Hilbert-space structure of the theory is unchanged from Everettian quantum mechanics: dynamics are still represented by unitary transformations, and measurements by POVMs. As such the framework which I have developed so far—the framework of quantum decision problems—can also be applied to operational theories of quantum mechanics. (It would be more normal to use a framework in which POVMs are primitive rather than derived from unitary dynamics and a fixed decoherence basis; however, even in an operational interpretation of quantum mechanics measurements are ultimately measurements in the decoherence basis, so my framework ought still to be applicable.)

The main difference between an operational approach and the Everett interpretation is that in the latter, the system under study (and, ultimately, the Universe) is regarded as having a real physical state, albeit one which is not necessarily known. Hilbert space is thus a state space in the sense that its rays represent different physically possible real states that the system might be in.[15] In operational quantum mechanics, on the other hand,

---

[15] In Ch. 10 I suggest that we might do better to take the set of density operators, not the set of rays, as the real state space; nothing here hangs on that distinction.

Hilbert-space rays do not represent physical states at all: it is the operators on Hilbert space which have direct physical significance as representing physical operations. (Indeed, it is possible to formulate operational quantum mechanics purely in terms of the algebra of these operators, and eschew mention of Hilbert space altogether.[16])

So what is the state vector supposed to represent in operational quantum mechanics, if not the physical state of the system? The answer (as given in various forms by e.g. Griffiths 2002, Peres 1993, or Fuchs and Peres 2000a) is that it represents the probability assigned by an experimenter to each possible experimental outcome. By appeal to Gleason's theorem (or to its POVM variant), operationalists argue that any distribution of probabilities over outcomes represented by quantum operators must be represented by a density operator. (And state vectors, or rather projectors onto state vectors, are just special cases of density operators.)

Now, conceptually speaking, I do not really think any of this makes sense, for unoriginal reasons (which I reviewed briefly in Chapter 1). But my purpose in this section is not to press such conceptual objections but to study the mathematics of the approach, and in fact the quantum decision theory framework fits that approach quite naturally. The noncontextual Inference theorem establishes that an approach to probability based on decision theory, if it is noncontextual, will lead to an agent's preferences being expressed via a density operator.

The fly in the ointment, of course, is the requirement for noncontextuality: it is not at all clear why noncontextuality is justified in the operationalist approach. Hemmo and Pitowsky (2007) put the point very clearly (ironically in the context of a criticism of Everettian quantum mechanics):

So the crucial question before us now is this: Can the non-contextuality of probabilities, in one or another formulation, be a principle of rationality? The easiest way to see why not is to consider two classical 'experiments': a toss of a coin and a throw of a die. A priori there are two logical possibilities.

1. The limit of relative frequency of heads in a sequence of coin tosses remains the same (say, converges to 0.5) whether the die is thrown or not ...

---

[16] For detailed discussion of this algebraic approach, see Landsman (2009) and references therein.

2. The limit of relative frequency of heads in the sequence of coin tosses converges to [say] 0.5 when the die in not thrown. However, when the die is cast, the limiting frequency of heads goes [say] down to 0.3.

The fact that case 2 does not occur in our world is contingent and not a priori.

The same objection, of course, would remain in a setup where measurements were formalized directly via POVMs rather than in my decision-theoretic framework. In either case, the point is: if it is just primitive that measurement operations are represented formally by certain operators, we will have no way of arguing that certain different sets of operators should be treated as representing decision-theoretically equivalent measurements.

Now, it lies outside the scope of this discussion to consider what other justifications operationalists might offer for noncontextuality (they could just take it as a bare postulate, for instance). But it should be clear that the justification used in this chapter is not available to operationalists, for (stripping away the technical detail) the Everettian defence of noncontextuality is that two processes are decision-theoretically equivalent if they have the same effect on the physical state of the system, whatever it is. Since operationalists deny that there *is* a physical state of the system, this route is closed to them.

In fact, there is nothing especially quantum-mechanical about the discussion so far, as can be seen if we attempt the same set of arguments in a classical context. Let us define a *classical* decision problem as a quadruple $\langle S, \mathcal{E}, \mathcal{U}, \mathcal{R} \rangle$, where S is a topological space, $\mathcal{E}$ (the event space) is a complete Boolean algebra of subsets of that space, $\mathcal{U}$ (the set of acts) is a collection of continuous maps from events into S which associates to each event a set of acts available at that event, and $\mathcal{R}$ is a set of disjoint elements of $\mathcal{E}$ whose union is S. (We need to require that the acts obey axioms of composition, restriction and the like, just as in the quantum case; a more formal account is given in Appendix C.5.)

We can define a history in a classical decision problem in exactly the same way as for the quantum decision problem: it is a sequence of acts and events such that each event is in the range of the preceding act and each act is available at the preceding event. To a history $h = \langle U_1, E_1, \ldots, U_n, E_n \rangle$ can be associated the partial map $C_h : \text{start}(h) \to S$ defined by

$$C_h(x) = U_n \cdots U_2 \cdot U_1(x) \qquad (6.43)$$

whenever it exists.

The definition of a solution of the classical decision problem, and most of the rationality constraints on such solutions, can be taken over directly from the quantum case (with the proviso that a solution is allowed to have any event as a starting point: there is no analogue of macrostates).[17]

There is also a classical analogue of representation by density operators. If E is the starting point of a solution $\succ_h$, and ρ is a (Borel) probability measure on E, then we define ρ_h recursively by

$$\rho_{\emptyset} = \rho \qquad (6.44)$$

(where $\emptyset$ is the empty history), and

$$\rho_{\langle E, U \rangle \cdot h}(X) = \frac{\rho_h(U^{-1}(X \cap E \cap \mathcal{O}_U))}{\rho_h(E \cap \mathcal{O}_U)} \qquad (6.45)$$

whenever the right hand side is defined (here $\mathcal{O}_U$ is the smallest event containing the range of U, as in the quantum case). And we say that ρ represents $\succ_h$ provided that $U \succ_h V$ iff $\mathrm{EU}_{\rho_h}(U) > \mathrm{EU}_{\rho_h}(V)$, where

$$\mathrm{EU}_\rho(U) = \sum_{r \in \mathcal{R}} \rho(U^{-1}(r) \cup \mathcal{O}_U) \times \mathcal{U}(r) \qquad (6.46)$$

for some real function $\mathcal{U}$ of $\mathcal{R}$, unique up to positive affine transformations.

It is possible to take an operational approach to classical mechanics just as with quantum mechanics, and thus to deny that the state space of classical mechanics (represented here by $\mathcal{S}$) *is* a state space in the physical sense. In this framework, we could choose to represent an agent's preferences via a probability measure over events; we can even, given certain constraints of problem richness and solution rationality, prove that an agent's preferences can be so represented.[18] But to justify the framework of classical inference (and in particular, classical statistical mechanics) requires more: it requires that the agent's preferences can be represented by some probability measure on the state space.

This in turn, it turns out, can be justified if we assume

---

[17] And both classical and quantum solutions are special cases of a solution to a general diachronic decision problem, as presented in Appendix B.

[18] Strictly speaking, it is only possible to prove this if we rule out the possibility of infinitesimal but nonzero probabilities and utilities; this technical caveat is discussed in more detail in Appendix B.

Classical noncontextuality. For any two acts U, V available at end(h), if $U(x)$ and $V(x)$ lie in the same reward as each other for any x then $U \sim_h V$.

I prove in Appendix C.5 that (given certain other rationality and richness assumptions) a solution to a classical decision problem is classically representable iff it is classically noncontextual.

In turn, the principle of classical noncontextuality is easy to justify *if* we assume that the state space really is a space of physical states. The structure of the argument is closely analogous to the quantum case, in fact (a formal statement and proof may be found in Appendix C.5):

1. We define a 'state-dependent solution' to a classical decision problem as an assignment, to each event E and each $x \in E$, of a two-place relation $\succ^x$ between acts available at E. (As in the quantum case, this is intended to represent an agent's preference between acts conditional on the current state being x.)
2. We assume a state-supervenience principle according to which, if $U(x) = V(x)$, then $U \sim^x V$.
3. By an erasure argument similar to the one used in the quantum case, we prove that if U_1 and U_2 are available at E and $U_1(x)$ and $U_2(x)$ lie in the same reward for each x, then for any $x \in E$ there are acts U_1' and U_2', available at $\mathcal{O}_{U_1}$ and $\mathcal{O}_{U_2}$ respectively, such that $1_{\mathcal{O}_{U_i}} \sim^{U_i(x)} U_i'$ and $U_1'U_1(x) = U_2'U_2(x)$. It follows from this that $U_1 \sim^x U_2$ for all $x \in E$.
4. We declare that $\succ_h$ is compatible with a state-dependent solution $\succ^x$ under the same conditions as we imposed in the quantum case: that is, if $U \succeq^x V$ for all $x \in$ end(h) then $U \succeq_h V$. It follows immediately that a solution compatible with some state-dependent solution must be noncontextual.

It follows that in the case of classical mechanics, as long as we treat the state space as a space of physical possibilities (and give this assumption decision-theoretic force by requiring state supervenience and compatibility between $\succ_h$ and $\succ^x$), any preference order must be represented in the usual way by a probability measure on the state space. This might be called a *classical realist inference theorem*, by analogy to the Everettian Inference theory (as I have argued throughout this book, to be an Everettian is just to be a realist about unitary quantum mechanics). In classical as in quantum mechanics, though, failure to make such a realist assumption about the state space means that the mathematical structure of the decision problem is largely inert.

There is another similarity between the classical and quantum accounts which is worth noting. In both cases, the actual microstate of the system drops out of view for the purposes of statistical inference: the relevant mathematical entities are, respectively, the probability measure and the density operator which represent the experimenter's assignment of probabilities. This can give the impression that the microstate itself is irrelevant (an impression perhaps stronger in the quantum case, where the same mathematical object can be used for both purposes, than in the classical case); in fact, though, the microstate plays a central role both conceptually (in making sense of what the theory itself actually means, and in particular what it means to represent a certain physical operation by a mathematical map) and technically (in deriving the required non-contextuality assumption).

There is, however, a crucial disanalogy between the classical and Everettian versions of the inference theory. The former provides no robust, agent-independent notion of objective probability: all probabilities are derived from the agent's initial choice of probability measure. In the Everett interpretation, on the other hand, it is easy to set up experiments where irrespective of their choice of density operator, agents must agree on the objective probability of each outcome: any act represented by the POVM $\{0.5\hat{1}, 0.5\hat{1}\}$, for instance, will be an act where all agents agree that the probability of each outcome is 0.5. Ultimately the reason for the difference is the profoundly different mathematical structure of the two theories: in quantum but not in classical mechanics, an act can take a microstate into a superposition of macroscopically definite outcomes.

What of hidden-variable theories? Recall that a hidden-variable theory preserves the full mathematical structure of quantum mechanics, but adds to it a set of 'hidden variables'. These variables follow dynamical trajectories which are influenced by, but do not influence, the evolution of the quantum state; in an important sense the word 'hidden' is a misnomer, as it is the hidden variables which determine the results of our observations.[19]

---

[19] In some interpretations of hidden-variable theories, such as that of Valentini (Valentini 1996; 2001; 2004; Valentini and Westman 2004), it is the hidden variables and the quantum state which jointly determine the results of observations; for simplicity, though,

For this reason, the notion of a state-dependent solution to a quantum decision theory does not make sense if there are hidden variables around—or at least, does not make sense unless an appropriate probability distribution over hidden variables has already been assumed. Instead, the appropriate notion of 'state-dependent solution' should be a preference order conditional both on the quantum state and on the hidden variables. A theory of this form, if formalized, would just be another classical theory: it should be possible to establish that an agent's preferences are represented by a joint probability distribution on quantum states and hidden variables, but this is insufficient to derive the Born Rule.

6.10 Theorems in context: the recent literature

In this section, I provide some context for the decision theoretic results of this and the last chapter, by showing how they fit in both with the classical (i.e. nonbranching) philosophy of probability and with other recent arguments (including some earlier ones of my own) for how the Probability problem could be resolved.

6.10.1 Classical probability: overview

To begin with the classical situation: Fig. 6.1 summarizes 'Lewisian dualism', the dual theory of objective and personal probability sketched in section 4.8, described in detail in section 6.4, and widely adopted in contemporary philosophy. Three components—the rationality principle that represents an agent's synchronic preferences by a personal probability measure, the Principal Principle which connects that personal probability with the objective probabilities of physics, and the Bayesian updating principle that tells us how to update personal probability in the light of evidence—are (it is claimed) jointly sufficient to give us the 'complete functional story', i.e. a complete functional account of how probability works.

I confine myself to the simpler position according to which the hidden variables alone suffice to do this. A particularly clear statement of this view may be found in Allori et al. (2008); in Brown and Wallace (2005), Harvey Brown and I argue that it is question-begging.

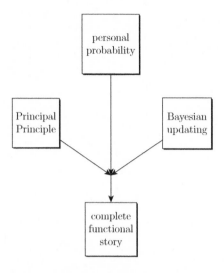

Figure 6.1. Classical inference (e.g. Lewis 1980)

6.10.2 Classical probability: the personal-probability side

Two of these principles—Bayesian updating and personal probability—appeal only to an agent's beliefs and preferences, without reference to objective probability. Probably the most common justification for these principles (advocated by e.g. Lewis 1997) is the one given in Fig. 6.2: if we assume the 'Dutch book principle'—i.e. the principle that rational agents are not vulnerable to Dutch books—then both principles can be derived.

It is more satisfactory, though, to derive both from more plausible constraints on rationality. This is generally understood in two stages: first

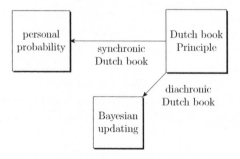

Figure 6.2. Dutch book arguments

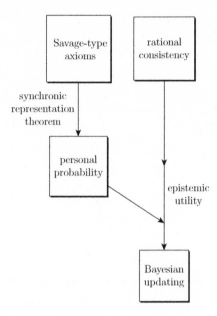

Figure 6.3. Decision-theoretic justification of personal probability and Bayesian updating

the existence of personal probability is derived from synchronic rationality axioms such as those of Savage (1972) or Jeffrey (1983), then consideration is given to the constraints rationality places on how personal probability is to be updated. Figure 6.3 is illustrative of this; by 'epistemic utility' I mean the kind of considerations used in Greaves and Wallace (2006) and Greaves and Myrvold (2010), which basically derive an agent's belief-updating strategy by requiring a consistency of his actions over time—specifically, that an agent adopts the updating strategy which leads to maximum expected utility as calculated via pre-updated personal probabilities. I use 'rational consistency' as a general label for such principles (I tighten its meaning later).

However, as Appendix B shows, this decision-theoretic framework can be unified to some extent. The Savage axioms[20] can be derived in a diachronic decision-making framework once we assume diachronic

---

[20] Actually, the axioms used in Appendix B are not *quite* the Savage axioms: for simplicity, I assume a slightly richer space of possibilities than Savage.

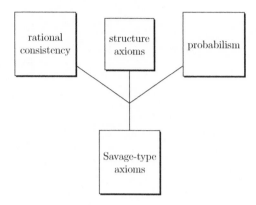

Figure 6.4. Deriving the Savage axioms in a diachronic framework (Appendix B)

consistency, certain 'structure axioms' which force probabilities and utilities to be real-valued, and the 'probabilism' assumption which guarantees that agents' preferences define a well-defined qualitative likelihood ordering on possible outcomes of their actions (and once we require that decision-making framework to be sufficiently rich). (Fig. 6.4.) With this framework in mind, from here on I sharpen the idea of 'rational consistency' to mean the combination of synchronic consistency (i.e. an agent's preferences define an ordering) and diachronic consistency.

The appearance of rational consistency as an ingredient of both the derivation of personal probability and the justification of Bayesian updating suggests that a unified treatment should be possible. And indeed it is: this is Appendix B's Diachronic Representation Theorem (Fig. 6.5).

6.10.3 Classical probability: the objective-probability side

So much for the personal-probability side of Fig. 6.1. The objective probability side—the Principal Principle—is, as we have seen, in much worse shape in the absence of quantum branching. To derive the Principal Principle from the physical facts and from decision theory, we first have to extract objective probabilities themselves from the underlying physics. Doing this by bare posit may seem unsatisfactory, but alternatives—notably frequentism and its sophisticated cousin, the Best-Systems analysis—are at best difficult. Secondly, we have to connect those probabilities with rational action—either by showing that they fit into the Principal Principle in the standard way, or at least by deriving a maximum-expected-utility rule

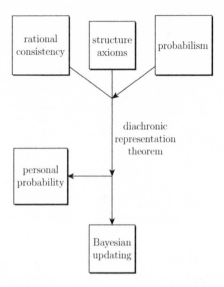

Figure 6.5. The Diachronic Representation Theorem (Appendix B)

for rational agents' actions conditional on their knowing the values of the objective probabilities (I called this latter result the minimal decision-theoretic link in section 6.5; in this section, I abbreviate it to 'decision-theoretic link'). In fact, the utility equivalence lemma introduced in section 6.5 (and proved in Appendix D) establishes the Principal Principle from the decision-theoretic link, so this apparently weaker result will suffice—but weaker or no, we have no idea how to establish the link in nonbranching physics except by positing it directly. Fig. 6.6 summarizes this situation; combining it with the account given by Fig. 6.5 gives (one analysis of) the best that can be done to understand probability in the absence of quantum-mechanical branching, and is illustrated in Fig. 6.7.

6.10.4 Quantum probability: subjective uncertainty and decision theory

What about the quantum situation? The decoherent-histories program alone suffices to extract the formal structure of probability from unitary quantum mechanics; in view of the difficulties in doing even this in nonbranching physics, it would be possible to regard this alone as sufficient to put quantum probability on, if not a firm footing, then a footing that is just as firm as in the non-quantum case (Fig. 6.8). Indeed, Papineau (1996)

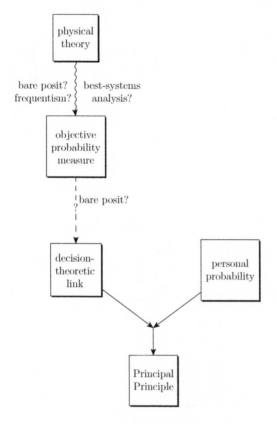

Figure 6.6. Objective probability in nonbranching physics

recommended just this strategy (though he did not stress the relevance of decoherence, in part because he adopted the then-popular language of 'many minds').

Saunders (1995; 1998) made explicit the role of decoherence, and then added a new ingredient: in his framework, while there was no more need in quantum than in classical physics to prove that some quantity defined by the theory was probability, there *was* need to prove in quantum physics what was obvious in classical physics: that probabilities were defined over alternative possibilities despite the objectively deterministic nature of quantum mechanics (what I later (Wallace 2002a) called 'subjective uncertainty'). Saunders did so by means of a thought experiment (critiqued

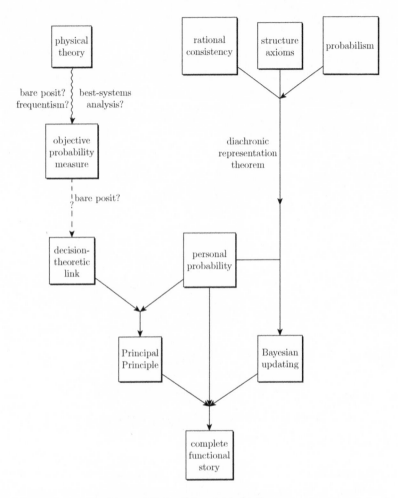

Figure 6.7. Justifying classical inference

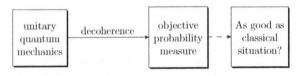

Figure 6.8. Minimal approach to quantum probability (Papineau 1996)

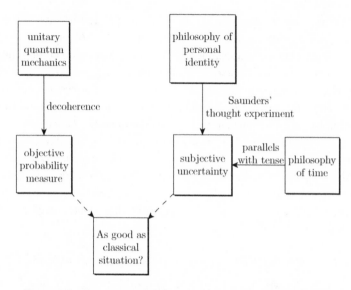

Figure 6.9. Quantum probability according to Saunders (1995; 1998)

in Greaves 2004 and Wallace 2006a) and by drawing parallels with the treatment of tense in special and general relativity; his overall framework is illustrated in Fig. 6.9.

Deutsch's introduction of decision theory into the debate Deutsch (1999) was quite minimalist in its scope (if not its significance): in my framework, 'Deutsch's theorem' (i.e. the central result proved in Deutsch 1999; see Wallace 2003b for further discussion of the argument) amounts simply to a proof of the decision-theoretic link between objective probability and action. Deutsch's paper, then, has the structure of Fig. 6.10 (though the role of decoherence in providing a quasi-classical basis is at most tacit in Deutsch's paper).

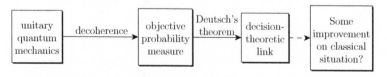

Figure 6.10. Quantum probability according to Deutsch (1999)

238   PROBABILITY IN A BRANCHING UNIVERSE

My own earlier approach to the problem (Wallace 2002a; 2003b; 2006a; 2007) combined Deutsch's and Saunders' ideas. My approach had a significant philosophical component, because I had regarded subjective uncertainty as essential for a solution of the probability problem, because (I thought—and argued in Wallace 2002a; 2006a) only if probabilities quantified alternative possibilities would it be possible to understand not just how probability functioned in an Everettian universe but how the Everett interpretation itself could be evidentially supported by the observed data (the 'epistemic problem').[21] I defended subjective uncertainty initially (Wallace 2002a) by appeal to Saunders' thought experiment, later (Wallace, 2005; 2006a) by the philosophy-of-language argument presented in Chapter 7 of this book (see also Saunders and Wallace 2008); my approach is summarized in Fig. 6.11.

6.10.5 Quantum probability: formal solutions to the Epistemic problem

Greaves (2007) explicitly addressed the question of what *additional* work would be required to solve the Epistemic problem, on the assumption that the practical problem (of why we should allow quantum probabilities to guide our actions) had been solved. She observed that (i) given Everett-compatible versions of personal probabilities, the Principal Principal and standard rules for updating personal probability, the Epistemic problem would be solved, and (ii) the classical arguments for those various positions went across *mutatis mutandis* to Everettian quantum mechanics. In the particular case of the Principal Principle, Greaves stated that the Principal Principle was arguably established by Deutsch's and my work, but was in any case content to postulate it. (For the reasons I discussed in section 6.5, I think Greaves is slightly too generous here: what Deutsch and I had proved was not quite the Principal Principle, but just the decision-theoretic link; section 6.5's utility equivalence lemma is required to go from one to the other.)

Greaves and Myrvold (2010) then went on to turn Greaves' informal arguments into a formal representation theorem; combining this with my

---

[21] At one point (Wallace 2002a; 2003b) I also thought subjective uncertainty was necessary to justify 'measurement neutrality', the tacit premises I identified in Deutsch (1999); in Wallace (2007), however, I developed the 'erasure' argument which is central to the Born Rule theorem, and which obviated the need for measurement neutrality.

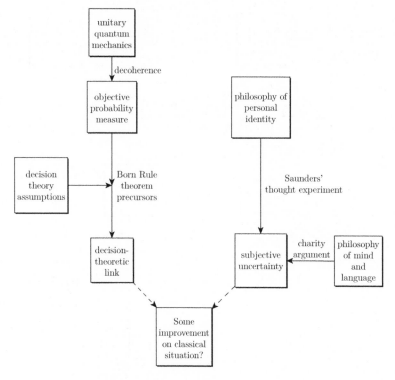

Figure 6.11. Quantum probability according to Wallace (2002a; 2003b; 2006a; 2007)

earlier work gives a picture like Fig. 6.12 (with Greaves and Myrvold being agnostic about whether the left-hand side is actually correct).

Greaves and Myrvold in fact develop the story in a slightly different way from Greaves (2007). Instead of using the Principal Principle, they adapt an older result ('de Finetti's theorem'; cf. references in their paper) to the effect that an observer who believes that the individual experiments in a sequence of experiments are 'exchangeable'—meaning, loosely, that he is indifferent as to swapping around which experiment(s) he bets on—will behave *as if* he assigns various *personal* probabilities to various hypotheses about the *objective* probabilities of each measurement outcome. Greaves and Myrvold then argue that *if* it is the case that an agent who accepts the Everett interpretation would accept the Born Rule (i.e. if the decision-theoretic link holds with respect to the branch weights),

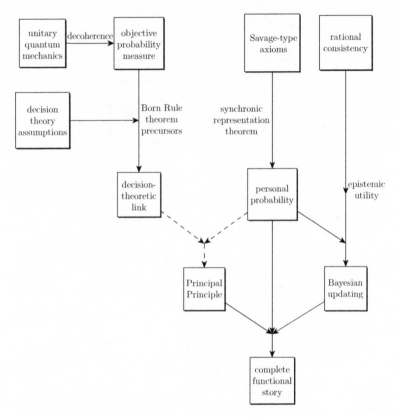

Figure 6.12. Quantum probability according to Greaves (2007), as formalized by Greaves and Myrvold (2010)

then their result establishes that an agent can confirm the hypothesis that Everettian quantum mechanics is correct by measurements of long-run frequencies. This approach is summarized in Fig. 6.13 (again, Greaves and Myrvold are agnostic about whether the derivation of the decision-theoretic link is valid or whether the link must be taken as a bare posit).

6.10.6 Quantum probability: a unified approach

As I noted at the start of section 6.6, if the probability problem is solved by combining Greaves and Myrvold's approach to the epistemic problem with Deutsch's and my approach to the practical problem, certain complexities arise because decision-theoretic considerations are being

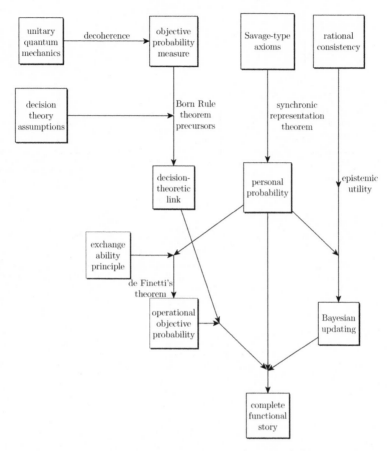

Figure 6.13. Quantum probability according to Greaves and Myrvold (2010)

deployed twice: once to establish the decision-theoretic link, and once to derive the apparatus of personal probabilities. This can be sharpened if we (i) replace my earlier, informal derivations of the Born Rule with Chapter 5's Born Rule theorem; (ii) replace Greaves and Myrvold's derivation of Bayesian updating and personal probability with Appendix B's Diachronic Representation theorem; (iii) fill in the connection between the decision-theoretic link and the Principal Principle using section 6.5's utility equivalence lemma. The result (which is basically the approach I described in section 6.5) is illustrated in Fig. 6.14; as can be seen, rational consistency is being used twice. As promised, I can now also tighten

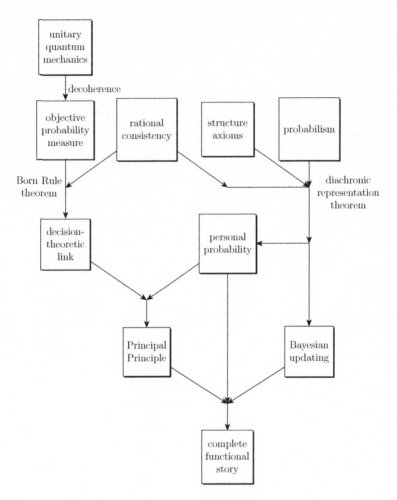

Figure 6.14. Quantum probability according to section 6.5

the definition of rational consistency: in this context, it means *only* the conjunction of ordering (i.e. the assumption that an agent's synchronic preferences form a total order) and diachronic consistency.

The results of sections 6.6–6.8 remove this dual aspect, and in doing so significantly tidy up the situation. If we consider a rational agent willing to presuppose Everettian quantum mechanics, all that is required in addition is the assumption of rational consistency (Fig. 6.15); if we are interested in

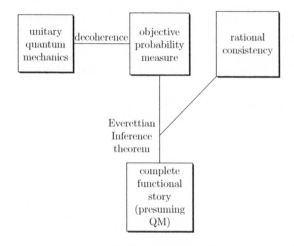

Figure 6.15. The Everettian Inference theorem (section 6.6)

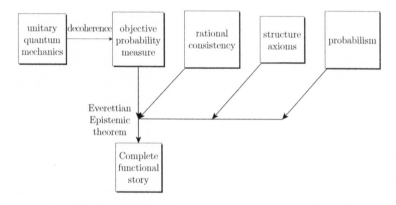

Figure 6.16. The Everettian Epistemic theorem (section 6.8)

the problem of how to gain evidence for Everettian quantum mechanics itself, we cannot rely on quantum mechanics to give us the structure required to set up decision theory and so we have to add additional assumptions: probabilism, and some structural axioms (Fig. 6.16). Either way, we have a complete solution to the problem of probability, going substantially beyond what can be achieved in the non-quantum framework. Far from being the Achilles' heel of the Everett interpretation, probability turns out to be a major philosophical advantage for Everettian quantum mechanics.

CHAPTER 6. Irrespective of the detailed analysis—Fisherian, Bayesian, or formal-decision-theoretic—made of statistical inference, Everettian quantum mechanics allows a fully satisfactory account to be given of how inference works, whose conclusion in each case vindicates existing forms of inference. As such, the Everett interpretation has no residual problem of probability; indeed, it allows us to improve significantly on our pre-quantum understanding of probability.

CHAPTER 7. Do the ordinary notions of 'uncertainty' and 'possibility' have any home in Everettian quantum mechanics?

Second Interlude

Do I dare
Disturb the universe?
In a minute there is time
For decisions and revisions which a minute will reverse.

from *The Love Song of J. Alfred Prufrock*, by T. S. Eliot

Double standards

SCEPTIC: You seem very ready to fall back on the idea that probability, however much or little sense it makes in the context of the Everett interpretation, makes no more sense in non-Everettian contexts.[1]

AUTHOR: Yes. I actually think we can get a *better* understanding of probability when we consider (Everettian) quantum mechanics, but that's a bonus, not a requirement.

SCEPTIC: But look: there's something crazy about that. We're totally used to probability in *non*-Everettian contexts. Okay, it might be *philosophically* a bit puzzling, but those puzzles don't really matter from the point of view of physics: practically speaking, we've got a sufficiently solid grip on probability to do science. In the Everett interpretation, though, we're worried that the whole idea of probability makes no sense at all. Failure to understand probability in a satisfactory way in the Everett interpretation isn't just problematic or puzzling: it's fatal.

AUTHOR: What makes you say we're used to probability in non-Everettian contexts?

---

[1] This section is loosely based on my recollection of a debate with David Albert in 2009; I'm grateful for David for the inspiration, but I make no claim to be accurately representing his views, then or now.

SCEPTIC: We use it all the time in science.

AUTHOR: We use *probability* all the time in science—but who's to say that we're not thereby using it in an Everettian context? After all, if Everettian quantum mechanics is true, then pretty much all appeals to (objective) probability in science are actually, directly or indirectly, appeals to probability in an Everettian context.

SCEPTIC: But our whole use of probability *presumes* the non-Everettian context! If we couldn't assume that, we'd have no reason to use it at all.

AUTHOR: It doesn't presume it—that's the whole point, in fact. If we had a satisfactory philosophical account of probability which made sense in the non-Everettian context but we didn't have one which worked in the Everettian context, I'd take your point. But that's not our situation. We know, practically speaking, how to use probability, but we don't understand it.

... Put it this way. Suppose you fall into a coma for twenty years, and when you awaken, to your horror everyone accepts Everettian quantum mechanics. You try to persuade them of the virtues of (say) a stochastic, dynamical-collapse theory, and they object as follows:

ANTI-SCEPTIC: We're totally used to probability in Everettian contexts. Okay, it might be *philosophically* a bit puzzling, but those puzzles don't really matter from the point of view of physics: practically speaking, we've got a sufficiently solid grip on probability to do science. In the single-world interpretation, though, we're worried that the whole idea of probability makes no sense at all. Failure to understand probability in a satisfactory way in the single-world interpretation isn't just problematic or puzzling: it's fatal.

Decision theory and the measurement of probability

SCEPTIC: I don't understand what all this decision theory has to do with physics. All this talk of 'betting' doesn't seem at all relevant to scientific practice.

AUTHOR: The decision theory is a necessary part of the explanation of how we measure probability. The 'bets' are just a convenient (if slightly stylized) framework for thinking about that explanation.

SCEPTIC: But there's no mystery about how we measure probability. We just make lots and lots of repeated measurements, and set the probabilities equal to the measured frequencies.

AUTHOR: That's what we should do if Everettian quantum theory is true, too.

SCEPTIC: But that method of measuring doesn't make sense in the Everett interpretation—all the outcomes occur, just on different branches.

AUTHOR: Now you've moved the discussion on from 'How do we measure probability?' to 'Why is our method of measuring probability justified?' And we don't actually have an answer to that question in the non-Everettian context.

SCEPTIC: Why can't we just say that probability is another word for long-run relative frequency?

AUTHOR: Well, for a start we run into technical problems. Our theories, after all, predict not that long-run relative frequency=probability, but that long-run relative frequency$\simeq$probability, *probably*. So there's a circularity—just the same circularity that occurred, in the Everettian context, in Everett's own frequency-based solution to the probability problem. There are *possible* ways around that technical problem—Lewis's 'best-system analysis', in particular—but they're pretty embryonic. And even if there is some satisfactory way to define probability in terms of long-run frequency, it really is going to have to be *long-run* frequency—overwhelmingly longer-run than we measure in the lab. So we don't really explain why our methods of measuring probability work just by analysing probability in frequency terms.

SCEPTIC: I feel many physicists will find this overly philosophical. We're just idealizing a long but finite run of experiments as an infinite run.

AUTHOR: That idealization works just as well (or as badly) in the Everettian context: in the infinite-experiment limit, the quantum state is an eigenstate of the relative-frequency operator.[2] There are severe problems—both conceptual and technical—with that, but they're basically the same problems that arise in the non-Everettian context.

[2] See Farhi, Goldstone, and Gutmann (1989).

Everything happens

SCEPTIC: A lot of your argument relies on this claim that probability makes no more sense in the non-Everettian context. But that's not true: there's a crucial difference. If I measure the relative frequency of some probabilistic process, if I believe the Everett interpretation to be true then I have to believe that *every outcome occurs*. How is that compatible with assigning probabilities to the various outcomes?

AUTHOR: Because (to simplify) it's rationally required to neglect the branches with extremely low weight. And if the probabilistic process is iterated enough times, then the collective weight of the branches in which the probability doesn't approximately equal the relative frequency will indeed be extremely low.

SCEPTIC: So what? They're still real. Why is it rational to neglect them, if they nonetheless exist?

AUTHOR: Because either (i) (the strong position) we can rationally prove it (see Chapter 5), or (ii) even if you don't accept we can prove it, it's no more unreasonable than (and, indeed, structurally isomorphic to) the assumption we make in non-Everettian contexts that low-probability outcomes should be neglected.

SCEPTIC: The difference is that in the Everettian context, the 'low-probability outcomes' still exist.

AUTHOR: Why is that difference salient?

SCEPTIC: It's—

AUTHOR: Please don't tell me 'It's intuitively obvious that it's salient'. The right time to stop reading on grounds of violated intuitions was five chapters back.

SCEPTIC: It's unreasonable to regard a decision-theoretic strategy as rational if we can guarantee that some of those who follow it (in this case, the experimenter's successors in the low-weight branches) will do badly as a result.

AUTHOR: Not only are you still not giving a positive argument for that view, it's clearly wrong for reasons independent of Everettian quantum mechanics. It's incompatible with the spatial infinitude of the (classical) Universe, for a start; more mundanely, there are plenty of sensible strategies which, on a planet with six billion inhabitants, are guaranteed to work out wrongly for someone. (See the discussion on pages 195–6 for an example.)

Subjective probability

SCEPTIC: Do you really find it acceptable to regard quantum probabilities as defined via decision theory? Shouldn't things like the decay rate of tritium be objective, rather than defined via how much we're prepared to bet on them?

AUTHOR: The quantum probabilities are objective. In fact, it's clearer in the Everettian context what those objective things are: they're relative branch weights. They'd be unchanged even if there were no bets being made.

SCEPTIC: In that case, what's the point of the decision theory?

AUTHOR: Although branch weights are objective, what makes it true that branch weight=probability is the way in which branch weights figure in the actions of (ideally) rational agents. By analogy, a dollar bill is a perfectly physical object. The economy of the world could cease to exist—all life could cease to exist—without needing to affect that physical object. But what makes that physical object deserve the name 'dollar bill' nonetheless depends on how that object plugs into the behaviour of human agents. (The analogy isn't perfect: there's an aspect of pure convention about our use of dollars as currency, whereas the Born Rule theorem tells us that any rational agent has to treat branch weight as probability.)

Diachronic consistency

SCEPTIC: So the idea is that, from supposed 'axioms of rationality', we derive the conclusion that rational agents should follow the Born Rule.

AUTHOR: Basically, yes. (I'm hedging because that's only the situation for the special case of rational action when the quantum state is known; cf. Chapter 6.)

SCEPTIC: But one of these 'axioms of rationality'—diachronic consistency—isn't a mainstream axiom at all, is it?

AUTHOR: Fair point. As I acknowledge, the classical decision theory in Appendix B isn't entirely mainstream. The reason for that is that while there has been lots of work on representation theorems in *synchronic* decision theory (i.e. the kind of decision theory that applies when an agent has just one decision to make), and lots of work on how to justify Bayesian updating *given* that synchronic preferences should be represented by probabilities, there's been little or no attempt to do both in a unified way. I think that's unfortunate: it leads to a distorted picture of decision-making as this instantaneous action whose rewards are instantaneous events, whereas actually any reasonable decision-making process is just that: a process.

SCEPTIC: But notwithstanding all that, all I have to do to reject your representation theorem is to deny diachronic consistency.

AUTHOR: I wouldn't say 'all'! I do think diachronic consistency—or something akin to it, like the diachronic Dutch book—is basically at the heart of pretty much any persuasive argument in favour of Bayesian updating. So denying diachronic consistency is going to cause trouble for probability and decision theory outside the Everettian context too. And I also think it's basically impossible to think about the process of rational behaviour—which, as I say, is a *process*—without accepting some kind of diachronic consistency. (By analogy: we take the idea of *synchronic* consistency as basically constitutive of rational behaviour—we can't apply the concept of agency to a physical system that can't be regarded as fairly accurately observing synchronic consistency. I think something pretty similar is true for diachronic consistency.)

SCEPTIC: But you'd admit that there's lots of philosophical work that needs to be done here.

AUTHOR: Sure. It's work in mainstream decision theory, though, and it doesn't really concern Everettian quantum mechanics except indirectly, so it didn't seem to belong in this book. (I take some small steps towards it in Wallace 2010a.)

SCEPTIC: But pending that work, the Everett interpretation is still in trouble.

AUTHOR: No: remember, the decision-theoretic representation theorems take us *beyond* what's possible in the non-Everettian context. Even if they're completely unsustainable, the Everettian still has the fallback position of maintaining that probability is *no less* mysterious in non-Everettian contexts.

The Born Rule theorem

SCEPTIC: I just don't find it plausible that we can't simply write down alternatives to the Born Rule.

AUTHOR: It isn't plausible, prima facie. That's why Deutsch's original result was so surprising.

SCEPTIC: But can't I just ...

AUTHOR: If we carry on like this, we're just going to reiterate section 5.8. In that section, I went through every 'alternative' I've seen anyone try to come up with and showed how those alternatives violate the rationality axioms. Not that I really needed to: ultimately, a theorem's a theorem.

SCEPTIC: I'm pretty unmoved by that. If some otherwise sensible strategy violates one of your so-called rationality axioms, so much the worse for that axiom.

AUTHOR: I'd be more moved by that objection if there were any such 'otherwise sensible strategies'. But isn't it interesting that no one seems to have come up with one that can even be properly defined in anything other than incredibly overstylised toy situations? (Not that I think they're justifiable even in those toy situations, but that's another matter.) In realistic, complicated, temporally extended situations, no one seems to be able to formulate any remotely tractable alternative to the Born Rule.

SCEPTIC: 'Remotely tractable' is a weasel phrase. Why couldn't the real rational strategy be *in*tractable? Couldn't we be unlucky enough to live in a universe where rational action isn't in principle possible?

AUTHOR: Not under any sensible interpretation of 'rational action', no. After all, following the Born Rule is at least—pending arguments to the contrary—rationally *permitted*. That is, someone who acts in accordance with the Born rule isn't behaving *irrationally*. All that has to be done is to establish that no *other* behavioural strategy is rationally permitted. (And notice that this is exactly how classical decision theory works, too. We prove that no nonprobabilistic decision-making strategy is rationally permissible, but we don't then provide further arguments as to why a probabilistic strategy *is* permissible, over and above showing that it satisfies the qualitative rationality axioms.)

Rationality, evolution, and anthropic reasoning

SCEPTIC: Okay: suppose, just for the sake of argument, that you can show that rational agents conform to the Born Rule. It doesn't ultimately help unless you can explain why our branch is one where rational agents exist.

AUTHOR: I don't see why quantum mechanics, any more than (say) general relativity, needs to answer that question. But in any case, here goes: we have a general framework, natural selection, for explaining why organisms in general are adapted to their environments. The fact that humans aren't such bad approximations to rational systems can be expected to be a special case of that explanation. And that explanation works just as well in the Everett interpretation.

SCEPTIC: I'm not so sure it does, actually. Natural selection tells us that we should expect to find organisms adapted to their environments in all but a very low-weight collection of branches. But why shouldn't our branch be one of those low-weight branches?

AUTHOR: Because the decision-theoretic representation theorems tell us (roughly) that we should reject hypotheses according to which our own branch has a low weight.

SCEPTIC: This brings up something that's bothered me throughout. In this case, just like all your other decision-theoretic cases, you're construing the problem as one about how we accept and reject hypotheses as data comes in. That's fine in a non-branching framework. But the whole epistemic game is different in the Everett interpretation, isn't it? It becomes a case of anthropic reasoning:[3] where in the branching multiverse *am I*? (Or, equivalently: which of the myriad qualitatively identical Sceptics *am I*?) That doesn't seem to show up at all in this decision-theoretic edifice you've created.

---

[3] Here the sceptic uses 'anthropic reasoning' in the physicists' sense, to mean any reasoning about the location of the observer in physical reality. Philosophers tend to use the term more narrowly: they should read 'anthropic reasoning' as 'reasoning about self-location'.

AUTHOR: Indeed not. Anthropic reasoning is fascinating, to be sure, and perhaps the Everettian context throws up some interesting examples of it. (I look at this in more detail in Chapter 10). But the ordinary, mundane, non-cosmological experiments we do in the lab to test quantum mechanics don't have to be thought about in an anthropic way.

SCEPTIC: It's question-begging to assume that.

AUTHOR: I'm not really *assuming* it. All I'm doing is asking: *if* we carry out the sort of experiments we do to test quantum theory, and *if* we get the kind of results we actually do get, is it rational to take them as evidence in favour of quantum theory—even if quantum theory is to be interpreted *à la* Everett? And I've shown that it is. Maybe—who knows?—we can reinterpret all that experimental work in an anthropic-reasoning format. But we don't have to. (No more does the hypothesis that the universe is spatially infinite have radical consequences for ordinary lab work, despite the infinitude of qualitatively identical replicas of the lab worker that hypothesis entails.)

Circularity

SCEPTIC: Isn't there something a bit circular about your whole position here? First you appeal to decoherence theory to argue that a branching structure is *approximately* realized—which is to say, is realised to within errors of small mod-squared amplitude. Then you appeal to decision theory, or symmetry, or whatever, to explain why the mod-squared amplitudes in that branching structure are probabilities. But surely the only thing that justifies regarding an error as small if it's of small mod-squared amplitude is the interpretation of mod-squared amplitude as probability.[4]

AUTHOR: Not really. We can think of the significance of the Hilbert space metric as telling us when some emergent structure really is robustly present, and when it's just a 'trick of the light' that goes away when we slightly perturb the microphysics. (Remember that the Hilbert space

[4] Here, the sceptic is presenting an objection first made in print by Baker (2007); see also Kent (2010).

norm is a perfectly objective feature of the physics, prior to any considerations of probability.)

SCEPTIC: But that's just the same thing again. What makes perturbations that are small in Hilbert-space norm 'slight', if it's not the probability interpretation of them?

AUTHOR: Lots of dynamical features of the theory. Small changes in the energy eigenvalues of the Hamiltonian, in particular, lead to small changes in quantum state after some period of evolution. Sufficiently small displacements of a wavepacket lead to small changes in quantum state too. Ultimately, the Hilbert-space norm is just a natural measure of state perturbations in Hilbert space, and that naturalness follows from considerations of the microphysical dynamics, independent of higher-level issues of probability.

SCEPTIC: It sounds awfully inconclusive and hand-waving, frankly.

AUTHOR: Well, I'll concede that there's plenty of work to be done here by philosophers of science with an interest in emergence. But, again—

SCEPTIC: *Please* don't tell me that you're going to claim *again* that there's nothing Everett-specific about the problem.

AUTHOR: Sorry!—but there isn't. Really, I don't think there's any profound difference here between the role of the Hilbert space metric in quantum physics and, say, the spatial metric in classical physics. Instantiation is always approximate, and we measure that approximation using the natural distance measures of the instantiating theory. There's a need for self-consistency—that distance measure had better appear natural from the perspective of the instantiated theory. But that self-consistency requirement doesn't make the whole enterprise viciously circular—it can't, on pain of undermining science in general, not just Everettian quantum theory.

PART III

Quantum Mechanics, Everett Style

7

Uncertainty, Possibility, and Identity

Words have no function save as they play a role in sentences: their semantic features are abstracted from the semantic features of sentences, just as the semantic features of sentences are abstracted from their part in helping people achieve goals or realize intentions.

Donald Davidson[1]

Do I contradict myself?
Very well then ... I contradict myself.
I am large ... I contain multitudes.

Walt Whitman, *Leaves of Grass*[2]

7.1 Everett and the future: the problem

In this chapter, I take a rather philosophical detour. My purpose here is to consider what, if anything, are the consequences of the Everett interpretation for some of our most everyday notions: those relating to our sense that we are beings who experience a single, but often unpredictable future. On the face of it, it looks as if we may have to abandon these notions if Everettian quantum mechanics is true. For how can it make sense, for instance, to continue saying things like 'The result of the experiment will be spin-up or spin-down, but not both' when quantum mechanics tells us that in fact there are branches in our future when each outcome occurs? And how can we continue to anticipate a single, but uncertain, future when we both know that there will be multiple future versions of ourselves

---

[1] Davidson (2001: 220).
[2] Whitman (1855).

and are able to predict what the experiences are that they will, collectively, have?

In a bit more detail: suppose an experimenter, asked what he believes the result of the experiment will be, says, 'Either outcome A or outcome B will occur, but not both, and I don't know which.' This statement (a perfectly normal response for an experimenter who accepts quantum mechanics) can be construed as the assertion of

- 'A will occur or B will occur', and
- 'Not both A and B will occur'

together with a refusal to assert either

- 'A will occur' or
- 'B will occur'.

However, if quantum mechanics is supposed to mean that the world branches, with A occurring in one set of branches and B in the other, it's unclear how this can be a reasonable position to adopt. Do we take A to occur if it occurs *in some branch* (i.e. somewhere in the branching Universe)? Then it cannot be true that not both A and B will occur, for each *will* occur in some branch. Nor can it make sense to plead ignorance as to which will occur, for both will. Or should we take A to occur if it is *the* outcome of the experiment, not just the-outcome-in-some-branch? If so, neither A nor B occurs; rather, some alien branching event occurs.

Similarly, ask the experimenter, 'What experiences will you have after the experiment?' and we would expect him to answer: 'I don't know: I'll see one of outcome A and outcome B, but not both; and I don't know which.' But if he accepts the Everett interpretation, this seems the wrong thing to say. For in his future are two (or two sets of) future versions of himself. Some see outcome A, some see outcome B. So it is not clear what justification there can be for his expecting one but not both outcomes: it seems that he should either expect *both* (because future versions of him see A and other future versions see B) or else *neither* (because, in the face of this kind of branching, the very idea of personal identity breaks down).

It might seem, then, that the Everett interpretation requires us to make a radical reassessment of our nature and our place in the world. Some critics of Everettian quantum mechanics (e.g. Albert and Loewer 1988 and Barrett 1999) have worried that the reassessment is *so* radical as to call into doubt the whole enterprise. And some advocates (such as Tappenden 2008; 2011 and possibly Greaves 2004) have embraced the idea that so radical a theory *ought* to have comparably radical consequences for our everyday worldview. My purpose in this chapter is to defend a more conservative position: I will argue that the ways in which we use talk of uncertainty, identity, the future and so forth will remain fully justified in the event that we come to accept Everettian quantum mechanics as true. We will have to abandon not our actual use of these concepts, but only (some of) our metaphysical and semantic theories as to how those concepts work. And this is not at all surprising, for those theories are indeed built on explicit assumptions about the structure of the world which Everett contradicts.

Some readers may be finding all of this unnecessarily metaphysical; they may wish to skip to the next chapter. Others, though, may feel that these issues should have been given a great deal more urgency: how can we possibly claim that Everettian quantum mechanics is empirically adequate without resolving them? And indeed, this would have been one way to proceed: I could have begun by presenting this chapter's arguments to the effect that our ordinary concepts of identity and uncertainty survive the Everett interpretation, and then developed the quantitative analysis of Part II from that starting point. (Indeed, in previous work—Wallace 2006a— I argued that this was the only way in which a satisfactory Everettian theory of probability and evidence could be constructed.[3]) However, the virtue of our present approach is that by treating probability purely in operational and functional terms—and thus reducing the problem to whether we should act *as if* branch weight were probability—it allows a solution to the probability problem which works independently of this chapter's considerations.

The purpose of this chapter, then, is not to form part of an argument for the Everett interpretation, but to be an investigation of some of its consequences. We shall find that, in this arena as well as that of physics, Everettian quantum mechanics is much more conservative than meets the eye.

---

[3] I was persuaded otherwise by Hilary Greaves and Wayne Myrvold, who have developed concrete theories (Greaves 2007; Greaves and Myrvold 2010) of confirmation appropriate to an agent contemplating the Everett interpretation; these results were a significant inspiration for my results in Ch. 6.

My starting point is the linguistic question: how can a deterministically branching future be reconciled with our ordinary talk of future uncertainty, and of alternative possibilities? And to answer this, I will need first to review briefly how such talk is generally understood in the philosophy of language on the understanding that the world does not, after all, branch.

7.2 How language works: a simple model

What is it to say that two sentences (or, if preferred, two acts of speaking) mean the same thing; that is, that they have the same meaning? What is it to say, for instance, that 'snow is white' (in English) means the same as 'la neige est blanche' (in French)? Saying that the one is a correct translation of the other seems[4] merely to beg the question, for we generally take the criterion of a good translation to be that it preserves meaning. At least on the face of it, to say that two sentences A and B have the same meaning is just to say that sentence A has meaning M and sentence B has meaning N and $M = N$.

Let us take this on-the-face-of-it idea seriously, at least for the moment. We can then ask: what is it to specify the meaning of a sentence? In a sense, this is *the* foundational question underlying philosophy of language, and much of twentieth-century philosophy more broadly, and many answers have been given, but I concentrate here on the most famous, having its origins in Frege's work (1892) and still widely supported: that to give the meaning of a sentence is to specify the conditions under which it is true. One understands 'snow is white' (says the truth-conditional theory) once one understands just what features of the world would have to change in order for it to be false. (I do not believe that the main conclusions of this chapter would fail on other mainstream approaches to a theory of meaning; to establish this fully would go far beyond the scope of this book, however.)

There is a useful (if on occasion dangerous) philosophical tool for getting precise about this: the concept of *possible worlds*. A 'possible world', or sometimes, 'a possible way the world could be' is generally taken to be some kind of abstract entity—a mathematical object, usually—representing a way that the world could be. What 'possible' and

---

[4] This is not to say that it is *uncontroversial* that there is more to meaning than translation; Quine, in particular, famously thought otherwise (Quine 1960).

'could' mean in full generality is (again) very controversial, but for present purposes we can restrict to a certain kind of physical possibility, and use the mathematical framework of physics to get a handle on them. (Here I follow Quine 1968 and Taylor and Dennett 2001; see also Dennett 2003: 63–9).

In a deterministic setting, in particular, we can take our physical theory to specify some state space S. Points in the state space, as usual, represent ways the world could be at an instant. So we could take possible worlds to be points in this space, but in fact it is more common to take them to be entire possible *histories* of the world (not least, as we'll see, because past- and future-tensed sentences are hard to handle otherwise). At the most fine-grained level, histories in physics are typically represented as trajectories through S: maps from some set of instants of time (discrete or continuous) into S. By *kinematically possible* trajectories, we normally mean any such maps (perhaps satisfying some kind of continuity or smoothness requirement); by *dynamically possible* trajectories, we mean only those trajectories satisfying the deterministic equations of the theory.

What is the point of this framework? Just this: if to specify the meaning of a sentence is to specify its truth conditions, then we can identify its truth *conditions* with the set of possible worlds *in which it is true*. That is, we can try *defining* the meaning of a sentence as a set of possible worlds; a sentence is then true if the actual world is (represented by) one of the possible worlds in the set of possible worlds which is its meaning. (This suggests that we should take the set of *kinematically* possible worlds of our physical theory to stand in for the philosopher's possible worlds, since we can generally make sense of a sentence being true or false in a possible world where the laws of physics are violated. Indeed, one can imagine sentences ('Spacetime is seven-dimensional'?) whose truth conditions require an even larger set of possible worlds; for our purposes, however, this set will be large enough.)

This simple theory of meaning is *too* simple even for our purposes. Consider, for instance, the banal sentence 'It is raining'. It doesn't make sense to ask, of this sentence alone, whether it is true or false (let alone, what its truth conditions are); what *does* make sense is to ask whether it is true or false when uttered at a given time, at a given place. Indeed, it makes sense to ask about truth *conditions* in that context: if Cecily utters the sentence at 6:05 pm on 11 August 2010 in Oxford, we can say that it is false but that it would have been true if such-and-such facts had been different.

This suggests a slightly more complicated theory of meaning: the meaning of a sentence is not a set of possible worlds: it is a map from contexts of utterance to sets of possible worlds.[5] The map tells us, if this sentence were to be uttered in this context, these are the conditions under which it would be true. A little philosophical terminology may help: the meaning of a sentence is a rule that assigns to each context of utterance a *proposition*, the proposition *expressed* by the sentence if uttered in that context. In our simple model, propositions are just sets of possible worlds (and, definitionally, a proposition P is true at a possible world w if $w \in P$); in a more complicated model, a proposition might have other structure too, but it would at least determine some set of possible worlds at which it is true.

For our purposes, we will be particularly interested in utterances of tensed sentences, such as 'The Democratic Party controls the US Congress', 'The Second World War ended twenty years ago', and 'The next president of the US will be a woman'. Whatever other contextual facts might be relevant to the truth conditions of utterances of these sentences, at least one is crucial: *when* it was uttered.

So much for the truth *conditions* of tensed utterence. Whether a given utterance is actually *true* is another matter. As of 2010, if I utter the sentence, 'The next US president will be a woman', I know perfectly well what its truth conditions are: it will be true if Barack Obama's successor is female, false otherwise. But is it *actually* true or false? Aristotle famously remarked that, since the future is open, such sentences are neither true nor false; Prior (1957; 1967) and Thomason (1970) developed this into a fully tensed concept of logic (cf. also McCall 1984; 1994 and Belnap 1992; 2002). The majority position in philosophy at present, however, is probably that all meaningful (declarative) utterances do not just have truth conditions, but truth *values*—i.e. each such utterance is either either true or false. The truth value of an utterance depends on future facts about the world, and even if those facts are unknowable even in principle, they still determine whether the utterance is true or false.

This theory of meaning appears deficient in many ways (for one, it does not really handle logically true claims: $2 + 2 = 4$ and $1 + 1 = 2$ are not synonymous, but are true in the same possible worlds, viz. all of them).

---

[5] Here I basically follow Kaplan (1989), though I have simplified somewhat for expositionary purposes.

The most promising way to make it more sophisticated is to recognize that to understand a sentence is not *just* to understand its truth conditions, but also requires an understanding of how those truth conditions follow from its components (this effectively makes a 'proposition' into something more structured than a mere set of possible worlds). For our purposes, though, the theory captures enough of the content of modern philosophy of language to allow us to explore how that content is affected by the move to quantum mechanics.

7.3 Using language: deterministic and indeterministic cases

What sort of sentences ought a rational person aim to utter, or to affirm (i.e. accept or agree to) when uttered by others? Here's a very crude, first-attempt model: they should aim only to utter, and to affirm, sentences that are *true*.

Let's acknowledge right away that this is unrealistically simple. A more sophisticated model would recognize that the point of language is com- munication, so that truths are *uttered* only so that those truths can come to be *believed* by those who hear the utterance; a more sophisticated model still would recognize that (alas) we sometimes have ulterior motives, and wish those who hear our utterances to believe things other than the truth.

Nonetheless, at least *other things being equal*, it seems that we normally aim to utter, and to affirm, sentences which are true: that is, in the possible-worlds jargon, to utter and to affirm sentences whose truth conditions evaluate to 'true' on that possible world which represents the actual world. And we can at least set out to model, and to analyse, *truthful* language use in this model, setting other uses aside.

To what extent is this a realistic goal? It depends on our aspirations *other* than truthfulness. If we are very unambitious in these aspirations, truthfulness is straightforward: we need only utter platitudes, things that are immediately obvious to our observations, or (if we want to be even more careful), things which follow by pure logic. But normally we aim to be more ambitious than this—to communicate more content. And that seems to involve trying to say things that are rather more specific: that is, which are true on rather more restricted subsets of the set of possible worlds than are the platitudes. In the ideal (though obviously unrealistic) case, we might aim to make utterances true in *exactly one* possible world: the actual world. That cannot be done, of course (not least because language is nothing like specific enough even in principle); still, it will be instructive to see what other constraints of principle restrict our approach to that ideal.

One constraint, of course, is an agent's knowledge of present-day facts. There are many true statements about the present (questions of how many sheep there are in Tibet, say) about which I know nothing—which is to say that I cannot pick out the actual world among a set of possible worlds which differ among themselves as to the number of Tibetan sheep. But let us idealize ourselves as present-day-omniscient, to the point that (at any rate, in the unrealistic limiting case) we know the actual present-day microstate.

In the event that the laws of physics (of this hypothetical model) are deterministic, knowing the present-day microstate suffices to know every-thing, because to each microstate corresponds exactly one dynamically possible trajectory. Put another way: in a deterministic physical theory, to every microstate corresponds exactly one possible world which is *physically* possible.

If the laws of physics are *stochastic*, we are not so lucky. In this situation, the laws of physics assign a probability to each trajectory, conditional on a given present-day microstate, and (by assumption) even knowing every-thing there is to know about the state of the present-day universe would not suffice to pick out the actual world from the set of possible worlds compatible with that present-day state. In general, in fact, almost any future is likely to have some (ridiculously small, but *nonzero*) probability of occurring, given the present-day state of the universe, so if we want to be *sure* that we are speaking truth, we are once again reduced to platitude. In practice, what we do is compromise: we accept some risk of error, and aspire not to utter sentences which are sure to be true, but only sentences which have a very high probability (according to the laws of physics) of being true.

And in fact, even if the world were deterministic, our situation would not be so different. For *exact* knowledge of the present is unattainable in practice, and *approximate* knowledge of the present does not in general fix the future, even in a deterministic universe; furthermore, even if we knew the present exactly, solving the relevant equations exactly is usually beyond our ability. In practice, we revert to higher-level, approximate approaches to the physics (typically, in physics, we obtain these from statistical mechanics), which are generally indeterministic. Indeed, insofar as we are analysing ordinary language talk (which is not in general sensitive to the fine details of microphysics), the right space of possible worlds is really not the space of kinematically possible trajectories in our microphysics, but the space of such trajectories in our emergent macrophysics.

It is important to note that *truth*—whatever its *theoretical* significance for explaining and justifying normal language use—has now essentially dropped out of the picture as far as *describing* normal language use, at least as far as future-directed utterances are concerned. Indeterminism—whether fundamental or emergent—has forced us to fall back on the goal of making *high-probability* utterances, where a high-probability utterance is one whose truth conditions pick out a set of possible worlds with high objective probability (as evaluated, relative to the present state, according to the appropriate fundamental or emergent probabilistic dynamics). If we want to say that a high-probability sentence is one with a high probability *of being true*, we can do so; but doing so doesn't actually play any role in determining what utterances are appropriate.

It is possible to place all of this in a familiar mathematical framework. Recall that S is our space of possible microstates (according to whatever background physics we are using). Let $\mathcal{E}$ be some Boolean algebra of subsets of S: in this context, the idea is that elements of $\mathcal{E}$ are possible truth conditions for present-tense sentences. (Following usage in Part II, I call $\mathcal{E}$ an *event space*.)

Discretize time (as usual, this is purely for convenience) into discrete moments $t_1, \ldots t_n$, and define a *history* as a time-indexed sequence of elements of $\mathcal{E}$. The histories stand in for sets of possible worlds, so that the truth conditions of an utterance (in some given context) is given by a history (histories, put another way, play the role of propositions). Histories themselves inherit Boolean structure from $\mathcal{E}$.

In this model, the *dynamics* are given by a probability measure Pr on the space of histories. This measure in turn defines a probability measure at each time over $\mathcal{E}$, and the probability of a history h *given* some initial history h_0 is given be

$$\Pr(h|h_0) = \frac{\Pr(h \& h_0)}{\Pr(h_0)}. \tag{7.1}$$

If the probabilities are past-deterministic (i.e. if for any given time t_i, two nonzero-probability histories which agree at t_i must agree at all past times), then we can speak of the probability of a history given some current event. A context of utterance picks out (perhaps amongst other things) some present event E. Conditional on that event, every history is assigned a probability, and so (since the truth conditions of utterances are represented by histories) every utterance is assigned a probability. The aspiration to truthfulness, in the model, is simply an aspiration to utter only high-probability sentences. No more lofty ambition is possible: no amount of present-day knowledge (in practice given determinism, in principle given stochasticity) would allow more.

Before considering quantum mechanics, let us apply this framework to the simple example I gave in the introduction: suppose some random physical process gives outcome A with probability ~ 0.5 and outcome B with probability ~ 0.5. An agent who aspires to truthfulness will not be willing to utter (or to affirm) 'A will occur' or 'B will occur', since neither utterance has probability ~ 1; he will be willing to utter 'either A will occur or B will, but not both', since the probability that one or the other occurs is ~ 1 (and he reasonably neglects the very small probability that neither will occur) and the probability that both will occur is ~ 0. That is, such an agent behaves as if uncertain as to which of A and B will occur.

7.4 Using language: branching case

How does this simple model of language use carry over to the Everett interpretation? To apply it, we need to work out what the possible worlds are on the assumption of Everettian background physics. There are basically two candidates:

1. Possible worlds are branches in a decoherent-history space: that is, sets of possible worlds are (more or less coarse-grained) histories within the emergent branching structure defined by decoherence.

2. Possible worlds are trajectories through Hilbert space: that is, they specify, at each time, the physical state of the entire Universe, and each possible world (in which decoherence occurs) is an entire branching structure.

Alternatively, we could express these two alternatives in the formal framework given previously:

1. The event space is a Boolean algebra of subspaces of Hilbert space; histories are sequences of such subspaces, inheriting a Boolean structure from the event space; the dynamics is given by the measure defined by the Schrödinger equation and by some given initial state; on the assumption that the set of histories is decoherent, this measure is indeed a probability measure and (by the Branching-Decoherence Theorem) it is past-deterministic.

2. The event space is a Boolean algebra of subsets of Hilbert space; histories are sequences of such subsets; the dynamics is given by the Schrödinger equation applied to some given initial state, and it assigns to each history either probability 1 (if the actual dynamical trajectory through Hilbert space lies in that history at all times) or probability 0 (otherwise).

Which of these models (on the assumption that Everettian quantum mechanics is correct) best gives the correct semantics for English and other natural languages?

I should acknowledge up front that the answer need not be uniform. For *some* of our most technical talk—in particular, for (some of) our talk about the Everett interpretation itself—the second model is fairly clearly correct. When I say, for instance, that the universe is constantly branching into multiple dynamically independent branches, I'm clearly saying something whose truth conditions should be evaluated relative to possible worlds that represent physical reality as a whole, not just some part of it: in other words, when I say things like that, what I'm saying needs to be interpreted according to the second model.

But even philosophers of physics don't spend most of their time talking about the Everett interpretation; most people don't spend any of their time talking about it. Our main interest here is in more mundane uses of language, whether in experimental physics or in everyday life, and here it is far less clear, at least to begin with, which model is correct. To resolve this, we need to consider what the truth conditions of ordinary utterances would be on each model.

As long as we confine our attention to utterances about the present and the past, the two models plausibly differ little. If (at time *t*) I say 'It is now raining in Oxford', then on the first model, my utterance is true in exactly those branches in which it is raining in Oxford at time t. In particular, the utterance is straightforwardly *true* if uttered in a branch in which it is raining in Oxford at time t.

On the second model, of course there are indefinitely many Oxfords in each possible world, in some of which it is raining at time t and in some of which it is not. But the truth conditions of 'It is now raining in Oxford', in this model, plausibly depend on more features of the context of utterance than just the time of utterance. In particular, it seems reasonable to suppose that the truth conditions depend on the branch in which the sentence is uttered as well as the time at which it is uttered, so that an utterance of 'It is now raining in Oxford' is true in exactly those possible worlds—i.e. those possible branching quantum universes— in which it is raining in Oxford at the time of utterance and in the branch of utterance. This is effectively the same set of truth conditions as for the first model.

Things are very different when we consider utterances about the future. If (at time t, just before carrying out a quantum measurement of spin on a spin-half particle) I say 'The measurement will give spin up', then on the first model, the utterance will be true in just those branches in which the result of the measurement after t is indeed up (with 'the measurement' being picked out by other contextual factors about the context of utterance, of course), and likewise for 'The measurement will give spin down'. Similarly, an utterance of 'The measurement will give either spin up or spin down' will be true in just those branches in which the measurement actually takes place and in which some freak event doesn't cause it to go wrong; 'The measurement will give both spin up and spin down' will be true in no branches at all; 'The measurement will give one, but not both, of spin up and spin down' will likewise be true in all branches where the measurement process works as intended. On the first model, then, other things being equal an agent will

- affirm 'Either spin up or spin down will occur';
- affirm 'At most one of spin up and spin down will occur';
- reject 'Spin up will occur';
- reject 'Spin down will occur'.

On the second model, certainly spin up will occur in some parts of pretty much any possible world (i.e. pretty much any quantum-mechanical branching structure); so will spin down. Indeed, both spin up and spin down results will occur in some of the branches which branch off from any instance in which the measurement is correctly performed. So the truth conditions of future-directed sentences appear to tell us that 'X will occur', uttered at time t, is true in all those possible worlds which assign nonzero branch weight to X occurring conditional on the context of utterance (that is: to X occurring in some branch futurewards of the event of utterance). On the second model, then, other things being equal an agent will

- affirm 'Either spin up or spin down will occur';
- reject 'At most one of spin up and spin down will occur';
- affirm 'Spin up will occur';
- affirm 'Spin down will occur'.

In sum: the first and second models more or less agree about the truth conditions of sentences which pertain only to events in the past and present of their being uttered, but they wildly disagree about the truth conditions—and hence, the rules of utterance—for sentences which pertain to events in the future of their being uttered.

Furthermore, it should be clear that on the first model, users of English would use it pretty much the way it would be used if quantum mechanics really were a single-Universe stochastic theory: in particular, being uncertain about the result of some quantum measurement would continue to make sense even for someone who was familiar with the Everett interpretation. The second model, however, requires English to be used in a sharply different fashion in quantum mechanics than in a formally similar stochastic single-Universe theory. According to the second model, someone about to perform a quantum experiment should not be uncertain about the result at all; rather, they should be certain that both possible results will occur. Indeed, someone who accepts Everettian quantum mechanics should no longer wonder about the weather next week: rather, they should be confident that it will be both rainy and sunny. If the second model is correct, then, adopting the Everett interpretation requires us to sharply revise our ordinary way of talking and thinking about the future.

To be sure, the revisionism of the second model can be blunted some-what. Even if 'Either spin up or spin down will occur, but not both' should not be affirmed, 'spin up and spin down will not occur in the same branch' can be affirmed, as can 'In each branch, either spin up or spin down will occur, but not both'. But this still requires us to bring talk of branching explicitly into our ordinary discourse. Ultimately—on the second model—we are still forced into recognizing that our old way of talking is just no longer usable in quantum mechanics.

This being the case, let us rename the models. Call the first model the *Conservative View*, and the second the *Radical View*, of language in an Everettian universe. On the Conservative View, the truth conditions of sentences are given by decoherent histories (i.e. possible worlds correspond to individual branches), and most of our ordinary thought and talk about the future remains coherent and reasonable in the light of the Everett interpretation. On the Radical View, the truth conditions of sentences are given by quantum-mechanical histories of the whole Universe (i.e. possible worlds correspond to entire branching structures), and accepting the Everett interpretation requires us to accept that most of our ordinary thought and talk about the future is really incoherent and meaningless. (Furthermore, I believe that this choice between views is not for the most part dependent on the detailed (and simplistic) theory of meaning which I used to motivate it.)

If there really were a conflict between our everyday concepts and our best theory of physics, so much the worse for our everyday concepts. But fortunately this is not our situation, because there are overwhelming reasons to regard the Conservative View as correct and the conflict as illusory. To set out those reasons, though, will require a further detour into the philosophy of language: we need to ask just what physical facts *make it the case* that a given theory of meaning is really true.

7.5 Meaning, use, and charity

It's tempting to think that *what makes it true* that a word, or a sentence, has a given meaning is just that speakers of the language *define* the word, or the sentence, to have that meaning. In a slogan: words mean what the dictionary says they mean. A moment's thought, though, shows that this cannot be true in general: while we do sometimes explain the meaning of new words using old words (especially when we explicitly define a term), on pain of infinite regress words cannot in general get their meaning this way. In a counter-slogan: dictionaries cannot explain meaning in general, because dictionaries themselves are written in words. Ultimately, a language must have the theory of meaning it has because of some nonlinguistic facts about the world.

In the light of modern neuroscience, it is tempting to look for those facts by looking inward: presumably (one might think) a sufficiently mature science of the brain could tell us just what the content of some utterance is simply by decoding the neural impulses from which it originates. But this just displaces the problem itself inwards: by virtue of what is it true that a given neural impulse has a given meaning? (If, for instance (as advocated by Fodor 1975) there is a 'language of thought' in the brain, and the meaning of a given utterance is its translation into the language of thought, what do sentences of that language itself mean?) Merely knowing the syntax of the language will not serve to fix its semantics in any algorithmic way; and even the syntax of a language—even the fact that it *is* a language—will not follow in a straightforwardly reductionist way from its microphysical details.[6]

As the reader of Chapter 2 will doubtless have recognized by now, we have here one more species of functionalist emergence: what *makes it true* that a given language has a certain semantics (or, better, what makes it true that a given pattern of behaviour is actually language-use with a certain semantics) is that by making that assumption, we are able to predict and explain structures in the phenomena which would otherwise be inexplicable.[7]

Indeed, this view, in one of a great many variants (see e.g. Quine 1960; Wittgenstein 1953; Lewis 1974; 1983a; Davidson 1973; or Dennett 1987b), is the dominant[8] view of semantics to emerge from twentieth-century philosophy of language. The common theme is that to understand the content of language—and, by the same token, to understand the content of thought—we need to make those ascriptions of content which make the most sense of the speakers. Quine (1960) referred to this as a *Principle of Charity*. In its simplest form, it enjoins us to ascribe to a community of speakers that interpretation of their language which maximizes the fraction of utterances which are true. In more

---

[6] For elaboration on this theme, see Dennett (1987b: 130–51).

[7] Arguably, we also need to require that no other semantics is clearly better at so predicting and explaining.

[8] Though not the only view; see Fodor (1990) and Dummett (1993: 1–116) for two very different dissenting positions.

sophisticated forms, it requires us to attribute to that community the combination of linguistic content and mental content (perhaps subject to constraints like simplicity, grammatical structure, and the like) which most makes sense of their actions as rational agents. Maximizing truthful utterances will play some role in more sophisticated schemes of this form, but untruths can be incorporated, up to a point, when they represent some rational form of error.

Fully explicating—let alone defending—this position lies far beyond the scope of this book; for the remainder of this section, I simply take it as read. I should, however, warn the reader against a common misconception. The view is not merely practical advice to translators: it is not simply the obvious theory that the most likely theory of semantics, other things being equal, is the one that makes most sense of a speaker's actions. Rather, it is the theory that there is *no more* to the question of what makes a given semantic theory correct for a language-using community than the goodness of fit of that theory to the way the community actually use their language. (What counts as 'goodness of fit', to be sure, remains controversial.)

One particular consequence may serve to illustrate this. Quine (1960: 27) called it *indeterminacy of translation*: in situations where two possible interpretations of a language are 'tied for first place'—i.e. equally good fits to facts about language use—there may be no fact of the matter about which is correct. To borrow an example from Dennett (1987a), suppose a tribe living in a methane swamp use the phrase 'glug' for explosions of methane gas. If one of those tribesmen, taken from his home, encounters an identical-looking explosion of some other combustible gas and exclaims 'glug!', has he erroneously identified the explosion as methane, or does 'glug' really mean 'explosive gas' rather than just 'methane'? Behavioural facts will not settle the matter; nor will any other plausible facts.[9] (In most cases, though, the cryptographic difficulties of finding multiple fits to a given set of behaviour suggest that such indeterminacy remains localized; cf. Dennett 2000: 343–8.)

---

[9] Philosophers tend to appeal to a doctrine of 'natural kinds' or 'natural properties' to remove ambiguities in the referents of theoretical terms (see Kripke 1981: lecture III and Putnam 1973). While this may suffice to rule out gerrymandered theories of meaning (as advocated by Lewis 1983b), it seems implausible to suppose that it helps to resolve problems like the tribesman's—at least if the resultant theory of semantics is supposed to do any actual explanatory work. (In any case, there are reasons to be sceptical that our intuitions about natural kinds really pick out any fundamental categories in Nature; cf. Dennett 1991a: 381.)

Returning to the Everett interpretation, the moral should be fairly clear. We have two views of how to understand the semantics of English and other natural languages given the Everett interpretation. According to the Conservative View, most of our everyday utterances remain reasonable even given the Everett interpretation. According to the Radical View, insofar as those utterances pertain to the future, they are by and large rendered meaningless by the Everett interpretation. On any plausible theory of how to ascribe meaning to utterances based on how well that theory explicates the action of the language-using organisms, the Conservative View comes out as a clear winner.[10]

This is not to say that the Conservative View entirely succeeds in preventing any of our commonplace utterances from being rendered false or meaningless by the Everett interpretation. In the next section—partly to illustrate further these issues of interpretative charity—I consider an often-cited example where this appears to occur.

First, though, a guest appearance.

SCEPTIC: Isn't it pretty absurd to apply the principle of interpretive charity to language users in this situation? After all, *ex hypothesi*, they're living in an Everettian universe, so their picture of the world is wildly wrong in a huge number of important respects.[11]

AUTHOR: Well, first of all, unless you've got an alternative theory of linguistic meaning available—one which doesn't make any use of charity-type arguments—then there isn't any choice but to apply the principle. (Unless you just want to say that linguistic communication is a priori impossible in a branching universe, I suppose; but that sounds pretty unmotivated.)

But in any case, who says that people living in an Everettian Universe are 'wildly wrong in a huge number of important respects' about the world? Sure, the Universe is a lot bigger than they thought it was;

[10] Interestingly, David Lewis—despite advocating an approach to semantics very much along the lines given here—claimed influentially that inhabitants of a branching Universe should (in effect) be understood according to the Radical rather than the Conservative View. ('The unfortunate inhabitants of such a world, if they think of 'the future' as we do, are of course sorely deceived, and their peculiar circumstances do make nonsense of how they ordinarily think': Lewis 1986b: 209.) However, Lewis gives no argument for this position, which he in any case defends in the context of philosophically, rather than scientifically, motivated reasons for considering a branching universe.

[11] Here the sceptic is channelling Lewis (2007b).

sure, there are lots of copies of themselves in causally removed bits of reality. Those are philosophically and scientifically important discoveries; maybe (I don't know) they're even theologically important. But do they matter to ordinary, banal thought, action and language? Friendship is still friendship. Boredom is still boredom. Sex is still sex.

SCEPTIC: But uncertainty isn't still uncertainty, and the future isn't still *the* future: it's many futures.

AUTHOR: Isn't whether 'uncertainty is still uncertainty' exactly the point at issue? You can't use the claim that our picture of the world has to be wildly modified in such-and-such a way as a premise in an argument against the claim that it doesn't have to be modified in such-and-such a way. (Well, I suppose you can, but it's not exactly going to convince anyone who's not already convinced.)

As for 'the' future: sure, on the Everett interpretation I have many possible futures; that was true already. And (arguably; cf section 7.9) more than one of those futures is physically realized. But again, that seems of (meta)physical significance rather than practical significance. (And after all, ordinary discourse about the future is committed to a metaphysics that's at best inchoate and at worst actively committed to the claim that it doesn't make sense to call future-directed claims true or false. Arguably, the Everett interpretation does more justice to people's intuitions about the openness of the future than does the classical view of a single fixed spacetime... not that I think that's any kind of reason to adopt it.)

7.6 Case study: what are we uncertain about?

I have argued that charity considerations require us to adopt the Conservative View of semantics and thus to take an attitude of uncertainty towards the results of experiments which have different results in different branches. But how can it even *make sense* to 'take an attitude of uncertainty' towards the future when we know all the relevant facts?

This objection is frequently brought up in informal discussions of the Everett interpretation; it has been made explicitly by Greaves (2004):

I can feel uncertain over P only if I think that there is a fact of the matter regarding P of which I am ignorant.

Of course, we have already seen a similar objection—to the use of probability rather than uncertainty—in Part II. And it was the central claim of Chapters 5 and 6 that as long as probability talk is understood operationally, the Everett intepretation is actually *better off* than nonbranching theories in making sense of that talk. An essentially similar strategy is available here: we can give a purely operational theory of uncertainty that is derivative on the theory of probability we have already constructed, according to which to be uncertain of something is to ascribe both it and its negation nonzero probability.

But the objection can also be responded to on its own terms, by noting that it is ambiguous between a truism of ordinary language and a more technical claim. As the former, it is indeed plausible to say that if a supposed concept of 'uncertainty' says that we can be uncertain about something without there being something to be uncertain about, then it does not deserve the name 'uncertainty'. However, there *is* something about which I am uncertain if I am uncertain about (say) whether the experiment will display 'spin up': trivially, that 'something' is *whether the experiment will display 'spin up'*.

The less trivial reading of Greaves' comment is that it asserts that we cannot be uncertain about something unless there is a third-party fact—a fact expressible via descriptions of the whole quantum universe, not just branch-relative descriptions—about which to be uncertain. But now it is far from clear what motivates Greaves, and others, to accept this technical claim.

To me, the most natural reading of the claim is that it is a *theory*—not something which is part of the structure of our ordinary pattern of language use or our ordinary conceptual scheme, but something theorized by philosophers as part of an attempt to analyse that language use and that conceptual scheme. The reading is strengthened when we note that most philosophers do not accept the unmodified claim, even in the absence of branching. For the philosophy literature is filled with examples of what is called 'indexical uncertainty': in this situation, an agent is uncertain not of what the world as a whole is like, but of where he is located within it. For a simple example (within the Everett intepretation), consider the state of an agent *after* a spin measurement has been made, but before he gets the result of the measurement. He knows all the salient facts about the Universe: there are two sets of branches, and in one set the result is spin up; in both sets, his epistemic state is the same. He is clearly ignorant of something, but his ignorance is not about the state of the Universe but about which branch *he is located in*.[12] (For non-Everettian examples, and for philosophical discussion, see Perry 1979 and Lewis 1979; there are also similarities with anthropic reasoning, for which see Barrow and Tipler 1986 and Bostrom 2002.)

This certainly *looks as if* it's a case of philosophers, entirely properly, (a) constructing a theory about the word 'uncertain' (or, if preferred, the concept of uncertainty), and then (b) modifying that theory when it turns out to conflict with actual usage. But in that case, it seems entirely right that the same philosophers, or those who share their methodology, ought once again to modify their theory when they recognize that it fails to fit other examples of normal usage: namely, *most* such examples, given that we live in a branching quantum universe.[13] To declare instead that ordinary people are wrong or confused if they persist in being uncertain in the face of branching seems no better motivated than to declare that ordinary people are wrong or confused if they so persist in the face of indexical uncertainty.

But, for the sake of argument (and against the possibility that there are better examples), let's grant that Greaves' claim, on the nontrivial reading (and, perhaps, modified to allow for indexical uncertainty) really is part of our ordinary use of the concept of uncertainty. In that case, it becomes impossible to apply interpretive charity in a way which makes sense of *everything* we do with the concept. On the Radical View, a large part of our ordinary usage (the part concerned with uncertainty about the future) becomes false or meaningless, but Greaves' claim remains valid. On the Conservative View, the converse is the case.

This situation would be an example of a more general issue in theories of meaning. Some language-users may have no theory of meaning which perfectly fits their actual pattern of use: the errors in their conceptual scheme mean that *any* such theory fails to make full sense of their behaviour in terms of rational action. It follows that there may be something a little holistic about where to localize their error: they must harbor some false or confused beliefs, and utter some false or confused

---

[12] Such examples, in the Everettian context, were first given by Vaidman (1998); I discuss them further in Wallace (2006a).

[13] Saunders (1995; 1996) makes a similar point about philosophers' construction of theories about tense and becoming, given that we live in a 4-dimensional universe.

statements, but exactly where the falsehood or confusion lies may have no precise answer.

Nevertheless, as a practical matter there can be—and in our situation, there *will be*—enormously strong reasons why one scheme rather than another will be correct. Namely: one scheme rather than another will ascribe to the language users enormously less confusion. In our case, on the Conservative View we have to accept as wrong or confused only those parts of ordinary language usage which come from a fairly theoretical analysis (so theoretical, indeed, that in fact I seriously doubt whether it deserves to be called *ordinary usage* at all.) By contrast, in accepting the Radical View we preserve the coherence of these theoretical claims, but at the cost of abandoning the coherence of most ordinary conversation about the future. Interpretative charity strongly favours the former move.

A famous example due to Putnam (1973) serves to illustrate this point further. Suppose (ignoring the scientific implausibility for the moment) that there exists a world called Twin-Earth, identical to our own world except that the liquid in the lakes and seas is not H_2O, but some other chemical—'XYZ'—which is indistinguishable from H_2O on casual inspection. There is a general consensus in philosophy of language that, if transported to Twin-Earth and given a glass of XYZ, I'm incorrect to say 'that's water', and correct to say 'water is H_2O'. We interpret my language so that 'water' refers to H_2O—even at the cost of making most of my post-arrival statements about water come out false—because this is the most explanatory interpretation of my linguistic community's overall language use.[14]

But now suppose that I never left Earth at all, but that in fact the stuff in the rivers and seas has *always* been XYZ. Only the tireless efforts of an International Conspiracy of Chemists has kept the public—even quite educated members of the public, such as those with chemistry degrees—duped into thinking that this stuff is H_2O. When investigative journalists expose the Conspiracy, how should we react? Surely not by clinging to our technical, theoretical belief that water is H_2O—thus preserving a few technical truths at the cost of making almost all the population's beliefs about water turn out false—but rather by acknowledging

---

[14] Putnam draws some relatively strong conclusions about semantics from this example—conclusions I do not necessarily wish to endorse here.

that water was *XYZ* all along and accepting that the technical connection between water and H_2O is simply false.

So too with claims like Greaves'. Our ordinary conceptual scheme—the one tacitly instantiated in our ordinary talk and our ordinary behaviour—copes perfectly well with the idea that uncertainty about the future always incorporated the possibility of branching. All that needs to be changed is our metaphysical theorizing about uncertainty; the concept itself survives just fine.

This concludes my general defence of conservatism in this chapter. In the remaining sections, I consider some more theoretical questions in metaphysics and philosophy of language from an Everettian perspective; readers whose interests lie elsewhere may wish to skip this part of the discussion.

7.7 Do sentences have truth values?

The (Conservative View) semantics offered above assigns bivalent truth conditions to any utterance of a sentence: that is, for each utterance, and each possible world, the utterance is true or false in that possible world. But traditionally, we are concerned not only with the truth *conditions* of an utterance, but with whether it is *true*. Can this be incorporated in an Everett-compatible semantics?

Short answer: up to a point. Suppose we are evaluating the truth value of an utterance, and we are doing so at some point x in the branching structure. (This need not be the moment at which the utterance was made; it's perfectly coherent to ask, for instance, 'Was what Bloggs said yesterday true?', and it's not automatically the case that the answer should be the same as to 'Is what Bloggs said true?', asked yesterday.) Because Everettian branches branch into the future but not into the past (a topic to be addressed in more depth in Chapter 9), all branches incorporating x are identical in its past. So if the utterance is concerned only with facts in the present or past of x, it will be true or false either on all branches incorporating x, or on none of them, and we might as well just call the utterance *true* in the former case, false in the latter.

This suggests a general theory of truth for Everettian semantics: an utterance is true at x if it is true on all branches in which x, false if it is true in no branches. But of course, a great many utterances about events futurewards of *x* will come out neither true nor false at *x* on this semantics: indeed, given the ubiquity of branching, virtually no contentful utterance about the future of *x* will be assigned a truth value at *x*.

Does this lack of future truth-values matter? My suspicion is that it does not, for a variety of reasons:

- There is no systematic presumption *at the level of ordinary usage* that ascribing truth *values* (as opposed to truth *conditions*) to claims about the future is coherent, much less necessary. (Again, consider that the contrary view has defenders all the way back to Aristotle; in more recent philosophy, authors such as Prior (1957; 1967); McCall (1984; 1994); and Belnap (1992; 2002) have defended a branching-time semantics on purely philosophical grounds, whilst Dudman (1985; 1992) has argued that English does not even have a genuine future tense in the sense of containing truth-evaluable utterances about the future.) So any requirement for future truth-values must be via theoretical considerations, not via considerations of ordinary usage.

- The semantics of our language itself remains bivalent, in the sense that a sentence is determinately true or false in any given possible world. There are therefore no problems with, for instance, understanding the semantics of composite sentences.

- We have seen that in a stochastic world (whether one which is fundamentally stochastic, or one like the Everett universe which is only emergently so) the goal of speaking *truths* about the future cannot even in an idealized sense be understood as a realizable normative ideal; the best substitute is to say *high-probability* things about the future. This goal is in no way undermined by the failure of future utterances to have determinate truth values.

Developing the details of any such semantics, though, lies beyond the scope of this book.

Suppose, though, that for whatever theoretical reasons it *does* remain necessary—or at least highly desirable—to attribute determinate truth values to all utterances, even those about the future. There is in fact at least[15] one proposal for doing so without giving up on overall conservatism

---

[15] A rather different method is developed by Ismael (2003).

about language in an Everettian universe. The best way to understand this proposal is to turn the discussion from matters of *language* to matters of *identity*: this will be my theme in the next section.

7.8 The nature of identity

As in the case of linguistic meaning, it will be helpful to review the question of identity, briefly, in a non-Everettian context.[16] Suppose that I possess some antique pot (call it P_2), and suppose that an antique dealer tells me that P_2 is *the same pot* as P_1, a famous pot owned by the emperor Tiberius in 30 AD. What is the content of this claim?

We can distinguish a metaphysical/semantic aspect from a more practical one. The physical facts that underlie the antique dealer's claim[17] are something like this: there is a four-dimensional tube P in spacetime, extending from the location of P_1 in 30 AD to the location of P_2 in 2009 AD, and the matter in that tube has certain structural and dynamical connections running along the tube. Formally speaking, if we write $P(t)$ for the contents of the tube at time t, the antique dealer's claim is underwritten by the existence of some structural-dynamical relation R holding, for each t, between $P(t)$ and $P(t + \delta t)$. In modern metaphysics it is standard to refer to the various $P(t)$ as the *stages* of the pot. (The details of R will not matter for our purposes; nor will the technical complications required by the need for δt, strictly speaking, to be infinitesimal.)

Given P, and given the R relation, there are then two basic *philosophical* analyses of the claim that P_1 and P_2 are *the same pot*, which Sider (2001) refers to as the Worm View and the Stage View. On the Worm View (espoused by Lewis 1983c), an object like a pot is a four-dimensional continuum. That is, P is the pot. It follows that 'P_1' and 'P_2' are just different names for the pot, and so, when we say 'P_1 is the same pot as P_2', we literally mean

$$P_1 = P_2.$$

---

[16] For more detailed discussions, see Sider (2001) and references therein.

[17] Metaphysicians will recognize that I assume a perdurantist rather than an endurantist picture of identity (cf. Lewis's definitions in Lewis 1983c, and also Sider's discussion in Sider 2001). My reasons are unoriginal: I find endurantism largely unintelligible, and insofar as I do understand it, it seems in conflict with physics.

On the Stage View (advocated by Sider himself), a pot is instead an instantaneous three-dimensional object, without temporal extent. That is,

$$P_1 = P(25 \text{ AD}); \quad P_2 = P(2009 \text{ AD}).$$

When we say 'P_1 is the same pot as P_2', on this view, we do not assert literal identity; rather, we say something like

P_2 is linked to P_1 by a continuous chain of R-related pots.

All of this assumes, of course, that pots don't branch, which is equivalent to requiring that the R relation is always (at most) one-to-one. If we drop this assumption—say, by shifting from pots to amoebae—and allow R to incorporate branching, the Worm View becomes ambiguous. Do we take an amoeba to be a maximal chain of R-related amoeba-stages, or do we take it to be any maximal set of amoeba-stages linked by R-related amoeba-stages? On the former view (which I call the Lewisian View, since it was proposed in Lewis 1983c), then amoebas overlap, in the sense that some amoeba stages are parts of many amoebas. On the latter view (which, following Tappenden 2000, I call the Hydra View[18]), then amoeba-stages separated by great expanses of space are nonetheless part of the same hydra.

There is an alternative way of seeing the distinction between Lewisian and Hydra Views. Instead of taking an amoeba-stage instantaneous section of an amoeba-tube, we could (a little unnaturally, perhaps) have defined it as a section of *all* amoeba-tubes which overlap at some point in the past. (Call this the *disconnected* version of an amoeba stage.) From this perspective, both the Lewisian and Hydra Views regard an amoeba as a single chain of amoeba stages; they just disagree as to what constitutes an amoeba-stage.

From this perspective, we can see that there is an fourth alternative available in the presence of branching, which stands to the Stage View rather as the Hydra View does to the Worm View. Namely, we could take an amoeba to be a disconnected amoeba stage; call this the Disconnected View of identity. The four options might be summarized thus:

---

[18] Tappenden's more recent use of this 'Hydra' terminology in Tappenden (2011) seems to fit less clearly with the 'Hydra View' presented here, and Tappenden states (in private conversation) that he does not currently accept that view. In any case, for my purposes I am simply adopting Tappenden's very evocative name; my arguments stand or fall independent of how they conform to Tappenden's past or present views.

	Stages are connected	Stages are disconnected
Objects are stages	Stage View	Disconnected View
Objects are sequences of stages	Lewisian View	Hydra View

In applying this framework to Everettian quantum mechanics, it is helpful to focus at first not on the theory of identity for individual macroscopic objects, but on the theory of identity for worlds themselves. For the story we have told as to how to find macro-objects in quantum physics is essentially (and ignoring relativistic considerations) a story of how to find entire quasi-classical worlds in quantum physics. Once we understand what it is for two worlds to be the same, it should be relatively straightforward to specialize this to objects within a world.

So, we have four options available to us (see Fig. 7.1). On the Hydra View, a 'world' is an entire branching structure: 'the world' is really the entire quantum Universe, or at least the entire macroscopic reality

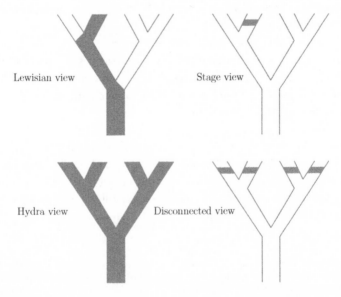

Figure 7.1. Four views of branching and identity (shaded area picks out one entity)

emergent from that Universe. On the Disconnected View, a 'world' is an instantangeous cross-section of the entire branching structure. On the Lewisian View, a 'world' is an entire decoherent history; on the Stage View, it is an instantaneous section of such a history.

Applying the frameworks now to individuals:[19] on the Hydra View, an object—a pot, say—is itself a massively branching structure. On the Disconnected View, a pot is a cross-section of such a structure. On the Lewisian and Stage Views, a pot is much as it was on the classical-universe version of those views. In each case, though, there is a caveat: 'pot stages' should not be understood as instantiated directly in the microphysics of the branching state. Rather, they are higher-order ontology, instantiated as features of the patterns that define entire quasiclassical worlds. (The reason this matters is that a given spacetime region will typically be occupied (at least on the Lewisian and Stage views) by a great many pots, each part of a different quasiclassical world. The fact that there are many pots there, in fact, does not in general supervene locally on that spacetime region; it depends on the nonlocal entanglement connections between that region and others—a matter to which I return in Chapter 8.)

7.9 Identity and language

It should now be obvious, I hope, that if something like the Conservative View of branching-Universe *language* is correct, the Hydra and Discon-nected Views of branching-Universe *identity* must be rejected. Indeed, these views of identity, committed as they are to a disconnected notion of world stage, are exactly the sort of view that would be appropriate to the Radical, rather than the Conservative, view (cf. section 7.5), committed as they are to the falsehood or meaninglessness of claims like

> I am not currently seeing both experimental outcome A and experi-mental outcome B.

and

> Once the experiment is performed, I will see one, but not both, of experimental outcome A and experimental outcome B.

---

[19] Here the impression may be given that the correct theory of identity for ordinary objects is the same as that for worlds. This need not be the case, though: a Lewisian View of worlds, for instance, is compatible with a Stage View of objects within each such world. (The converse looks more problematic.)

In addition, these two views seem to fail even on their own terms. For the correct identity relation for macroscopic objects, and indeed for quasi-classical worlds, must answer to the same considerations as does any candidate for higher-level ontology: that is, it must pick out an explanatorily important structure in the world. And clearly, disconnected stages cannot do so: they are mere conglomerateas of mutually non-interacting structures.

This leaves the Lewisian and Stage Views as possible candidates for the correct theory of identity in Everettian quantum physics. It is possibly unrealistic to expect to decide between them: after all, in mainstream metaphysics the Worm and Stage Views seem fairly deadlocked, and it is not even clear (to me, at any rate) whether there is anything really substantive to the debate, there or here. Certainly, considerations like Ockham's razor seem irrelevant: it is not in question that we can coherently talk about both world-stages and sequences of world-stages (i.e. decoherent histories); all that is in question is which of them is to be identified with the word 'world'.

Nonetheless (and here I return to the semantic question which brought us to this topic), there are semantic implications to each position. On the Stage View, we seem[20] committed to the theory of truth which I developed in section 7.7, in which statements about the future do not in general have truth values. By contrast, on the Lewisian View semantics can proceed pretty much as they do in nonbranching theories: everything is relativized to a world, understood as a decoherent history. So utterances of sentences take place in worlds, and 'The result of the experiment will be A' comes out true in all worlds where the outcome is indeed A (and where something picked out by 'the experiment' occurs after the utterance). An utterance which bears on facts in the future of some point x is determinately true or false for each history containing x, and so is determinately true or false for each *person* who, at x, is considering its truth

[20] In previous work (Wallace 2005) I proposed an alternative view of semantics (suggested to me by Simon Saunders) which I called *ambiguity semantics*, in which the semantic content of an utterance like 'X will happen' was a claim about the whole branching structure, of form 'X will happen in branch h'—but in the case of assertions about the future, it was unavoidably and irreducibly ambiguous which claim was being expressed (i.e. which h was meant). I am now (tentatively) rather more sceptical about the viability of ambiguity semantics, since it does not straightforwardly seem to have the resources to deal with situations where the context from which we *evaluate* an utterance's truth value differs from the context in which it is uttered.

value; but it will in general be in principle impossible for that person to determine the truth value, except by waiting.

Detailed philosophical considerations are likely to bear on which of these alternatives is preferable. For instance, the Lewisian View seems overall more conservative, in the sense that (for better or worse) it preserves more of the standard philosophical picture of language and identity. However, this is not universally the case: for instance, whilst it gives a very clear *formula* for reference (an utterance can be regarded as represented by an ordered pair of a microphysical event x and a complete history h, and a referential expression in the utterance refers to whatever thing in h it would have referred to on our conventional view of semantics if we regarded h as the only history), it is unclear what *causal* story could be told as to how this reference occurs (unless we wish to regard causation as itself relativized to branches), since nothing intrinsic about the microphysical x ties it to a particular history.[21]

Conversely, the Lewisian View tells a rather more conventional story than the Stage View about the metaphysical underpinnings of uncertainty claims. On its account, *my world* does not literally branch at all when (say) a quantum measurement is made. Rather, the referent of 'my world' is a non-branching, four-dimensional continuum, the pre-measurement part of which qualitatively identical with—indeed, supervenient on the same microphysical facts as—many other worlds up until the time of measurement. By definition, then, there is a determinate (if rather trivial) matter of fact as to whether I will see (say) outcome A or outcome B: I will see A if the outcome in my world is A, and B if the outcome in my world is B. My uncertainty about which I see is mere indexical uncertainty—uncertainty about my place in the world—but it is an uncertainty which cannot, even in principle, be removed prior to the branching.

According to the Stage View, on the other hand, I, and *my world*, undergo real branching, in the sense that there is no determinate answer as to what will happen to my world in the future. My uncertainty about the future, therefore, is not indexical on this account: it is a new kind

---

[21] It is for these sort of reasons that Lewis himself claimed (1983c: 73–6) that neither one of a pair of individuals who will be undergoing fission can refer to himself and not his twin until the fission occurs. As such, he would in effect reject the semantics I have read into the Lewisian View (I am grateful to Oliver Pooley for this observation). Saunders and Wallace (2008) presents an argument that Lewis was wrong to do so, or at least that he would be wrong to do so in a branching-worlds context.

of uncertainty, one which has no analogue in non-branching situations. The Lewisian View, therefore, is much more conservative with respect to current philosophical theorizing about uncertainty than is the Stage View.

Now, as noted in section 7.6, I personally ascribe essentially no weight to *this* kind of conservatism, conservatism, that is, about the current theoretical position in philosophy, especially insofar as it rests on a tacit rejection of the Everett interpretation, not about everyday non-theoretical discourse. Those that disagree, however, may thus find reason to prefer the Lewisian view (this issue is developed further in Saunders and Wallace 2008).

Overall, I do not wish to attempt here to adjudicate between the various semantic proposals that I have suggested: further discussion would involve too many controversies and would go far beyond the scope of this book. (I briefly discuss (and dismiss) one particular metaphysical puzzle in Box 7.1.) I rather suspect that there may not really even be any fact of the matter as to which is correct: here perhaps, even if nowhere else, we really do find *widespread* indeterminacy of translation, as Quine predicted.

Box 7.1. Overlap vs divergent worlds

In the conventional framework of possible worlds, if two possible worlds are identical up to some time *t* and if both contain (say) Abraham Lincoln, it appears that we can coherently ask whether the Abraham Lincoln in world *A* is the same as the one in world *B*, or whether they are numerically distinct but qualitatively indistinguishable. Insofar as possible worlds are mathematical abstracta (the position advocated in this chapter), this is a meaningless or at any rate a pretty shallow question, akin to asking whether two qualitatively described mathematical objects which are partially isomorphic are *identical* in their isomorphic parts or merely *indistinguishable*. But David Lewis (for reasons that lie beyond the scope of this discussion; cf. Lewis 1986b) took worlds to be concrete particulars (so that the actual world *is*, and is not merely *represented by*, some possible world), and furthermore regarded a possible world as the mereological sum of the objects contained within it. From this perspective the question is well-posed; Lewis answered that a given object is a part of exactly one world.

It is superficially tempting to suppose that an analogous question applies in Everettian quantum mechanics: given two histories (identical to our own until 1950, say), it seems that we can coherently ask whether the Abraham Lincoln in the first history is *identical to* or merely *qualitatively indistinguishable from* the one in the second history.

But the temptation should be resisted: the question is not meaningful, at any rate not without further precisification. The reason is straightforward: Everettian histories are not in any helpful sense mereological sums of their parts. (In fact, in general I see very little reason to think that the part–whole relation embodied in mereology finds much, if any, application in science.) To be sure, they are *represented* by sequences of time-indexed projectors; but the question of whether two such sequences have the *same* initial members or merely *qualitatively indistinguishable* ones is fairly contentless (after all, if it were convenient to add a label to each projector, we could do so without affecting their representational usefulness), and in any case, representation (whatever it is) is not identity (no one claims that a world literally *is* a sequence, i.e. a map from the natural numbers into some space of projectors).

We can perfectly well (albeit only in principle, and highly nonlocally) pick out a set of expectation values of spacetime operators, and note that facts about Lincoln supervene on these expectation values; we can furthermore observe that facts about Lincoln in history A supervene on the same set of expectation values as facts about Lincoln in history B. That much is conceptually unproblematic (if technically difficult). If there is some further metaphysical question to be asked, in my view it has yet to be unproblematically formulated.

For dissenting views, however, see Saunders (2010) and Wilson (2010a; 2010b).

7.10 Possibilities

I have used the concept of *possible worlds* throughout this discussion, due to the clarity they provide in discussions of *semantics*. However, the possible-worlds machinery is used at least as often to give some clarity to discussions of *modality*—i.e. of possibility and necessity—and in this section I wish to consider briefly what impact Everettian quantum mechanics has on these issues.

When I say that some future event is possible, sometimes I speak simply of epistemic possibility, so that 'possibly P' means just 'I do not know that it isn't the case that P'. As we have seen, interpretative charity tells us:

- that if I know that it is true that P on some non-negligible-probability[22] history containing my current location in the branching structure, then I should assent to 'possibly P';
- that if I know that it is true that P on no such history then I should reject 'possibly P'.

Sometimes, however, we seem to use possibility in a non-epistemic sense (as might be the case for physical, historical, logical possibility, etc). It is fairly clear that if P is true of some history through x, then 'possibly P' must also be true at x: that which cannot in principle be ruled out should not be impossible. However, nothing in the interpretive-charity argument entails that the converse holds: that for 'possibly P' to be true at x, P must be true at some history through x.

This is not to deny that some philosophically relevant notion of possibility *could* be defined via

'possibly P' is true at x iff P is true in some history containing x;

indeed, as far as I can see anyone tempted by the purely philosophical arguments for branching time should embrace this notion of possibility, for it seems to provide all that the advocates of such arguments desire of 'objective possibility'.

But it is also not to say that it is *required* that this particular notion of possibility has conceptual significance. As I mentioned previously, I am myself sceptical that free will, causation, etc. require a branching-time picture. This raises the question of how those analyses of possibility which traditionally have no notion of branching time fare under the branching-world hypothesis.

---

[22] There is in fact a considerable literature (see Wheeler 2007 and references therein) on how to allow for the fact that I can know that P is the case even while knowing that there is a nonzero (but very small) probability that P is false. This debate, so far as I am aware, carries over to Everettian quantum mechanics *mutatis mutandis*.

In this context, it is worth considering issues of possibility in the possible-worlds framework. Lewis (1986b) defined possibility relative to some pre-defined class X of possible worlds (such as the physically permitted worlds, or the physically permitted worlds compatible with some data x, etc.): P is possible relative to this class iff it is true for some world w in the class. (See also Taylor and Dennett 2001 or Dennett 2003: ch. 3 for a development of this position.) For future-directed possibility statements (i.e. for statements of the form 'It is possible that Athens will host the Olympics in 2500 AD', rather than of the form 'It is possible that Hitler won the Second World War'), it is a requirement that the actual world is in X.

Branching-time semantics makes only a slight alteration to this framework. In such a semantics, through a given moment of time x there is no unique 'actual history' but rather a set of such histories: all those through x. Any future-directed notion of possibility, defined via a set X of possible worlds (i.e. histories) and assessed at some point x, is coherent only if the entire set of histories through x is contained within X. As far as I can see, this alteration does not have profound effects for the possible-worlds analyses of possibility. (Also note that on a Lewisian theory of identity, everything is almost exactly as it is in the absence of branching, with exactly one world 'actual' relative to a given utterance.)

One final note: I have assumed, throughout this discussion, that possible worlds are *abstracta*, akin to mathematical objects. There is an alternative tradition, though—*modal realism* (Lewis 1986b), according to which possible worlds are as real as the actual world. It has sometimes been suggested that modal realism should be combined with the Everett interpretation, so that possible worlds should be identified with physically present Everettian branches: David Deutsch endorses the proposal, at least in the sense of viewing Everettian branches as capable of doing the philosophical work that Lewis's possible worlds are supposed to do (Deutsch 1997: 340–41; 2010: 543) and Alastair Wilson has explored it extensively (Wilson 2010a; 2010b). I'm personally somewhat sceptical of the proposal: for unoriginal reasons (cf. Kripke 1981 and Stalnaker 2003), it seems to me that possible worlds *should* be abstract rather than concrete, and those reasons don't seem appreciably changed by Everettian quantum mechanics. But the issue clearly deserves further attention—attention, however, which lies far beyond the scope of this book.

7.11 Summary

If the only consistent ways to explain the semantics of tensed talk all assume a certain structural property of time, then that would indeed be a reason to assume that time does have that structure. This is not the case for the question of the linearity or otherwise of time: though I see no reason to doubt that nothing in our experience *rules out* linear time and linear semantics, I have argued here that if our universe actually has a branching structure, then the correct semantics must also be branching.

As such, the semantics of tense are more naturalistic than meets the eye. If the right semantics for tense depends on the structure of time, we can be no more certain of our semantic theory than of our theory of temporal structure. (Perhaps this shouldn't be too surprising: if linguistic meaning supervenes on linguistic behaviour and if the correct third-party description of linguistic behaviour depends on facts about the world about which we are ignorant, then our theory of linguistic meaning should inherit that ignorance.)

In particular, given Everettian quantum mechanics, *our Universe* is branching. In such a situation, we are required to assume that the semantics for our language is itself branching. The apparent conceptual paradoxes of the branching Universe are caused by insisting instead, without justification and against the principle of interpretive charity, on adopting linear-time semantics.

Once we recognize the semantic implications of the branching-Universe hypothesis, we realize that the statement 'Either A or B will occur, but not both, and I don't know which', uttered at some point x in the branching structure of the Universe, becomes a perfectly rational thing for someone to say if A occurs in some of the branches futurewards of x and B occurs in the others. A rational agent at x (and who knows the above facts) would affirm 'Either A or B will occur' (for in every branch one or the other occurs) and reject 'Both A and B will occur' (for in no non-negligible-probability branch do both occur) but will neither affirm nor reject 'A will occur' and 'B will occur', for each occurs in some but not all branches futurewards of x. As such, Everettian quantum mechanics has no difficulty in explaining why experimenters are entirely justified when, as they routinely do, they express ignorance as to the outcome of their experiments.

CHAPTER 7. Despite appearences, the branching structure of Everettian quantum mechanics has few or no consequences for our everyday beliefs and actions. Ordinary talk of the future is in general neither false nor meaningless if Everettian quantum mechanics is true.

CHAPTER 8. Is Everettian quantum mechanics a nonlocal theory? How are we to think about the relation between the quantum state and space-time; indeed, how are we to think about the quantum state, in realist terms, at all?

8

Spacetime and the Quantum State

[E]very theoretical physicist who is any good knows six or seven different theoretical representations for exactly the same physics.

Richard Feynman[1]

8.1 Concepts of locality

Part of my purpose in this chapter is to determine whether Everettian quantum mechanics is in a relevant sense *nonlocal*. Nonlocality, of course, is frequently held up as a characteristic feature of quantum mechanics; on the one hand, it is said to be responsible for such iconic quantum phenomena as violations of Bell's Inequality[2] and quantum teleportation;[3] on the other, it appears at least at first sight to be in flat contradiction both with our commonsense view of the world and (far more importantly) with the relativity of simultaneity and with relativistic covariance more generally. Absent a solution to the quantum measurement problem, it is not easy to make precise just what we mean by the claim that quantum physics is nonlocal; since the Everett interpretation provides a solution, though, we might hope that it allows just such precision.

I should begin by getting a bit more precise—independently of quantum theory—on just what 'nonlocality' is supposed to mean. In fact (and here I largely follow Healey 1991; 1994) we can usefully distinguish

---

[1] Feynman (1967: 162).

[2] Bell (1966); for further discussion, see e.g. Bell (1981a), Maudlin (2002), Butterfield (1992), or Redhead (1987).

[3] Bennett et al. (1993); see e.g. Vedral (2006: 46–7) or Benenti, Casati, and Strini (2004: 208–12) for a presentation, and Timpson (2006) for philosophical discussion.

SPACETIME AND THE QUANTUM STATE   293

two sorts of nonlocality, which are usually called *action at a distance* and *nonseparability*.

Action at a distance occurs when, given two systems A and B which are separated in space, a disturbance to A causes an immediate change in the state of B, without any intervening dynamical process connecting A and B. (Conversely, theories which do not involve action at a distance are said to satisfy *Local Action*.) Theories which allow *nonlocal signalling*—i.e. allow macroscopically readable *messages* to be transmitted instantaneously from A to B without propagating through the intervening space—clearly violate Local Action, but the converse is not true: if the microdynamics of a theory is sufficiently carefully arranged, it may be impossible to influence the microphysical action-at-a-distance processes in order to signal. (This loophole is exploited by the de Broglie–Bohm theory and other hidden-variable theories to reconcile action at a distance (which such theories do have) with the absence of macroscopic signalling (which the no-signalling theorem[4] rules out for any theory empirically equivalent to quantum mechanics[5]).

The most famous example of a physical theory involving action at a distance, of course, is Newtonian gravity, where every body in the universe *instantaneously* influences every other body. But precisely because such instantaneous influence seems to imply a preferred notion of simultaneity, relativistic physical theories (at least at the classical level) generally do satisfy Local Action: influences propagate from A to B at the speed of light or more slowly, and they do so through intervening fields.

Nonseparability is a matter, not of dynamics, but of ontology. A theory is nonseparable if, given two regions A and B, a complete specification of the states of A and B separately fails to fix the state of the combined system $A + B$. That is, there are additional facts—nonlocal facts, if we take A and B to be spatially separated—about the combined system, in addition to the facts about the two individual systems.

[4] See Redhead (1987: 113–16) and references therein.

[5] Antony Valentini (1996, 2001, 2004; cf. also Valentini and Westman 2004) has explored the possibility that some sort of hidden-variable theory is correct but that the no-signalling theorem only holds in some kind of approximate sense, roughly analogous to the assumption thermal equilibrium; if so, the theory would in principle be empirically inequivalent to quantum mechanics, and thereby testable. Of course, I doubt that any such tests will in fact vindicate Valentini; philosophically, though, his is exactly the right approach. If we are to modify or supplement the equations of quantum theory, we should expect empirically detectable consequences.

When we are dealing with something like a field theory—i.e. when to specify the state of some system is to specify all the fields within some region—there is an alternative way of defining separability, which roughly corresponds to what philosophers call *Humean supervenience*. (The term originates from David Lewis (1986c: ix–xvii), who named it after the Enlightenment philosopher David Hume; how accurately it describes Hume's own philosophical position is a moot point.) On this definition, a theory is separable if, for any collection $\mathcal{U}$ of open subsets of spacetime whose union is all of spacetime, specifying the states of each element of $\mathcal{U}$ separately suffices to specify the state of the entire Universe according to the theory.[6]

Nonseparability, too, can be found in nonquantum situations. In classical electromagnetism (expressed in terms of the vector potential), if A and B are simply connected spatial regions whose union forms an annulus around some solenoid, the electromagnetic states of regions A and B are each trivial (i.e. each is gauge-equivalent to a connection which is everywhere zero). Yet if there is a current in the solenoid, and thus magnetic flux through the hole in $A \cup B$, then it is well known that the electromagnetic state of $A \cup B$ is *not* trivial, and indeed is empirically detectable: this is the Aharonov–Bohm effect (Aharonov and Bohm 1959). See Healey (2007) and Belot (1998) for more detailed analysis and for more extensive defences of the view that the Aharonov–Bohm effect indicates that classical electromagnetism is nonseparable.[7]

Our project, then, might seem simple. Does Everettian quantum mechanics violate Local Action? And does it violate Separability? But things are not quite that easy: even saying what *is* the state of a given physical region in quantum theory requires us to have a more solid grasp of the physical reality that 'the quantum state' represents than is available

---

[6] Lewis's own definition of Humean supervenience requires that the individual properties of all the spacetime *points* sufficed to determine all the facts about that world; for technical reasons, this definition is ill-suited to modern field theories (see Butterfield 2006a).

[7] Arguably, the Aharonov–Bohm effect is quantum-mechanical: the empirical prediction it makes is that a certain set of electron interference fringes will be shifted by the solenoid. I call it classical for two reasons: firstly, the electromagnetic field is treated classically throughout the process; secondly (as I argue in Wallace 2009) the effect is fundamentally a result of classical field theory, and it is only the particular details of our world—in particular, the absence of a regime in which charged quantum fields behave like classical fields—that prevent us observing it at the classical level. (Note that the gravitational analogue of the Aharonov–Bohm effect (Ford and Vilenkin 1981) is unambiguously classical.)

from its abstract, Hilbert-space definition. And this brings me to the other purpose of this chapter: to get a firmer grip on just what the physical world is like according to Everettian quantum physics. We have already seen what the *macro*-world is like: it is a branching structure instantiated in the microphysical reality represented by the quantum state. But to talk sensibly about nonlocality, it will be helpful to find a way to talk sensibly about that microphysical reality itself; this will be my purpose for the next three sections. That done, I will use the account I develop to consider locality: I will argue that quantum mechanics is nonseparable but obeys Local Action, and I will analyse measurements of entangled states from an Everettian perspective. In the last two sections, I return to the question of how we should think about the quantum state, and explore alternatives to the picture I develop.

Much of this chapter draws heavily upon joint work with Chris Timpson (see in particular Wallace and Timpson 2010).

8.2 Thinking about physical states

A traditional view is that the quantum state should be understood as somehow being a description or encoding of various classically describable or measurable properties. According to this view, the potential properties of the system are represented by projectors on Hilbert space (although typically not all such projectors will correspond to bona fide classical properties), and the system determinately possesses a property if (and only if) its state is in the subspace projected onto by the associated projector, and determinately does not possess a property if (and only if) the state is orthogonal to that subspace. As such, many properties seem to be neither determinately possessed nor determinately not possessed.

This understanding is adequate for interpretations of quantum mechanics in which the theory is simply an algorithm for predicting measurement outcomes; it may also be the right starting point for interpretations which abandon classical logic (cf. section 1.6). For more traditionally realist interpretations—and in particular, for the Everett interpretation—it is (as it stands) hopeless. Everettian quantum mechanics reads the quantum state literally, as itself standing *directly* for a part of the ontology of the theory. To every different quantum state corresponds a different concrete way the world is, and the quantum state *completely* specifies the ontology.

From this perspective, regarding the state as encoding properties of the system in the traditional way is at best unhelpful and incomplete—many properties, like 'being in an entangled state' or 'being in some eigenstate of energy' or 'possessing an even number of zero amplitudes in configuration-space' cannot be expressed using the traditional approach.[8] Focusing on projectors to represent properties is too crude to capture all of the interesting properties of the world when the quantum state directly describes ways the world is.

In that case, how *should* we think about the quantum state? Mathematically, it is a vector (or, strictly, a ray) rotating in a very high-dimensional Hilbert space: is that what the Universe is like according to quantum physics? If so, it becomes hard to see what justifies speaking of the quantum state as 'highly structured': a ray in Hilbert space is just an unstructured line, and all rays seem alike.

This apparent paradox is partially resolved when we recall that a quantum theory can only be specified if we also give some particular set of (socalled) 'observables' (e.g. the position and momentum operators in one-particle QM, or the various field observables in QFT)—this breaks the symmetry of Hilbert space, allowing us to see how merely mathematically distinct vectors could represent physically distinct states of affairs. The state vector can now be seen as a way of codifying the various rich properties of the physical state: in one-particle quantum mechanics, for instance, to any self-adjoint function $f(\widehat{X}, \widehat{P})$ of the position and momentum operators it assigns an 'expectation value' $\langle \psi | f(\widehat{X}, \widehat{P}) | \psi \rangle$. Differences between states correspond to differing patterns of assignment of numbers to operators, and *those* patterns can be very highly structured.

Put another way: just as quantum mechanics can be expressed in a very elegant way as the theory of a ray evolving in Hilbert space, so classical mechanics can be expressed in a very elegant way as the theory of a point evolving in phase space. But quantum physics no more says that the physical universe is an unstructured ray in a high-dimensional space than classical mechanics says that the physical universe is an unstructured point in a high-dimensional space.

---

[8] E.g. the disjunctive property of being in some energy eigenstate or other would, in projectors language, be given by a sum of projectors onto a complete set of energy eigenstates, returning the projector onto the whole Hilbert space; thus every quantum state has this property. Clearly something has gone wrong.

But this answer is essentially negative. How *are* we to think of the reality represented by the quantum state, once we recognize that it should not be definitionally related to measurement, nor understood by too literal an appeal to the Hilbert-space formalism?

Again by analogy with classical physics, we should not expect a one-size-fits-all answer to the question. The framework of Hamiltonian mechanics can be used to describe essentially any classical dynamical theory, from the harmonic oscillator, through the electromagnetic field or the Milky Way's star distribution, to the large-scale dynamics of spacetime itself. So the question is not 'What kind of physical reality does the quantum state *in general* represent?' but rather 'what kind of physical reality do the quantum states of *particular physical theories* represent?'.

In particular, we should ask what physical reality is like, at the micro-level, according to our best currently extant quantum theories: namely, the quantum field theories. The proposal I present (essentially the proposal of Wallace and Timpson 2010) can be motivated by considering how we answer such questions in a similar but simpler case: that of classical field theory, and in particular, classical electromagnetism.

8.3 The ontology of electromagnetism

We are happy enough that electromagnetism does not present us with an irredeemably obscure picture of the world, yet it is not as if we really have an intuitive grasp of what an electric or a magnetic field is, other than indirectly and by means of instrumental considerations ('A test charge would be accelerated *thus*', for example). But in the case of field theory, we do at least understand that 'the electric field at spacetime point p' denotes a property *of point p,*[9] and this gives us substantial intuitive understanding of how to think about that field: namely, it describes certain, admittedly somewhat alien, properties *of spacetime regions*.

Does electromagnetism offer any further elucidation of its ontology, beyond an assignment of vectorial quantities to spacetime points? I think not: what could that understanding derive from? All we have to work with are the mathematical structure of the theory and the instrumental considerations alluded to above—and the latter, when one takes seriously the

---

[9] Strictly speaking, assigning a vectorial quantity to a point will also involve making reference to relations with neighbouring points; cf. Butterfield (2006a; 2006b).

idea of physics as universal, ultimately collapse into interactions between the electromagnetic field and other comparably alien entities. It may help here to recall that *matter*, in the sense of the macroscopic material bodies we observe, consist of, and manipulate in the lab, cannot ultimately be thought of as some non-electromagnetic entity on which electromagnetic forces act: modern physics makes clear that solid matter is made up of electromagnetic fields as well as fermionic matter. In any case, the latter is scarcely less alien than the former. Even if we naively regard fermions as classical point particles, 'point particles' are shapeless, colourless, texture-less entities with little in common with familiar macroscopic bodies, for all that schoolbook physics may have acclimatized us to them.[10]

Thus it seems that we gain a basic understanding of the electromagnetic field by seeing it as a property of spatial regions; and our further understanding must be mediated by reflecting on its role in the theory (including, importantly, the instrumental considerations). Beyond that there doesn't seem to be much further to be grasped. And the structural complexity of a given electromagnetic field is represented not in the properties of very small spacetime regions (indeed, in the limit as these regions become point-sized, the field's structure becomes almost trivial) but in the way in which those properties vary across spacetime. Furthermore, this general model is characteristic of pretty much any classical field theory, except that vector fields seem mathematically tame compared to the sorts of mathematical objects used to represent the field values of many classical field theories (the affine connections of general relativity, for instance, or the principle–bundle connections of gauge field theory, or the Grassman fields used in classical fermionic field theory). In the next section, I see to what extent we can take this model over to quantum theories, and in particular to quantum field theory.

8.4 Spacetime state realism

Suppose one were to assume that the Universe (or at least, the Universe according to some quantum theory) could naturally be divided into sub-

---

[10] Notwithstanding this, Allori et al. (2008) try to take exactly this line about the electromagnetic field: i.e. to interpret it entirely via its propensity to act on point particles. Their attitude to electromagnetism and to charged matter seems to me at best outdated, at worst appealing to a view of the subject which was never historically accepted; space, however, prevents more detailed consideration of their case. See Diacu and Holmes (1999) and Wilson (1998) for further discussion.

systems. We know what this means in mathematical terms: the Hilbert space of the theory can be decomposed as the tensor product of a number of distinct Hilbert spaces

$$\mathcal{H} = \bigotimes_{A_i} \mathcal{H}_{A_i} \qquad (8.1)$$

where the product ranges over all the subsystems, so that each subsystem is assigned a Hilbert space.

If the state of the Universe is $|\psi\rangle$, what is the state of some subsystem A? Mathematically speaking, there is only one candidate object: the partial trace of $|\psi\rangle$, over all components of (8.1) with the exception of the Hilbert space $\mathcal{H}_A$ corresponding to A itself.

Timpson's and my proposal (which we call *spacetime state realism*, for reasons which will become apparent) is very straightforward: just take the density operator of each subsystem to represent the intrinsic properties which that subsystem instantiates, just as the field values assigned to each spacetime point in electromagnetism represented the (electromagnetic) intrinsic properties which that point instantiated. While the property that having a given density operator represents may not be a familiar one, the case need be no different in principle from that of electromagnetism. Insofar as one can continue to press for the physical meaning of the density operator, the theory in which these objects are postulated must provide the answer. (We can say such things, for example, as: the property is not a scalar or vector one, in contrast with the (classical) Klein–Gordon or electromagnetic fields.)

To provide a simple model, imagine a Universe consisting of a great many interacting qubits whose spacetime trajectories we approximate as classical (cf. Deutsch and Hayden 2000). The qubits each bear the property or properties represented by their two-dimensional density operator; pairings of qubits bear properties represented by a four-dimensional operator; and so on. There need be no reason to blanch at an ontology merely because the basic properties are represented by such objects: we know of no rule of segregation which states that, for example, only those mathematical items to which one is introduced sufficiently early on in the schoolroom get to count as possible representatives of physical quantities!

Note that if, by contrast, we were to treat the Universe just as one big system, with no subsystem decomposition, then we would only have a single property bearer (the Universe as a whole) instantiating a single property (represented by the Universal density operator), and we would lack sufficient articulation to make clear physical meaning of what was presented (as with section 8.2's one-particle-in-high-dimensional-space view of classical mechanics, one would struggle to see the structure being imputed to the world in this case).

Now, our proposal becomes more concrete when one connects this system–subsystem analysis of the quantum state with spacetime. To do this we need some notion of the spatial location of a physical system. In nonrelativistic quantum mechanics (NRQM) this is somewhat complicated: the systems are naturally taken to be particles and assemblies of particles, and a particle's spatial location is one of its dynamical properties, not something to be specified *ab initio*.

It is somewhat simpler if we consider the Fock-space formalism of NRQM, where we allow the number of particles to vary.[11] On the one-particle Hilbert space of any nonrelativistic particle we can define projectors $\widehat{P}_\Delta$ projecting onto those states with wavefunctions having support in spatial region Δ. If $\{\Delta_i\}$ is a partition of real space into measurable subsets and if $\mathcal{H}_i = \widehat{P}_{\Delta_i}\mathcal{H}$ is the Hilbert space built from all states having support in Δ_i, then we have $\mathcal{H} = \oplus_i \mathcal{H}_i$ and moreover

$$\mathcal{F}(\mathcal{H}) = \mathcal{F}(\oplus_i \mathcal{H}_i) = \otimes_i \mathcal{F}(\mathcal{H}_i), \qquad (8.2)$$

so the Hilbert spaces $\mathcal{F}(\mathcal{H}_i)$ may be taken as representing the possible states of the subsystem in, or comprising, region Δ_i.[12] This means that

---

[11] Recall that if $\mathcal{H}$ is any Hilbert space, the associated symmetric (bosonic) and antisymmetric (fermionic) Fock spaces are

$$\mathcal{F}_S(\mathcal{H}) = \oplus_{i=0}^{\infty} \mathcal{S}(\otimes^i \mathcal{H})$$

and

$$\mathcal{F}_A(\mathcal{H}) = \oplus_{i=0}^{\infty} \mathcal{A}(\otimes^i \mathcal{H})$$

where $\mathcal{S}$ and $\mathcal{A}$ are symmetrization and antisymmetrization operators. The distinction between $\mathcal{F}_S(\mathcal{H})$ and $\mathcal{F}_A(\mathcal{H})$ plays no part in the argument, and will be ignored in the text.

[12] Intuitive support for the result expressed in eqn. (8.2) can be seen once we recognize that the Fock-space operation is a sort of 'exponential' of the Hilbert space (this is clearest from the power-series expansion of the exponential), so that we can write $\mathcal{F}(\mathcal{H}) = \exp(\mathcal{H})$. Then equation 8.2 becomes $\exp(\oplus_i \mathcal{H}_i) = \otimes_i \exp(\mathcal{H}_i)$. More formally, for each i let $\widehat{a}^\dagger_{i,k}$ be a set of creation operators for states in $\mathcal{H}_i$. Then it is easy to see (via the observation that creation operators on different Hilbert spaces commute) that states of form

$$\widehat{a}^\dagger_{i_1,k_1} \cdots \widehat{a}^\dagger_{i_n,k_n} |\Omega\rangle$$

(where $|\Omega\rangle$ is the vacuum) form a basis both for $\otimes_i \mathcal{F}(\mathcal{H}_i)$ and for $\mathcal{F}(\oplus_i \mathcal{H}_i)$.

we can take the regions of space (and their unions) as our basic bearers of properties: tensor products of states belonging to each region (in general, superpositions of such products) allow us to express the original total state of varying particle number.

Already our presentation is sounding somewhat field-theoretic. To illustrate what is going on, consider a particular region of space Δ_j. This region has a Fock space $\mathcal{F}(\mathcal{H}_j)$ whose (pure) basis states can be represented in the form $|n_1, n_2, \ldots\rangle$, where n_1, n_2 etc. represent the occupation numbers of what we can think of as the available modes of Δ_j, i.e. the number of excitations in each of some orthogonal set of states of $\mathcal{H}_j$. What we would normally think of in NRQM as a single particle localized in Δ_j will, in this setting, be represented by a singly excited state of $\mathcal{F}(\mathcal{H}_j)$, e.g. $|1, 0, 0 \ldots\rangle$, tensor product with the vacuum state $|0\rangle$ for all the other regions' Fock spaces. In general, then, a single particle (which usually *won't* be localized in some particular region) will be represented by an entangled state composed of a superposition of states each differing from the vacuum only in a small region Δ_i:

$$\ldots |1, 0, \ldots\rangle |0\rangle |0\rangle \ldots + \ldots |0\rangle |1, 0 \ldots\rangle |0\rangle \ldots + \ldots |0\rangle |0\rangle |1, 0, \ldots\rangle \ldots$$

and so on.

Things become simpler still when we move to full quantum field theory. In the algebraic formulation of QFT, we associate to each spacetime region $\mathcal{R}$ a C^*-algebra $\mathcal{A}(\mathcal{R})$ of operators, representing the dynamical variables associated to region $\mathcal{R}$. A state ρ of such a region is a positive linear functional on $\mathcal{A}(\mathcal{R})$ (often described in rather instrumentalist terms as giving the expectation value of each element of $\mathcal{A}(\mathcal{R})$). By the Gelfand–Naimark–Segal construction we can associate ρ with a state in a Hilbert space $\mathcal{H}_R$, and represent $\mathcal{A}(\mathcal{R})$ as an algebra of operators on $\mathcal{H}_R$ (see e.g. Haag 1996: 122–4 for the details). $\mathcal{H}_R$ can then be taken as the Hilbert space of the field in region $\mathcal{R}$.[13] If preferred, one can even remain at the

---

[13] In the standard presentations of AQFT, the algebra $\mathcal{A}(\mathcal{R})$ is infinite-dimensional (even though it is associated with a spatially finite region); as such it has multiple non-isomorphic representations, and so different states lead to different Hilbert spaces. This makes it less clear that we are licensed to talk about $\mathcal{H}_R$ as 'the' Hilbert space for region $\mathcal{R}$. On the modern (Wilsonian) understanding of renormalization, however, this is an artefact of the formalism, which disappears when we cut off the unphysically high-energy degrees of freedom; as such, we ignore this complication in the text. (See Wallace 2006b, and references therein, for further discussion.)

more abstract level, forgo the representation theorems, and just take the C^*-algebraic state itself as denoting the properties of a region.

This way of thinking is well-defined for any quantum theory with compositional structure, but in particular, it makes sense for any quantum field theory, and for any many-particle theory once it is expressed in field-theoretic terms. It has a certain simplicity: it respects the dynamical structure of QM, indicating no preference for Schrödinger over Heisenberg or interaction dynamics (as the state is just construed as a linear functional of the dynamical variables); it adds no additional interpretational structure (given that the compositional structure of the system is, *ex hypothesi*, already contained within the formalism); it prefers no basis for each individual system, in keeping with the physics norm of treating different bases as on a par; and it gives an appropriately central role to spacetime. Hence (and for want of a better one) our name for this position: spacetime state realism.[14]

8.5 Locality

Now that we have a way of understanding the microphysical reality represented by the quantum state, we can return to the questions of locality raised in section 8.1; it is, I think, a virtue of spacetime state realism that it gives clear, definite answers to these questions.

8.5.1 Does Everettian quantum mechanics display action at a distance?

No.

In a quantum field theory, the quantum state of any region depends only on the quantum state of some cross-section of the past light cone of that region. Disturbances cannot propagate into that light cone. This follows from the well-known fact that spacelike separated field operators commute: any disturbance outside the past light cone of a region $\mathcal{R}$ can be represented on the global quantum state as the action on that state of a unitary operator built out of some operators localized outside the past light cone of $\mathcal{R}$, and since those operators commute with all operators

---

[14] Although if the theory has a compositional structure not induced by spacetime regions—as in, say, an abstractly specified quantum computer, this name is a misnomer.

localized in $\mathcal{R}$, they have no effect on the expectation values of operators in $\mathcal{R}$, and so no effect on the physical state of $\mathcal{R}$.[15]

We can get more insight into this by once again considering a network of qubits and idealizing the dynamics as discretized: at each discrete time, some set of quantum gates (i.e. unitary transformations acting on some qubit or N-tuple of qubits) is applied to the qubits. The state of a given qubit Q at time t_{n+1} then clearly depends only on the state, at time t_n, of those qubits (including Q itself) which interacted with Q between n and $n + 1$.

8.5.2 Does Everettian quantum mechanics display nonseparability?

Yes.

Because of entanglement, knowing the density operators of regions A and B does not suffice to fix the density operator of $A \cup B$. Some of the properties of $A \cup B$ are genuinely nonlocal: they have local physical manifestations only if we arrange appropriate dynamics.

For instance, suppose that we have a long row of qubits $q_1, \ldots q_n$, and we simulate local *interactions* by only ever applying gates to adjacent pairs of qubits. The system might start in a state with no nonlocal properties: say, with each qubit in the state $|0\rangle\langle 0|$.[16] By applying a gate to the first two qubits, we can transform them so that their joint state is (the projector onto) any one of the four states

$$\{|X_\pm\rangle = (|0\rangle \otimes |0\rangle \pm |1\rangle \otimes |1\rangle), \quad |Y_\pm\rangle = (|0\rangle \otimes |1\rangle \pm |1\rangle \otimes |0\rangle)\}). \quad (8.3)$$

In each case, the states of qubit 1 and qubit 2 are each

$$\frac{1}{2}(|0\rangle\langle 0| + |1\rangle\langle 1|). \quad (8.4)$$

Now we can effectively transport qubit 2 along the chain by applying a sequence of swap operations between adjacent qubits. At the end of this

---

[15] In many presentations of QFT, the operators corresponding to fermions *anti*-commute. Full discussion of how to understand this lies beyond the scope of this book; however, there is good reason to think that the conclusion of this section is unaffected. Specifically, fermionic operators invariably appear in the Hamiltonian in QFT in pairs (i.e. as a product of two field operators), and the product of two fermionic field operators localized at x commutes with any fermionic or bosonic field operator localized somewhere spacelike separated from x.

[16] Strictly speaking, this is an unrealistic idealization: in relativistic quantum field theory, vacuum entanglement means that *all* states have some nonlocality. Vacuum entanglement is generally negligible on macroscopic scales, however.

local process, all qubits except q_1 and q_n are in the state $|0\rangle\langle 0|$, q_1 and q_n individually are in the state (8.4), and q_1 and q_n jointly are in one of the four states (8.3). However, the joint state of q_1 and q_n is not locally accessible: no local operation (i.e. no operation on adjacent qubits) leads to local results (i.e. states of individual qubits) which depend on the joint state of q_1 and q_n.

Finally, we can transport qubit 1 along the chain in the same manner. At the end of the process, q_{n-1} and q_n will be in the entangled state, and this can be determined by a local interaction.

Somewhat picturesquely, we can think of entanglement between states as a string connecting those states, representing the nonlocal relation between them. We can move either end of the string by local interactions, and we can cause the string to 'fray' at either end by entangling the system at each end with adjacent systems. But we cannot access the information content of the string—i.e. we cannot set up dynamical processes whose outcomes are dependent on the nonlocal properties represented by the string—without moving the two ends of the string until they coincide. In this way, nonseparability remains fully compatible with dynamical locality. (In this sense, quantum nonseparability and the nonseparability associated with the Aharonov–Bohm effect are rather similar.)

(An interesting corollary of quantum nonseparability in the relativistic regime has recently been discussed by David Albert (2007); see also the earlier discussions in Aharonov and Albert (1980) and Myrvold (2002).) We might expect that if we knew the complete state of the Universe on every element of some foliation of spacetime—the complete history of the world as expressed on that foliation, if you like—then this would suffice to fix all facts whatsoever about the Universe. But this is not so: there can, in principle, be entanglement between two spacelike separated spatial regions on different slices of the foliation which leaves no trace on the quantum states of each individual slice of the foliation. Albert calls this a failure of *narrativity*, of the principle that we can say everything about the world by narrating a story of how it began and how it changed over time. (Of course, all *local* facts are fixed by such a story; it is only the nonlocal facts that may not be.) For further discussion of narrativity failure (in the Everett interpretation, and also in non-unitary quantum mechanics where the failure is far more severe), see Wallace and Timpson (2010).

The overall story about locality in Everettian quantum physics, then, is this: the dynamics of the theory are local: there is no action at a distance, and no clash with relativistic covariance. But quantum entanglement means that a great deal of the information contained within the quantum state is nonlocal, associated with large spatial regions but not with any given subregions of those regions. As David Deutsch once put it, quantum theory is a theory of local interactions and nonlocal states.[17]

8.6 The true state and the relative state

The picture of the world given in the previous section is apt to look pretty alien, and pretty unlike the world we see around us. This is unsurprising, of course: Everettian quantum mechanics, after all, is a many-worlds theory, and so our own world constitutes just one small aspect of the universal quantum state. Nonetheless, it will be helpful to see how the decoherence account of Chapter 3 connects to what we have discussed.

In essence, the situation is this: the state ρ_A of some large region A will typically correspond to no single quasi-classical situation, but rather can be decomposed as a convex sum

$$\rho_A = \sum_i p_i \rho_A^i \qquad (8.5)$$

of states ρ_A^i which do correspond to quasi-classical situations.[18] Each quasi-classical component evolves independently of the others, so ρ_A encodes not one, but a great many simultaneously present quasi-classical situations. And the spatial extension of the quasi-classical goings on is encoded in the entanglement between states: if B is some largeish region adjacent to A, then the state of $A \cup B$ might be

$$\rho_{A \cup B} = \sum_i p_i \rho_A^i \otimes \rho_B^i; \qquad (8.6)$$

the local states of A and B each encode many simultaneously present quasi-classical situations, and the nonlocal information in $\rho_{A \cup B}$ encodes the connections between them that justifies the 'many-worlds' language.

---

[17] I heard this comment from Deutsch in a public lecture in Oxford around 1996; as I note below, he probably wouldn't put it this way any more.

[18] As I observed in section 3.11, the exact fineness of grain of this distribution is largely conventional: there is no well-defined notion of branch count in Everettian quantum mechanics.

This should all bring home the point that the true state of a spatial region is very far from being directly accessible to any realistic agent. An observer at A (present in the quasi-classical situation encoded by ρ_A^i, say) might well speak of *the* state of A being ρ_A^i and *the* state of B as being ρ_B^i, but these would be emergent, approximate notions (somewhat akin to Everett's original 'relative states'). The true, ontologically primary state of A would still be ρ_A.

It will be useful to see how all this plays out dynamically in the spacetime language. A simple model will be helpful: suppose now that A is some small, but macroscopic, spatial region, and that $B_1, B_2, \ldots$ are a series of concentric, non-overlapping shells of successively larger radius surrounding B. Suppose (somewhat unrealistically) that the quantum states of $B_1, \ldots B_n \ldots$ are all classical states with very little entanglement with one another: a thermal bath of radiation of some sort, say. And suppose that A is initially in some state corresponding to a macroscopic superposition such as Schrödinger's cat. Initially, then, A and the B_i each have states that are pure or close to pure, and so the state of the Universe as a whole is close to a product state:

$$|\psi(t_0)\rangle = (\alpha|X\rangle + \beta|Y\rangle) \otimes |B_1\rangle \otimes \cdots |B_n\rangle \cdots \qquad (8.7)$$

(where $|X\rangle$ and $|Y\rangle$ are the macroscopically definite states of A).

Very quickly—at the speed of whatever the dynamical interactions between them is, which for a radiation bath will usually be the speed of light—A will become entangled with B_1: B_1 will record the state of A, and the new global state will be

$$|\psi(t_1)\rangle = (\alpha|X\rangle \otimes |B_1^X\rangle + \beta|Y\rangle \otimes |B_1^Y\rangle) \otimes |B_2\rangle \otimes \cdots |B_n\rangle \cdots \qquad (8.8)$$

with $\langle B_1^X|B_1^Y\rangle \simeq 0$. The individual states of systems A and B_1 are now impure:

$$\rho_A = |\alpha|^2|X\rangle\langle X| + |\beta|^2|Y\rangle\langle Y|; \; \rho_{B1} = |\alpha|^2|B_1^X\rangle\langle B_1^X| + |\beta|^2|B_1^Y\rangle\langle B_2^Y| \qquad (8.9)$$

but the combined state of $A \cup B$ remains pure.

The decoherence propagates outwards, of course. Pretty soon, B_2 will be entangled with B_1 (and so, indirectly, with A); then B_3 will be entangled with B_2, and so forth. The state of A itself does not change at all in the process; what changes are the nonlocal states of successively larger regions including A. Where originally we had a quantum state with no nonlocality at all, in due course we will have a quantum state with structure

$$|\psi(t_N)\rangle = \left(\alpha|X\rangle|B_1^X\rangle\cdots|B_N^X\rangle + \beta|Y\rangle|B_1^Y\rangle\cdots|B_N^Y\rangle\right)|B_{N+1}\rangle\cdots \quad (8.10)$$

Assuming the decoherence interaction spreads at lightspeed, if C is any region contained entirely in the future lightcone of the initial preparation of the superposition at A, then the state of C will have the form

$$\rho_C = |\alpha|^2\rho^X + |\beta|^2\rho^Y \quad (8.11)$$

for some density operators ρ^X, ρ^Y. All the phase information about the superposition has been dispersed into nonlocal degrees of freedom inaccessible at C.

As we would expect from the absence of action at a distance, then, branching is not a global phenomenon. Rather, when some microscopic superposition is magnified up to macroscopic scales (by quantum measurement or by natural processes) it leads to a branching event which propagates outwards at the speed of whatever dynamical interaction is causing decoherence—in practice, it propagates out at the speed of light.

I illustrate this for our simple model, somewhat schematically, in Fig. 8.1. (Notice that the illustration does not show the true quantum state—which will already contain vast amounts of multiplicity—but some

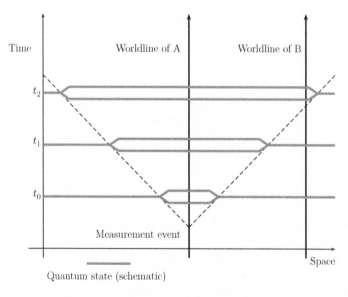

Figure 8.1. Schematic representation of spacetime branching

appropriate relative state.) From the point of view of an observer at B, between times t_0 and t_1 the state of an observer at A is indefinite—or, better, the state of region A is a nonclassical state instantiating two sets of observers with macroscopically different observations. (Because of the relativity of simultaneity, as long as the measurement event is spacelike separated from B, it is a matter of convention for the observer at B just when he deems A to have entered this nonclassical state.)

After t_1, though, the expanding branching event encompasses the observer at B, and he too branches; thereafter, for each version of the observer at B there is a unique version of the observer at A, and their observations match up.

8.7 Aspect-type experiments in Everettian quantum theory

Things get more interesting if instead two measurements are made. Suppose, then, that observers at A and B each prepare some spin-half particle in a superposition of spin-up and spin-down along some axis, and then measure its spin on that axis: suppose, say, that A measures

$$a_1 | \uparrow \rangle + a_2 | \downarrow \rangle \tag{8.12}$$

and B measures

$$b_1 | \uparrow \rangle + b_2 | \downarrow \rangle \tag{8.13}$$

and, for illustrative purposes, suppose that a third observer is at C, midway between A and B. Now there are two independent branching processes: one spreading out from the measurement at A, one from B. Any description of how this happens, therefore, will have to be frame-dependent; I illustrate, in one particular frame, in Fig. 8.2.

So, from C's point of view: at time t_0 region A has branched but region B has not. At time t_1, branching has occurred at B and the branching at A has spread almost as far as C. By time t_2, the observer at C has branched; for each such observer, A has a definite relative state but B remains branched. By time t_3, C has entered the light cone of both measurements and there are now four relevantly distinct sets of observers at C, for each of which both A and B have definite states. At this time, however, the observers at A have not entered the future light cone of the measurement at B, so this

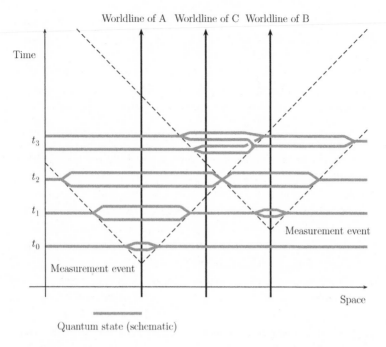

Figure 8.2. Schematic representation of spacelike separated measurements

is asymmetric: for these observers, the C observer's state is indefinite, and does not become definite until the worldline of A interects that future light cone.

What are the amplitudes of the various branches that the observer at C in due course splits into? At t_2, the branches in which the measurement result is 'up' at A have weight $|a_1|^2$ and the remaining branches have weights $|a_2|^2$. When these branches are themselves branched by interactions with signals from B, the A-up branches split into 'up-up' branches with relative weight $|b_1|^2$ and so overall weight $|a_1|^2|b_1|^2$, and similarly for the other three sets of branches. The overall weights, then, are

up-up	$|a_1 b_1|^2$
up-down	$|a_1 b_2|^2$
down-up	$|a_2 b_1|^2$
down-down	$|a_2 b_2|^2$

The amplitudes of the branches are determined locally: information about the weights of the splitting at A and at B propagates outward to C from A and B respectively.

Things are interestingly different if the measurements at A and B are performed not on independently prepared particles but on two entangled particles—say, on a pair of spin-half particles with joint state

$$c_{11} \mid \uparrow \rangle \otimes \mid \uparrow \rangle + c_{12} \mid \uparrow \rangle \otimes \mid \downarrow \rangle + c_{21} \mid \downarrow \rangle \otimes \mid \uparrow \rangle + c_{22} \mid \downarrow \rangle \otimes \mid \downarrow \rangle. \qquad (8.14)$$

In this case, the amplitudes of the four sets of branches into which C eventually branches are not determined simply by the separate weights of the branchings at A and at B. Nor is this to be expected: as I stressed previously, in Everettian quantum mechanics interactions are local but states are nonlocal. The entanglement between the particle at A and the particle at B is a nonlocal property of $A \cup B$. That property propagates outwards, becoming a nonlocal property of the forward light cone of A and that of B. Only in their intersection can it have locally determinable effects—and it does, giving rise to the branch weights which, in turn, give rise to the sorts of statistical results recorded in Aspect's experiments (Aspect, Grangier, and Roger 1982) and their successors: statistical results which violate Bell's inequality.

But in Everettian quantum mechanics, violations of Bell's inequality are relatively uninteresting. For Bell's theorem, though its conclusion arguably entails not only non-separability but action at a distance (cf. e.g. Bell 1981a, Maudlin 2002, Butterfield 1992, or Redhead 1987 for details), simply does not apply to the Everett interpretation. It assumes, tacitly, among its premises that experiments have unique, definite outcomes.[19]

From the perspective of a given experimenter, of course, her experiment *does* have a unique, definite outcome, even in the Everett interpretation. But Bell's theorem requires more: it requires that from her perspective, her distant colleague's experiment also has a definite outcome. This is not the case in Everettian quantum mechanics—not, at any rate, until that distant experiment enters her past light cone. And from the third-person perspective from which Bell's theorem is normally discussed, no experiment has any unique definite outcome at all.

---

[19] For a more detailed analysis of Bell's theorem in an Everettian context—which gives a slightly different account of why the Everett interpretation is an exception to Bell's result—see Timpson and Brown (2005).

The upshot, in any case, is that while Everettian quantum mechanics can be approximated within relatively small volumes as a theory of (emergent) branching of the whole world, really branching is a much more local phenomenon than this: branching *spacetimes*, rather than branching *times*, might be a more apt description.[20] It follows that the machinery of decoherent histories, when applied to systems large enough for relativistic travel times to be relevant, has a certain arbitrariness in how branches in distant regions are combined to create overall histories. This is only to be expected: the arbitrariness corresponds to the arbitrariness in the definition of distant simultaneity.

One more point should be stressed before I leave this topic. I have spoken, throughout, as if the branching events play out against a *fixed* background spacetime. And this is indeed what our current best theories of quantum mechanics—the quantum field theories—predict. Other branches, then, are located in precisely the same space and time as our own world; it is just that they are dynamically incapable of (significantly) affecting our world, and vice versa.

However, while this is true in our best *current* quantum theories, it is rather unlikely to survive the inclusion of gravity in those theories. According to general relativity, spacetime itself is dynamical, and influenced by the distribution of matter; superpositions of macroscopically different states of affairs, therefore, should become entangled with the spatial geometry, giving rise to superpositions of different geometry. (Were this not the case—if the curvature of spacetime were generated by, say the *expected* distribution of mass-energy—we would expect that the other worlds would, like dark matter, show themselves via their gravitational effects, something decisively ruled out by experiment.[21])

---

[20] In fact, philosophers have developed a formal framework for branching spacetimes (for totally different reasons); see e.g. Belnap (1992) or McCall (1994) for details, and Bacciagaluppi (2002) for an application of this framework to the Everett interpretation.

[21] See Page and Geilker (1981) for a direct laboratory test; more indirectly, since classically chaotic systems correspond to quantum-mechanical indeterminism, we have good reason to expect that the distribution of stars and galaxies varies wildly from branch to branch, and so we would expect that the observed gravitational field would bear little resemblance to the distribution of matter in *this* branch.

Of course, there is another possibility: perhaps the linear dynamics of quantum mechanics must themselves be modified in the presence of gravity, so that superpositions of geometry do not arise. Penrose has explored this proposal (see e.g. Penrose 1989 or 2004: 816–68), which as a bonus (for some!) would eliminate the 'many worlds' of quantum theory altogether; there is, however, no evidence whatever in its favour at this stage.

For this reason, I expect that the true quantum state is a superposition of geometries, but that matter and geometry are fairly entangled, so that with respect to a given macroscopically definite geometry, facts about the matter distribution are fairly definite too. But of course, we lack the knowledge to turn this qualitative description into a quantitative—still less a testable—mathematical theory. If you know how to do this, dear reader, I imagine that you are reading this while on a plane to Stockholm.

8.8 A digression on metaphysics

Spacetime state realism is not the only way to interpret the quantum state which has been proposed; it is not even the only way which is viable. In the next section I want to consider alternatives, and to give some arguments for and against them as ways of interpreting the ontology of quantum physics.

Before I do so, though, a caveat is in order. I have spoken, so far, mostly of the pragmatic virtues of spacetime state realism: I have advocated it because of what I see as its clarity in understanding issues of locality and of the relation between quantum theories and spacetime. But one might think that questions like this require more substantive and principled arguments: we wish to know if a given way of understanding the quantum state is *true*, not merely *convenient*.

Certainly, this is the predominant view in traditional metaphysics: if there are multiple ways of understanding a given physical theory *even given* that the theory is to be understood in realist terms, then at most one must be right, and various non-empirical considerations (generally involving simplicity, elegance, and perhaps intuitive appeal) need to be brought to bear to decide which is correct. A classic example is the interpretation of general relativity: granted that the spacetime metric is physically real, is it really *spacetime*, as defended by e.g. Misner et al. (1973), or just another field, as Rovelli (2004) argues? From the point of view of traditional metaphysics, both are coherent possibilities, but at most one can be correct.

This has not always been the standard view in philosophy. In the first half of the twentieth century, philosophers like Ayer, Carnap, and Schlick worried that questions like these, precisely because they were so divorced from anything empirical, did not really mean anything.[22] According to the frameworks of *logical positivism* and *logical empiricism* which they and others developed, for any two claims about matters of fact (as distinct from claims about mathematics or logic) to have different meanings, they must somewhere make different empirical claims. Indeed, claims about unobservable entities, for the logical positivist, were actually just disguised claims about the observable consequences of those unobservable entities, so that a claim about the unobservable which had *no* observable consequences was, quite literally, meaningless. (Logical empiricism took a somewhat more sophisticated line on such statements, but they were still regarded as essentially pragmatic tools, with no meaning of their own beyond their usefulness in describing the observable world.)

As I noted in passing in Chapter 1, logical positivism and logical empiricism are essentially defunct doctrines in modern philosophy. The reasons are many, but from our point of view, the clearest is that they rely on a sharp distinction between the language of observation and the language of theory, and no such distinction is really to be found in science or in the world, both because of the irreducibly holistic nature of evidence (so that the empirical consequences of a claim cannot be understood in isolation from all other claims that we make) and (relatedly) because the language we use to describe the observable is irreducibly imbued with theoretical claims. We have seen a fairly similar point ourselves in discussion of quantum mechanics: there, operationalist interpretations of quantum theory founder on the fact that the experimental processes which we wish to explain are themselves not black boxes, but are constructed and analysed using precisely the same physics of the unobservable that those interpretations wish to analyse away. When we say that a solenoid *generates a magnetic field*, for instance, or that a laser *produces coherent light at 450 nm*, we violate the supposed distinction between the language of observation and the language of theory, and doing so is an essential part of scientific discourse.

Over the decades following the demise of logical empiricism, modern analytic philosophy (or at least that branch of modern analytic philosophy known as analytic metaphysics) has become determinedly relaxed about

---

[22] For further information, see e.g. Hanfling (1996), Ray (2000), Salmon (2000), and references therein.

discussion of the nature of the unobservable—hence the default assumption that questions like 'Is the metric really spacetime?' have a substantive meaning, and have answers which go beyond analysis of what we mean by 'spacetime' and what the equations of our theory say.

But one can worry that in the flight from positivism, the baby has been thrown out with the bathwater.

The worry can be given an epistemic or a semantic flavour. On the epistemic side: what evidence could really bear on such questions? Elegance and simplicity may have been good guides to choice *between theories* (intuitive appeal, less so), but there is no evidence—there *can be* no evidence—that they cut any ice at all as guides to choice of ontology *within a theory*. On the semantic side: the positivists' positive(!) theory of meaning is certainly defunct—the meaning of a sentence cannot be identified with its observational consequences—but it does not automatically follow that every grammatical sentence which mentions unobservable entities says something meaningful about them. Wayne Myrvold (in conversation) has put it very well: the sceptic about some supposed difference between two observationally equivalent theories need not have a positive case that there is *no difference*; he merely need insist that his interlocutor *tell him what the difference is*.

This seems to suggest a gap in the market for some intermediate philosophical position, one which respects scepticism about overly 'metaphysical' claims while incorporating the impossibility of any coherent theory/observation divide. The gap is currently in the process of being filled by *structural realism*, a philosophical position whose advocates argue that science is concerned only with structural claims about the world, and that nonstructural claims about a physical theory are not worth making either because we have no access to nonstructural facts (Worrall 1989), or because there are no such facts (Ladyman 1998).[23] (See Ladyman and

---

[23] Note that to accept structural realism is not to reject wholesale questions like 'Does the metric represent spacetime?'; it is just to reject the idea that positive and negative answers to such questions are both coherent possibilities for a given, mathematically specified, physical theory. Eleanor Knox, for instance, building on the work of Brown (2005), has argued persuasively (2009a; 2009b) that what it is for a mathematical entity within a theory to represent spacetime structure is for it to have certain links to the inertial behaviour of matter; on these grounds, she concludes not only that the Einsteinian metric really represents spacetime structure, but that so does the Newtonian gravitational potential.

Ross 2007: esp. Ch. 2, and references therein, for the current state of the art in structural realism.)

For the structural realist, the quantum measurement problem is worth taking seriously, because operationalism about quantum mechanics is not an option, and so we need either to understand how quantum mechanics, taken literally, gives us the (approximately) true structure of the world, or to replace quantum mechanics with another theory which gives us that structure. And having accepted (let us suppose) an Everettian resolution to the measurement problem, structural realists can readily become interested in the question of how *best* to understand and describe the structure of the world according to quantum theory: that is, they can readily engage with the questions of this chapter. But unlike traditional realists, structural realists will be relaxed about using pragmatic considerations to decide how best to understand the structure of the quantum state: they do not accept that there is a *true* answer to be found.

My own view, for what it's worth, is that something like structural realism will turn out to be the right way to think about scientific theories—not least because it seems to do a far better job of respecting which distinctions do seem to be scientifically salient. But for my purposes in this chapter, I can remain agnostic about it. For the considerations that lead the structural realist to regard a given account of quantum structure as *helpful* or *perspicuous* are, in general, precisely the sort of considerations that lead the traditional realist to regard that account as *likely to be true*.

(For the broader purposes of this book, note that the structuralist approach to *higher-level ontology* advocated in Chapter 2 does not require structural realism about *micro-ontology* to be viable—though the two are natural bedfellows, of course. A traditional realist could coherently believe that the classical world is a structure multiply instantiated in the underlying quantum reality, whilst maintaining that there are determinate, empirically underdetermined nonstructural matters of fact about that underlying reality itself.)

8.9 Alternatives to spacetime state realism

Having got a bit clearer on just what is (and is not) at stake in questions about 'the right' way to understand the quantum state, let us consider three alternative proposals in the recent literature. (I make no claim to be exhaustive here.[24])

8.9.1 Wavefunction realism

If we take nonrelativistic quantum particle mechanics as our paradigm quantum theory, there is one fairly obvious possible way to make sense of the quantum state: it is typically represented as a wavefunction on configuration space, so just take that literally. On this reading (which in Wallace and Timpson 2010 we call *wavefunction realism*) quantum mechanics is not a theory of events in 3+1-dimensional spacetime at all: it is a theory of a complex field evolving in a very high-dimensional space. (If we could treat the whole observable universe non-relativistically, for instance, the space would have $\sim 10^{80}$ dimensions.)

Wavefunction realism was first explicitly proposed and discussed by Albert (1996) (though Barbour 1994; 1999 and Vaidman 2002 advocate similar positions). It has been criticized (notably by Monton 2002; 2006) on the grounds that it fails to provide the kind of ontology that can underwrite requisite facts about three-dimensional objects and their behaviour. From this book's point of view, though, such concerns miss the point: all that is needed is that we recover high-level ontology at a structural level (so that, for instance, we can find, within the wavefunction of the universe, components of that wavefunction which behave quasi-classically).

In Wallace and Timpson (2010) we respond in rather more detail to Monton's criticisms and defend the metaphysical coherence of the position. We do, however, criticise it on more technical grounds, which I briefly reprise here. First, wavefunction realism misrepresents the structure of quantum mechanics by singling out the position basis for special treatment. Secondly (though relatedly), it is difficult at best to extend it to quantum field theory, where no single basis seems to have the preferred status which the position basis might arguably be said to have in nonrelativistic quantum mechanics. Thirdly, it obscures the relations between spacetime and quantum theory: relativistic covariance and nonlocality, for instance, become hard to analyse. And fourthly, the phenomenon of narrativity failure (mentioned briefly above) is a curiosity from the

---

[24] For an alternative survey of the possibilities (in the context of Albert's 'narrativity' considerations, and from a less explicitly Everettian perspective) see Arntzenius (n.d.).

perspective of spacetime state realism, but a pathology for wavefunction realism.

A further reason is more conceptual than technical. We saw in section 1.2 that thinking about the quantum state as a ray in Hilbert space, or the classical state as a point in phase space, is unhelpful because it obscures the structure of the state—essentially by coding that structure in the position of the state in a highly structured, highly inhomogenous, highly nonsymmetric, high-dimensional space, rather than in the state itself. Something similar is going on in wavefunction realism: the space is lower-dimensional, but still very high-dimensional and very highly structured. Indeed, an entire quasi-classical world is represented by a single structureless wavepacket in such a space. The wavefunction represents the superposition structure over classical worlds but obscures the structure within each world.

For this reason, as well as the technical ones noted above, I see wave-function realism as in general an unhelpful way to think about the ontology of quantum mechanics.[25]

8.9.2 Quantum relationism

In teaching relativity to students in the modern fashion, one normally begins with the concept of an *event* and its spatiotemporal relations to other

---

[25] We can, however, use the present discussion to obtain a nice way of seeing just *how much more* structure can potentially be present in a quantum system than in the individual branches of that system. In wavefunction realism, each point in configuration space (more accurately, each localized wavepacket) represents the structure of an entire classical branch. In Hilbert space, though, each point represents the structure of the entire quantum Universe. So the difference in dimensionality between configuration space and Hilbert space gives us a measure of how much structure is potentially present in the quantum state over and above the structure of any given quasi-classical branch in that state.

Of course, the straightforward answer to the question is that while the configuration space of an N-particle theory is $3N$-dimensional, the Hilbert space of that theory is infinite-dimensional. But rather more insight can be gained by supposing that the particles are confined to a fixed volume V and to total energies below a fixed energy E, in which case a crude estimate of the dimensionality of the available region of Hilbert space is

$$\mathrm{Dim}\mathcal{H} \sim \left(V \times (2mE)^{3/2}/\hbar^3 \right)^N .$$

For the $\sim 10^{57}$ particles in the Sun, for instance, if we take the Sun's kinetic energy of $(k_B \times \sim 10^7 \mathrm{K} \times \sim 10^{57} =) \sim 10^{41}$ joules as a limit on the total energy of the system, and its volume of $\sim 10^{27}$ cubic metres as a limit on the total volume, we get a configuration space of dimension 3×10^{57} and a Hilbert space of dimension $\sim 10^{10^{59}}$.

events. Initially one appeals to intuitive characterizations of 'event' ('hands clapping', 'flashes of light', 'particles colliding'), but ultimately none is adequate: 'event' (understood as an instantaneous, pointlike happening) is just a primitive in the theory, and spacetime is the nexus of events and their relations with one another. What distinguishes one event from another is simply their different relations to all the *other* events.

When we make our account of spacetime more mathematically rigorous, this 'nexus of events' picture fades into the background, to be replaced by a framework in which the bare manifold is primary and the spacetime structure secondary. But at the physical level, it remains an attractive way to think about spacetime—one which goes back to Leibniz, and which has been championed in recent years by Barbour (1999). And it extends satisfactorily to dynamical spacetimes (at least at a conceptual level): specifying the spacetime distance between all the events suffices to recover the full metric.

Simon Saunders has proposed[26] a way of understanding the quantum state which is closely analogous to the nexus-of-events view of spacetime. (A similar proposal was made by Mermin 1998, though Mermin does not accept the Everettian implications of the position.) In Saunders' view (which, like spacetime state realism, is most naturally understood in the context of quantum field theory), an 'event' is not simply a spacetime point but a projector at spacetime points: a projector, that is, built from the quantum field operators at that spacetime point.[27] Any two events stand in some spatiotemporal relation in the usual way—but they also stand in what Saunders calls 'relations in norm': the relation in norm between projectors $\widehat{\Pi}_1(x)$ and $\widehat{\Pi}_2(y)$ is defined as

$$\langle \psi | \widehat{\Pi}_1(x) \widehat{\Pi}_2(y) | \psi \rangle \tag{8.15}$$

where $|\psi\rangle$ is the Heisenberg state. Furthermore, we can extend this notion to N-tuples of events straightforwardly, so that it makes sense to ask for the relation in norm between any finite set of events, however many members it has. Mathematically speaking, the crucial point is this:

---

[26] See Saunders (1995; 1997).

[27] Handling this in a mathematically satisfactory way—i.e. without running into the infinities of naive quantum field theory—requires us either to take some short-distance cutoff seriously or replace talk of spacetime *points* with talk of arbitrarily small spacetime *regions*. For an account of quantum field theory on this latter strategy, see Haag (1996); I defend the philosophical cogency of the former strategy in Wallace (2006b; 2011).

if we know all the relations in norm defined by a quantum state, we can discard the state itself without loss, because the relations in norm suffice to determine the state uniquely. (An elementary proof of this can be found in Mermin 1998.)

How are we to think about this 'relation in norm'? Mermin, rejecting the Everett interpretation as 'a cloud-cuckoo land of many worlds' (1998: 762) advises us to see it as a straightforward statistical correlation, no matter that there are no unique outcomes to *be* correlated. He calls this 'correlations without correlata'; I call it incoherent.[28]

Better, I think, is just to see it as a new physical relation in our theories. Consider: in classical field theory, we get by *largely* with non-relational properties of spacetime points: field values at spacetime points, for example, can be understood as properties of those points. But we do have to admit one irreducible relation to our theory: that of spacetime distance. In quantum mechanics on Saunders' proposal, we simply need to broaden our horizons, and admit a larger class of relations to our ontology. Put this way, I think, quantum relationism bears much the same relation(!) to spacetime state realism as the nexus-of-events view of spacetime does to the points-on-a-manifold view, and the structural realist's claim that is a non-question which is correct starts to look more attractive.

One virtue of quantum relationism is that it gives a natural language to discuss the emergence of quasi-classicality. Just as in classical physics we talk about 'events' with a nonzero spatiotemporal size, so in quantum physics we can extend the definition of event to include projectors localized in rather larger spacetime regions. In particular, among these projectors (given a dynamics, and a quantum state, which lead to decoherence) will be those which define decoherent, branching histories. The quasi-classical history structure can then be understood directly as a structure of relations in norm between events with the particular feature that it instantiates emergent quasi-classical dynamical behaviour. On the other hand, the link between the underlying quantum theory and the spacetime description of entanglement and branching is perhaps slightly less clear on the quantum relationism account.

---

[28] In terms of the taxonomy of interpretations I offer in Ch. 1, Mermin's 'Ithaca interpretation' is a 'vestiges of reality' interpretation, inasmuch as it does not explicitly abandon scientific realism, but it does abandon the idea that quantum mechanics can be understood in the same way as that of other scientific theories.

8.9.3 Operator-valued fields

One sometimes (mostly in conversation) hears the claim that quantum field theory is not about Hilbert-space states (as is the case in non-relativistic quantum mechanics), but is about fields—only, instead of being real, complex, or vector fields, they are *operator-valued* fields—fields which take self-adjoint operators as values.

As stated, this is pretty misleading: in particular, there is no relevant difference here between nonrelativistic quantum mechanics and quantum field theory. Quantum field theory uses operator-valued fields for exactly the same purpose that ordinary quantum mechanics uses position, momentum, and spin operators: the Hamiltonian of the theory is defined in terms of them, and the structural properties of the system are given by the expectation values of those operators with respect to the quantum state. And in quantum field theory, just as in quantum mechanics, we can formally speaking transfer the dynamics from states to operators, shifting to the Heisenberg picture. In this picture, the operators are time-dependent (and so, in QFT, *space*time-dependent), and the state does not evolve.

In neither case is the state vector redundant. Furthermore, it is generally held that the state represents the contingent physical details of the system in question: different states of affairs are represented by different state vectors. From this perspective, the operators—time-dependent or not—do not directly give the state of the system; they are simply part of the mathematical machinery which breaks the symmetries of Hilbert space and allows different rays in Hilbert space to represent different states of affairs.[29]

This seems to make the Heisenberg picture a poor choice from the point of view of conceptual understanding of quantum mechanics, for all its formal convenience in quantum field theory (a convenience which derives largely from the manifest covariance of Heisenberg-picture quantum field theory). But David Deutsch and Patrick Hayden (2000) have suggested a rather different reading of the Heisenberg picture. In their reading, the *state vector* is taken to be 'mere mathematical machinery': they hold it fixed throughout, writing it as $|0\rangle$ and fixing it once and for all, and independent

---

[29] The situation is very similar to formulations of classical mechanics (see e.g. Woodhouse 1997: 20–24) where the phase space is interpreted as a space of solutions rather than instantaneous states and the dynamics is represented by transformations of the observables.

of any contingent facts about the world, as some particular state. The operators corresponding to the dynamical variables, on the other hand, are taken perfectly seriously as the theory's ontology. Deutsch and Hayden develop their proposal largely in the context of quantum computation. In a computer consisting of N qubits, for instance, the state of the Mth qubit at time t is a triple of operators $\langle \widehat{S}_1^M(t) \widehat{S}_2^M(t), \widehat{S}_3^M(t)$ on a 2^N-dimensional Hilbert space, satisfying the constraint equations

$$\left[\widehat{S}_i^M(t), \widehat{S}_j^{M'}(t) \right] = i\delta_{M,M'} \epsilon_{ij}^k \widehat{S}_k^M(t) \tag{8.16}$$

but otherwise able to take arbitrary values. In the context of quantum field theory, their ontology is quite literally a set of operator-valued fields. The significance of the state vector $|0\rangle$ is that it gives the probability rules: the expected value of a measurement of the field $\widehat{\phi}(x, t)$, for instance, is

$$\langle 0 | \widehat{\phi}(x, t) | 0 \rangle.$$

Deutsch and Hayden's reading of quantum mechanics has a great advantage: not only does it display no action a distance, it is entirely separable! The state of some spatial region A at time t is completely specified by the (operator) values of the fields at the spacetime points within the region at that time, and the state of $A \cup B$ is specified uniquely once the states of A and B are separately known.[30] For this reason, Deutsch and Hayden claim that quantum mechanics is actually a completely local theory.

However, this advantage comes at a price (here I outline an argument given in more detail in Wallace and Timpson 2007). The alert reader will already have qualms at my account of the significance of the state vector: the probability rule it defines seems to be an interpretive postulate which fits uneasily into a purportedly realist account of quantum physics. What makes the properties defined by the expectation value of *that state* special? And what is the physical relation between operator-valued fields both of which give the same expectation values with respect to the preferred state?

The right response for Deutsch and Hayden to make to the latter question, in our view, is to regard any two such fields as representing the

---

[30] Here I assume, for simplicity, that the quantum field theory in question has no independent sources of nonlocality, as would arguably be the case for, say, the quantization of a classical gauge theory.

same *physical* state of affairs. Transformations which preserve the expectation value of all operators with respect to the preferred state should be understood as symmetry transformations, and it is widely accepted (see, e.g. Saunders 2003 and references therein) that two states differing by a symmetry transformation are just the same state, differently described. In this particular case (see Wallace and Timpson 2007 for the details), since the symmetries can be applied over arbitrarily short periods of time, it makes sense to think of them as a kind of gauge symmetry, so that the Deutsch–Hayden picture of quantum theory, like the four-potential picture of classical electromagnetism has surplus structure in the sense of Redhead (1975; 1980): not all of the mathematical structure of the operator-valued field corresponds to structure in the physical world.

But if we ask for a less redundant description of Deutsch and Hayden's picture than the one given by the operator fields, we run into an old friend: it is easy to show that equivalence classes of operator fields (at a given time) under these symmetry transformations correspond one-to-one to density operators at that time. That is: once we quotient out by the gauge symmetries, the Deutsch–Hayden picture reduces to spacetime state realism.

We have come full circle. Not only does quantum mechanics, like the electromagnetic field in the context of the Aharonov–Bohm effect, have non-separability without action at a distance; in both cases, we can find a separable formulation, but it comes at the price of gauge freedom.

CHAPTER 8. Physical reality, according to Everettian quantum field theory, should not be thought of as being a structureless point in Hilbert space, but can be understood in spatiotemporal terms just as in classical physics. Quantum entanglement means that the states of composite regions are not determined by the states of their subregions, but interactions nonetheless propagate locally. When this is applied to decoherence processes, we find that branching, too, occurs locally, spreading out, in general, at the speed of light. When two branching regions intersect, if the branchings were caused by decoherence acting on some previously entangled system whose components were at the respective branching centres, the nonlocal information associated with the entanglement propagates outward and serves to determine just how the branches intersect.

CHAPTER 9. The branching process which leads to emergent quasi-classicality is not time-reversal invariant: the time-reverse of decoherent branching is interference and recombination. How can this be squared with the time-reversibility of the underlying microdynamics, and what relation does it bear to better-known cases of irreversibility in physics?

9

The Directions of Branching and the Direction of Time

Most conventional physicists are concerned with predicting the future, rather than retrodicting the past. To them, if the second law of thermodynamics holds any mystery—or even if it merely falls marginally short of being rigorously understood—what would need to be understood would be: why it is that the entropy in an isolated system goes *up* in the *future*? To me, the mystery is a different one: why does the entropy go *down* in the *past*?

Roger Penrose[1]

9.1 The asymmetry of branching

Given decoherence, the unitarily evolving quantum state has a macroscopic structure that is branching, with quasi-classical histories diverging from one another but not reconverging. Such was the thesis of Chapters 2 and 3, and the Everett interpretation rests upon it.

But to say that histories 'diverge but not reconverge' is to make a distinction between past and future. For histories which agree before some time and disagree afterwards—that is, diverging histories—are the time reversals of histories which agree after some time and disagree before—that is, converging histories. Except in the trivial case where there is no branching at all, it only makes sense to say that the universe has emergent branching structure relative to some direction of time. If it has it in one direction, it cannot have it in the other. (The formal definition of

[1] Penrose (1994a: 218).

'branching' given in sections 3.8 and 3.9 should make this clear; by the branching-decoherence theorem, it follows that except in the trivial case of macroscopic determinism, if a set of histories is decoherent then its time reverse is not.)

For this reason, there is a momentary temptation to suppose that quantum mechanics builds in some arrow of time at the base level. But ultimately, the Schrödinger equation remains essentially[2] time-symmetric. The direction of time in the branching structure does not reflect any underlying asymmetry in the microdynamics: it is an emergent, not a fundamental, asymmetry.

This leaves the question, though, of where that emergent asymmetry comes from. Almost by definition, if a higher-level time-asymmetric state of affairs comes about despite time-symmetry at the fundamental level, it must be due to some asymmetry in the boundary conditions; but it is not prima facie clear where any such asymmetry was smuggled into Chapter 3's discussion of the dynamical processes that give rise to branching.

Put like this, the problem may start to sound familiar. Microdynamical time-reversibility, macrolevel irreversibility, appeal to special boundary conditions ... this is, of course, the familiar problem of how irreversible thermodynamic processes can be understood consistent with a lack of irreversibility at the microscopic level.

In an ideal world, then, the strategy would be clear: look at the accepted explanations for thermodynamic irreversibility and see whether they apply to decoherent branching. There would be three basic possibilities: either the exact same conditions that lead to other irreversibilities also lead to branching; or some analogous approach can be taken to making sense of branching; or the two phenomena are actually quite different.

But this particular branch of the multiverse, unfortunately, is not an ideal world. In fact, the foundations of statistical mechanics are if anything in more disarray than the philosophy of quantum mechanics. So to get clearer on why branching occurs, I will need to take an extended detour into the philosophy of statistical mechanics, in order to say something about not just Everettian variants of irreversibility, but macrolevel irreversibility in general. This will be the subject of this chapter: its conclusion is

---

[2] That is, putting aside CP violation in particle physics, which (a) appears very small, and (b) much more importantly, just requires us to replace 'time-symmetric' with 'CPT-symmetric', with no significant consequences.

that thermodynamic and decoherence-driven irreversibility are essentially consequences of the same assumptions. (As a corollary of this, situations in which the direction of branching and the direction of the Second Law are opposed cannot occur.)

A useful starting point will be the nearest we currently have to a consensus in foundations of statistical mechanics. In recent work, two claims have become popular.[3]

1. The tendency of systems' entropy to increase is basically just a consequence of the geometry of phase space. That region of phase space corresponding to a system being at equilibrium is so very large compared to the rest of phase space that unless either the dynamics or the initial state are (as Goldstein 2001 puts it) 'ridiculously special', then the system will in fairly short order end up in the equilibrium region.

2. The observed asymmetry in statistical mechanics—in particular, the tendency of entropy to increase rather than decrease—can be derived from time-symmetric microphysics provided we are willing to postulate that the entropy of the early universe is very low compared to the current entropy of the universe—the so-called Past Hypothesis (Albert 2000).

There is something rather puzzling about both views. Take the first: it seems to suggest that any given system, unless it is 'ridiculously special', will quickly end up in equilibrium. But of course, in the real world, we very frequently find systems far from equilibrium—indeed, life itself depends on it. And many of those systems, even when isolated from their surroundings, refuse to evolve into equilibrium. A room filled with a mixture of hydrogen and oxygen, at room temperature, can remain in that state for years or decades, yet one has only to strike a match in that room to be reminded (briefly) that it is not an equilibrium state. Indeed, a room filled with hydrogen at room temperature is not really at equilibrium: it is thermodynamically favourable for it to fuse into iron, but you would wait a long time for this to happen.

---

[3] Claims of roughly this form are espoused by e.g. Goldstein (2001), Lebowitz (2007), Callender (2009), Albert (2000), and Penrose (1989; 2004).

Furthermore, we have a detailed, quantitative understanding of exactly how quickly systems in various non-equilibrium states evolve towards equilibrium, We have already seen just these quantitative details in Chapter 3's discussion of decoherence, but of course, it is also available in less foundationally contentious areas of physics. In particular, chemists (whether of the ordinary or nuclear variety) have precise and thoroughly tested dynamical theories which predict, from the microdynamics, just how quickly systems complete their irreversible movement towards equilibrium. It is, at best, very difficult to see how these quantitative theories of the approach to equilibrium fit into the very general argument for equilibration given by Goldstein et al. and embodied in claim 1.

The Past Hypothesis is puzzling in a different way. It suggests, or seems to suggest, that our knowledge of the low entropy of the early Universe is somehow special: we are not supposed to know the Past Hypothesis in the way we usually know information about the past, but rather, we are justified in postulating it because without that postulate, all of our beliefs about the past would be unjustified. But there is something a little odd here: after all, we have (or think we have) rather detailed knowledge of the macroscopic state of the early Universe gained from cosmology, and we can calculate its entropy fairly accurately. It is also not clear (see in particular the trenchant criticisms of Earman 2006) exactly how imposing a low entropy at the beginning of time can lead to irreversible physics here and now.

And yet...for all that, there seems to be something to both views. There does seem to be some important sense in which irreversibility is connected with phase space volume and the behaviour of typical—i.e. not ridiculously special—systems. And I have already noted that, absent time asymmetry in microphysics, there must be some link between the boundary conditions of the Universe and the observed time asymmetry in macrophysics.

So my purpose in this chapter is to try to get as clear as possible on just how the logic of deriving macrophysical irreversibility from micro-dynamics-plus-past-hypothesis is supposed to go. My starting point is the observation above: that we actually have a large body of quantitative theory about irreversible physical processes, and any adequate account of irreversibility needs to explain the quantitative success of these theories and not just the qualitative observation that systems tend to equilibrium. So in sections 9.2–9.4 I set aside philosophical concerns and try to get as clear as possible on what the mathematical route is by which we derive empirically reliable irreversible macrodynamics from reversible microdynamics. In sections 9.5 and 9.6 I examine just when this mathematical route is physically justified, and conclude that a Past Hypothesis is indeed needed, but of a rather different character from what is usually argued.

I should note that one frequent theme in criticism of the two views given above has been their lack of mathematical rigour and care; see, in particular, Frigg's criticism of Goldstein (Frigg 2008) and Earman's objections (2006) to any assignment of entropy to the early Universe. By contrast, I am perfectly happy to allow their proponents to make whatever plausible-sounding mathematical conjectures they like (indeed, I will myself make several such in my own account). My concern, rather, is in understanding just what those conjectures are supposed to achieve, and why they can be expected to achieve it. The purpose of the philosopher of physics, it might be argued, is not to prove theorems but to see which theorems are worth proving.

Also, in this chapter I break from my general principle of discussing exclusively *quantum* mechanics: for the most part, I discuss the quantum and classical theories side by side. This is partly for the sake of making contact with the literature—most conceptual and technical work on the foundations of statistical mechanics is carried out in a classical framework—and partly because it helps clarify what, if anything, is distinctly quantum-mechanical about irreversible phenomenology.[4] I will have no qualms, however, about specializing to quantum mechanics whenever it seems necessary: classical mechanics, after all, is true of our world only insofar as it is a good approximation to quantum mechanics, so features of classical mechanics which do not go over to quantum mechanics are unphysical artefacts of the formalism.

---

[4] One might be inclined to reply that there *can't* be anything quantum-mechanical about at least thermodynamic irreversibility, because we have a perfectly satisfactory classical statistical mechanics. But of course, the issue here is precisely whether statistical mechanics, classical or quantum, really can account for macroscopic irreversibility in a satisfactory way. We have pretty good reason to think that quantum statistical mechanics can, independent of the detailed argument, because quantum statistical mechanics is our best extant theory of microphysics and because our world is demonstrably irreversible at the macro level. But we have no (independent) grounds to think this of classical statistical mechanics, since classical mechanics is false. See Wallace (2001b) for more on this point.

9.2 The macropredictions of microdynamics

For present purposes, classical and quantum mechanics have, essentially, a similar dynamical form. In both cases, we have

- a state space (phase space or (projective) Hilbert space);
- a deterministic rule determining how a point on that state space evolves over time (generated by the classical Hamiltonian and the symplectic structure, or the quantum Hamiltonian and the Hilbert-space structure, as appropriate);
- time reversibility, in the sense that given the state at time t, the dynamics is just as well suited to determine the state for times before t as for times after t.

I will also assume that, whatever particular version of each theory we are working with, both theories have something which can reasonably be called a 'time reversal' operator. This is a map τ from the state space to itself, such that if the t-second evolution of x is y then the t-second evolution of τy is τx; or, equivalently, if $x(t)$ solves the dynamical equations then so does $\tau x(-t)$. I'm not going to attempt a formal criterion for when something counts as a time reversal operator; in classical and quantum mechanics, we know it when we see it. (Though in quantum field theory, it is the transformation called CPT, and not the one usually called T, that deserves the name.)

Both theories also have what might be called, neutrally, an 'ensemble' or 'distributional' variant, though here they differ somewhat. In the classical case, the deterministic dynamics induces a deterministic rule to evolve functions over phase space, and not just points on phase space: if the dynamical law is given schematically by a function φ_t, so that $\varphi_t(x)$ is the t-second evolution of x, then $\varphi_{t*}\rho = \rho \cdot \varphi_t$. In more concrete and familiar terms, this takes us from the Hamiltonian equations of motion for individual systems to the Liouvillian equations for ensembles.

In the quantum case, we instead transfer the dynamics from pure to mixed states. If the t-second evolution takes state $|\psi\rangle$ to $\widehat{U}_t|\psi\rangle$, the distributional variant takes ρ to $\widehat{U}_t\rho\widehat{U}_t^{\dagger}$.

I stress: the existence of these distributional variants is a purely mathematical claim; no statement of their physical status has yet been made. The space of functions on, or density operators over, the state space can be thought of, mathematically speaking, as a state space in its own right, for the distributional variant of the theory.

In principle, the way we use these theories to make predictions ought to be simple: if we want to know the state of the system we're studying in t seconds' time, we just start with its state now and evolve it forward for t seconds under the microdynamics. And similarly, if we want to know its state t seconds ago, we just time-reverse it, evolve it forward for t seconds, and time-reverse it again. (Or equivalently, we just evolve it forwards for $-t$ seconds.)

And sometimes, that's what we do in practice too. When we use classical mechanics to predict the trajectory of a cannonball or the orbit of a planet, or when we apply quantum mechanics to some highly controlled situation (say, a quantum computer), we really are just evolving a known state under a known dynamics. But of course, in the great majority of situations this is not the case, and we have to apply approximation methods. Sometimes that's glossed as being because of our lack of knowledge of the initial state, or our inability to solve the dynamical equations exactly, but this is really only half the story. Even if we were able to calculate (say) the expansion of a gas in terms of the motions of all its myriad constituents, we would have missed important generalizations about the gas by peering too myopically at its microscopic state. We would, that is, have missed important, robust higher-level generalizations about the gas. And in quantum mechanics, the emergent behaviour is frequently the only one that physically realistic observers can have epistemic access to: decoherence strongly constrains our ability to see genuinely unitary dynamical processes, because it's too difficult to avoid getting entangled with those same processes.

The point is that in general we are not interested in all the microscopic details of the systems we study, but only in the behaviour of certain more coarse-grained details. It is possible (if, perhaps, slightly idealized) to give a rather general language in which to talk about this: suppose that $t_1, \ldots t_N$ is an increasing sequence of times, then a set of *macroproperties* for that sequence is an allocation, to each time t_i in the sequence, of either

(i) in the classical case, a Boolean algebra of subsets of the system's phase space whose union is the entire phase space; or

(ii) in the quantum case, a Boolean algebra of subspaces of the system's Hilbert space whose direct sum is the entire Hilbert space.

In both cases, it is normal to specify the macroproperties as being unions or direct sums (as appropriate) of macro*states*: a set of macrostates for a (classical/quantum) system is a set of mutually (disjoint/orthogonal) (subsets/subspaces) whose (union/direct sum) is the entire state space. Throughout this chapter, I will assume that any given set of macroproperties is indeed generated from some set of macrostates in this way. (And in most *practical* cases, the choice of macrostate is time-independent.) For the sake of a unified notation, I will use $\oplus$ to denote the union operation for classical sets and the direct sum operation for quantum subspaces, $\subseteq$ to denote the subset relation for classical sets and the subspace relation for quantum subspaces, and 'disjoint' to mean either set-theoretic disjointness or orthogonality, as appropriate.

The idea of this formalism is that knowing that a system has a given macroproperty at time t_i gives us some information about the system's properties at that time, but only of a somewhat coarse-grained kind. We define a *macrohistory* α of a system as a specification, at each time t_i, of a macroproperty $\alpha(t_i)$ for that time; the set of all macrohistories for a given set of macroproperties is the *macrohistory space* for that set. It should be fairly clear that given the macrohistory space of a given set of macroproperties, we can recover that set; hence I speak interchangably of a macrohistory space for a theory and a set of macroproperties for the same theory. For simplicity, I usually drop the 'macro' qualifier where this is not likely to cause confusion.

A few definitions: by a history of length K (where $K < N$) I mean a history which assigns the whole state space to all times t_i for $i > M$. Given histories α and β of lengths K and K' (with $K < K'$) then α is an *initial segment* of β if $\alpha(t_i) = \beta(t_i)$ for $i \leq M$. Given macrohistories α and β, we can say that α is a *coarsening* of β if $\beta(t_i) \subseteq \alpha(t_i)$ for each time t_i at which they are defined, and that α and β are *disjoint* if $\beta(t_i)$ and $\alpha(t_i)$ are disjoint at each t_i. A history β is the *sum* of a (countable) set of mutually disjoint histories $\{\alpha_j\}$ (write $\beta = \oplus_j \alpha_j$) if there is some decomposition of each $\beta(t_i)$ into disjoint properties β_i^j such that a history is in $\{\alpha_j\}$ iff it is a sequence of these properties, and, in particular, a set of disjoint histories is *complete* if their sum is the trivial history $\widehat{1}$ whose macroproperty at each time is just the whole state space. And a *probability measure* Pr for a given history space is a real function from histories to $[0, 1]$ such that

1. If $\{\alpha_j\}$ is a countable set of disjoint histories then $\Pr(\oplus_j \alpha_j) = \sum_j \Pr(\alpha_j)$, and

2. $\Pr(\widehat{1}) = 1$.

332 QUANTUM MECHANICS, EVERETT STYLE

The point of a probability measure over a history space is that it determines a (generally stochastic) dynamics: given two histories α and β where α is an initial segment of β, we can define the *transition probability* from α to β as $\Pr(\beta)/\Pr(\alpha)$. A *macrodynamics* for a (classical or quantum) system is then just a history space for that system, combined with a probability measure over that history space. A macrodynamics is *branching* iff whenever α and β agree after some time t_m but disagree at some earlier time, either $\Pr(\alpha) = 0$ or $\Pr(\beta) = 0$; it is *deterministic* if whenever α and β agree before some time t_m but disagree at some later time, either $\Pr(\alpha) = 0$ or $\Pr(\beta) = 0$. (The history formalism used here is essentially the same as that adopted in Chapter 3.)

With this formalism in place, we can consider how classical and quantum physics can actually induce macrodynamics: that is, when it will be true, given the known microdynamics, that the system's macroproperties obey a given macrodynamics. The simplest case is classical mechanics in its non-distributional form: any given point x in phase space will have a determinate macrostate at any given time, and so induces a deterministic macrodynamics: if $U(t) \cdot x$ is the t-second evolution of x under the classical microdynamics, then

$$\Pr_x(\alpha) = 1 \ (\text{if } U(t_n - t_1) \cdot x \in \alpha(t_n) \text{ for all } n)$$

$$\Pr_x(\alpha) = 0 \qquad \qquad (\text{otherwise}) \qquad \qquad (9.1)$$

To get stochastic dynamics from classical microdynamics, we need to consider the distributional version. Suppose that at time t_1 the probability of the system having state x is $\rho(x)$; then the probability at time t_n of it having state x is given by evolving ρ forward for a time $t_n - t_1$ under the distributional (Liouville) dynamics. Writing $L(t) \cdot \rho$ for the t-second evolution of ρ and $P(M) \cdot \rho$ for the restriction of ρ to the macrostate M, we define the *history super-operator* $H(\alpha)$ by

$$H(\alpha) \cdot \rho = P(\alpha(t_n)) \cdot L(t_n - t_{n-1}) \cdot P(\alpha(t_{n-1}))$$

$$L(t_{n-1} - t_{n-2}) \cdots L(t_2 - t_1) P(\alpha(t_1)) \cdot \rho. \qquad (9.2)$$

$H(\alpha) \cdot \rho$ is the distribution obtained by alternately evolving ρ forward and then restricting to the successive terms in α. So we have that the probability of history α given initial distribution ρ is

$$\mathrm{Pr}_\rho(\alpha) = \int H(\alpha) \cdot \rho \qquad (9.3)$$

where the integral is over all of phase space.

A formally similar expression can be written in quantum mechanics. There, we write ρ for the system's density operator at time t_1, $L(t) \cdot \rho$ for the t-second evolution of ρ under the unitary dynamics (so if $\widehat{U}(t)$ is the t-second unitary time translation operator, $L(t) \cdot \rho = \widehat{U}(t)\rho\widehat{U}^\dagger(t)$), and $P(M) \cdot \rho$ for the projection of ρ onto the subspace M (so that if $\widehat{\Pi}_M$ is the standard projection onto that subspace, $P(M) \cdot \rho = \widehat{\Pi}_M\rho\widehat{\Pi}_M$). Then (9.2) can be understood quantum-mechanically, and (9.3) becomes

$$\mathrm{Pr}_\rho(\alpha) = \mathsf{Tr}(H(\alpha) \cdot \rho). \qquad (9.4)$$

The resemblance is somewhat misleading, however. For one thing, in classical physics the macrodynamics are probabilistic because we put the probabilities in by hand, in the initial distribution ρ. But in quantum physics, (9.4) generates stochastic dynamics even for the pure-state version of quantum theory (relying on Part II to explain why the weights of histories deserve to be called 'probabilities'). And for another, (9.4) only defines a probability measure in special circumstances. For if (as in Chapter 3) we define the *history operator* $C(\alpha)$ by

$$\widehat{C}(\alpha) = \Pi_{\alpha_n}\widehat{U}(t_n - t_{n-1})\Pi_{\alpha_{n-1}} \cdots \widehat{U}(t_2 - t_1)\Pi(\alpha_1), \qquad (9.5)$$

we can express $H(\alpha)$ by

$$H(\alpha) \cdot \rho = \widehat{C}(\alpha)\rho\widehat{C}^\dagger(\alpha) \qquad (9.6)$$

and rewrite (9.4) as

$$\mathrm{Pr}_\rho(\alpha) = \mathsf{Tr}(\widehat{C}(\alpha)\rho\widehat{C}^\dagger(\alpha)), \qquad (9.7)$$

in which case

$$\mathrm{Pr}_\rho\left(\sum_j \alpha_j\right) = \sum_{j,k} \mathsf{Tr}(\widehat{C}(\alpha_j)\rho\widehat{C}^\dagger(\alpha_k)), \qquad (9.8)$$

which in general violates the requirement that $\mathrm{Pr}_\rho\left(\sum_j \alpha_j\right) = \sum_j \mathrm{Pr}_\rho(\alpha_j)$. As was shown in Chapter 3, to ensure that this requirement is satisfied, we need to require that the history space satisfies the decoherence condition: that the decoherence function

$$d_\rho(\alpha, \beta) \equiv \mathsf{Tr}(\widehat{C}(\alpha)\rho\widehat{C}^\dagger(\beta)) \tag{9.9}$$

vanishes unless α is a coarsening of β. (Recall from section 3.9 that a weaker requirement—that the real part of the decoherence functional vanishes—would be formally sufficient but seems to lack physical significance.)

So there already is a connection here between macrohistories and decoherence: macrodynamics can only arise for a decoherent set of histories. It follows from the branching-decoherence theorem (section 3.9) that any macrohistory space which gives rise to macrodynamics is a coarsening of a macrohistory space whose macrodynamics are branching.

Before moving on, I should stress that the entire concept of a history operator, as defined here, builds in a notion of time asymmetry: by construction, we have used the system's distribution at the initial time t_1 to generate a probability measure over histories defined at that and all subsequent times. However, we could equally well have defined histories running backwards in time—'antihistories', if you like—and used the same formalism to define probabilities over antihistories given a distribution at the final time for those antihistories.

9.3 Coarse-grained dynamics

The discussion so far has dealt entirely with how macroscopic dynamics can be extracted from the microscopic equations, assuming that the latter have been solved exactly. That is, the framework is essentially descriptive: it provides no shortcut to determining what the macrodynamics actually are. In reality, though, it is almost never the case that we have access to the exact microlevel solutions to a theory's dynamical equations; instead, we resort to certain approximation schemes both to make general claims about systems' macrodynamics and to produce closed-form equations for the macrodynamics of specific systems. In this section, I wish to set out what I believe to be *mathematically* going on in these approximation schemes, and what assumptions of a purely technical nature need to be made. For now, I set aside philosophical and conceptual questions, and ask the reader to do likewise.

The procedure we use is intended to allow for the fact that we are often significantly ignorant of, and significantly uninterested in, the microscopic details of the system, and instead wish to gain information of a more coarse-grained nature, and it seems to go like this.[5] Firstly, we identify a set of macroproperties (defined as above) in whose evolution we are interested. Secondly, we define a map $\mathcal{C}$—the *coarse-graining map*—which projects from the distribution space onto some subset S_C of the distributions. By 'projection' I mean that $\mathcal{C}^2 = \mathcal{C}$, so that the distributions in S_C—the 'coarse-grained' distributions—are unchanged by the map. It is essential to the idea of this map that it leaves the macroproperties (approximately) unchanged—or, more precisely, that the probability of any given macroproperty being possessed by the system is approximately unchanged by the coarse-graining map. In mathematical terms, this translates to the requirement that for any macroproperty M,

$$\int_M \mathcal{C}(\rho) = \int_M \rho \qquad (9.10)$$

in the classical case, and

$$\mathrm{Tr}(\Pi_M \mathcal{C}(\rho)) = \mathrm{Tr}(\Pi_M \rho) \qquad (9.11)$$

in the quantum case. I will also require that $\mathcal{C}$ commutes with the time reversal operation (so that the coarse-graining of a time-reversed distribution is the time-reverse of the coarse-graining of the distribution).

We then define the *forward dynamics induced by* $\mathcal{C}$—or the $\mathcal{C}+$ dynamics for short—as follows: take any distribution, coarse-grain it, time-evolve it forward (using the microdynamics) by some small time interval Δt, coarse-grain it again, time-evolve it for another Δt, and so on. (Strictly speaking, then, Δt ought to included in the specification of the forward dynamics. However, in practice, we are only interested in systems where (within some appropriate range) the induced dynamics are insensitive to the exact value of Δ_t.)

By a *forward dynamical trajectory induced by* $\mathcal{C}$, I mean a map from (t_i, ∞) into the coarse-grained distributions (for some t_i), such that the distribution at t_2 is obtained from the distribution at t_1 by applying the $\mathcal{C}+$ dynamics whenever $t_2 > t_1$. A *section* of this trajectory is just a restriction of this map to some finite interval $[t, t']$.

---

[5] Note added in proof: after the completion of the manuscript I discovered that a rather similar framework has been developed by Zwanzig (e.g. *J. Chem. Phys.* 33(1960) pp. 1338–41); see H. D. Zeh, *The Physical Basis of the Direction of Time* (Springer, 1989) and references therein for details.

What is the coarse-graining map? It varies from case to case, but some of the most common examples are:

The coarse-grained exemplar rule. Construct equivalence classes of distributions: two distributions are equivalent if they generate the same probability function over macroproperties. Pick one element in each equivalence class, and let the coarse-graining map take all elements of the equivalence class onto that element. This defines a coarse-graining rule in classical or quantum physics; in practice, however, although it is often used in foundational discussions, rather few actual applications make use of it.

The measurement rule. Replace the distribution with the distribution obtained by a nonselective measurement of the macrostate: that is, apply

$$\rho \to \sum_M \Pi_M \rho \Pi_M \qquad (9.12)$$

where the sum ranges over macrostates.[6] (This obviously only counts as a coarse-graining in quantum mechanics; the analogous classical version, where ρ is replaced by the sum of its restrictions to the macrostates, would be trivial.)

The correlation–discard rule: Decompose the system's state space into either the Cartesian product (in classical physics) or the tensor product (in quantum physics) of state spaces of subsystems. Replace the distribution with that distribution obtained by discarding the correlations between subsystems (by replacing the distribution with the product of its marginals or the tensor product of its partial traces, as appropriate).

One note of caution: the correlation–discard rule, though very commonly used in physics, will fail to properly define a coarse-graining map if the probability distribution over macroproperties itself contains nontrivial correlations between subsystems. In practice this only leads to problems if the system does not behave deterministically at the macroscopic level, so that such correlations can develop from initially

---

[6] To avoid problems with the quantum Zeno effect (Misra and Sudarshan 1977; see Home and Whitaker 1997 for a review) for very small δt, the measurement rule strictly speaking ought to be slightly unsharpened (e.g. by using some POVM formalism rather than sharp projections onto subspaces); the details of this do not matter for our purposes.

uncorrelated starting states. Where this occurs, the correlation-discard rule needs generalizing: decompose the distribution into its projections onto macrostates, discard correlations of these macrostates individually, and re-sum. Note, though, that in quantum mechanics this means that two coarse-grainings are being applied: to 'decompose the distribution into its projections onto macrostates and then re-sum' is just to perform a nonselective measurement on it—i.e. to apply the measurement rule for coarse-grainings.

Another example is again often used in foundational discussions of statistical mechanics, but turns up rather less often in practical applications:

The smearing rule. Blur the fine structure of the distribution by the map

$$\rho' = \int dq' \, dp' f(q', p') T(q', p') \cdot \rho \qquad (9.13)$$

where $T(q', p')$ is translation by $(q'p')$ in phase space and f is some function satisfying $\int f = 1$ and whose macroscopic spread is small. A simple choice, for instance, would be to take f to be a suitably normalized Gaussian function, so that

$$\rho' = \mathcal{N} \int dq' dp' \, \exp[-(q - q')^2/(\Delta q^2)] \exp[-(p - p')^2/(\Delta p^2)]\rho(q, p)$$
$$(9.14)$$

where ρ is to be read as either the phase-space probability distribution (classical case) or the Wigner-function representation of the density operator (quantum case).

For a given system of $\mathcal{C}+$ dynamics, I will call a distribution *stationary* if its forward time evolution, for all times, is itself. (So stationary distributions are always coarse-grained.) Classic examples of stationary distributions are the (classical or quantum) canonical and microcanonical ensembles. Distributions involving energy flow (such as those used to describe stars) look stationary, but generally aren't, as the energy eventually runs out.

How do we generate empirical predictions from the coarse-grained dynamics? In many cases this is straightforward, because those dynamics are deterministic at the macroscopic level ('macrodeterministic'): if we begin with a coarse-grained distribution localized in one macrostate, the $\mathcal{C}+$ dynamics carries it into a coarse-grained distribution still localized in one (possibly different) macrostate. More generally, though, what we want to know is: how probable is any given sequence of macrostates? That is, we need to apply the history framework used in the previous section. All this requires is for us to replace the (in practice impossible to calculate) macrodynamics induced by the microdynamics with the coarse-grained dynamics: if $L^{C+}(t) \cdot \rho$ is the t-second evolution of ρ under the $C+$-dynamics, and $P(M) \cdot \rho$ is again projection of ρ onto the macroproperty M, then we can construct the coarse-grained history super-operator

$$H^{C+}(\alpha) = P(\alpha(t_n)) \cdot L^{C+}(t_n - t_{n-1}) \cdot P(\alpha(t_{n-1})) \cdot L^{C+}(t_{n-1} - t_{n-2}) \cdots$$
$$L^{C+}(t_2 - t_1) \cdot P(\alpha(t_1)). \tag{9.15}$$

(It should be pointed out for clarity that each $L^{C+}(t_k - t_{k-1})$ typically involves the successive application of many coarse-graining operators, alternating with evolution under the fine-grained dynamics; put another way, typically $t_k - t_{k-1} \gg \Delta t$. Even for the process to be well-defined, we have to have $t_k - t_{k-1} \geq \Delta t$; in the limiting case where $t_k - t_{k-1} = \Delta t$, we obtain $H^{C+}(\alpha)$ by alternately applying *three* operations: evolve, coarse-grain, project.)

We can then define the probability of a history by

$$\mathrm{Pr}_{\rho}^{C+}(\alpha) = \int H^{C+}(\alpha) \cdot \rho \tag{9.16}$$

in the classical case and

$$\mathrm{Pr}_{\rho}^{C+}(\alpha) = \mathrm{Tr}(H^{C+}(\alpha) \cdot \rho) \tag{9.17}$$

in the quantum case. The classical expression automatically determines a (generally stochastic) macrodynamics (i.e. a probability measure over histories); the quantum expression does so provided that all the coarse-grained distributions are diagonalized by projection onto the macrostates: that is, provided that

$$C \cdot \rho = \sum_{M} P(M) \cdot C \cdot \rho \tag{9.18}$$

where the sum ranges over macrostates. This condition is satisfied automatically by the measurement and correlation-discard rules (the latter rules, recall, build in the former); it will be satisfied by the coarse-grained exemplar rules provided the exemplars are chosen appropriately; it will be satisfied approximately by the smearing rules given that the smearing function is small on macroscopic scales. Examples in physics where this process is used to generate a macrodynamics include:

Boltzmann's derivation of the H theorem. Boltzmann's 'proof' that a classical gas approached the Maxwell–Boltzmann distribution requires the *Stosszahlansatz*—the assumption that the momenta of gas molecules are uncorrelated with their positions. This assumption is in general very unlikely to be true (cf. the discussion in Sklar 1993: 224–7), but we can reinterpret Boltzmann's derivation as the forward dynamics induced by the coarse-graining process of simply discarding those correlations.

More general attempts to derive the approach to equilibrium. As was already noted, the kind of mathematics generally used to explore the approach of classical systems to equilibrium proceeds by partitioning phase space into cells and applying a smoothing process to each cell. (See Sklar 1993: 212–14 for a discussion of such methods; I emphasize once again that at this stage of the discussion I make no defence of their conceptual motivation.)

Kinetic theory and the Boltzmann equation. Pretty much all of non-equilibrium kinetic theory operates, much as in the case of the H theorem, by discarding the correlations between different particles' velocities. Methods of this kind are used in weakly interacting gases, as well as in the study of galactic dynamics (Binney and Tremaine 2008). The BBGKY hierarchy of successive improvements of the Boltzmann equation (cf. Sklar 1993: 207–10 and references therein) can be thought of as introducing successively more sophisticated coarse-grainings which preserve N-body correlations up to some finite N but not beyond.

Environment-induced decoherence and the master equation. Crucially, given our goal of understanding the asymmetry of quantum branching, quantitative results for environment-induced decoherence are generally derived by (in effect) alternating unitary (and entangling) interactions of system and environment with a coarse-graining defined by replacing the entangled state of system and environment with the product of their reduced states (derived for each system by tracing over the other system). The models discussed in section 3.5, in particular, are derived by this method.

Local thermal equilibrium. In pretty much all treatments of heat transport (in e.g. oceans or stars) we proceed by breaking the system up into regions large enough to contain many particles, small enough to treat properties such as density or pressure as constant across them. We then take each system to be at instantaneous thermal equilibrium at each time, and study their interactions.

In most of the above examples, the coarse-graining process leads to deterministic macrodynamics. Some (rather theoretical) examples where it does not are:

Rolling dice. We don't normally do an explicit simulation of the dynamics that justifies our allocation of probability 1/6 to each possible outcome of rolling a die. But qualitatively speaking, what is going on is that (i) symmetry considerations tell us that the region of phase space corresponding to initial conditions that lead to any given outcome has Liouville volume 1/6 of the total initial-condition volume; (ii) because the dynamics are highly random, any reasonably large and reasonably Liouville-smooth probability distribution over the initial conditions will therefore overlap to degree 1/6 with the region corresponding to each outcome; (iii) any coarse-graining process that delivers coarse-grained states which are reasonably large and reasonably Liouville-smooth will therefore have probability 1/6 of each outcome.

Local thermal equilibrium for a self-gravitating system. Given a self-gravitating gas, the methods of local thermal equilibrium can be applied, but (at least in theory) we need to allow for the fact that a distribution which initially is fairly sharply peaked on a spatially uniform (and so, unclumped) state will in due course evolve through gravitational clumping into a sum of distributions peaked on very non-uniform states. In this situation, the macrodynamics will be highly non-deterministic, and so if we want to coarse-grain by discarding long-range correlations, we first need to decompose the distribution into macroscopically definite components.

Decoherence of a system with significant system–environment energy transfer. If we have a quantum system being decohered by its environment, and if there are state-dependent processes that will transfer energy between the system and environment, then macrolevel correlations between, say, system centre-of-mass position and environment temperature may develop, and tracing these out will be inappropriate. Again, we need to decompose the system into components with fairly definite macroproperties before performing the partial trace.

9.4 Time reversibility in coarse-grained dynamics

The process used to define forward dynamics—as the name suggests—is explicitly time-asymmetric, and this makes it at least possible that the forward dynamics are themselves time-irreversible. In fact, that possibility is in general fully realized, as we shall see in this section.

Given a dynamical trajectory of the microdynamics, we know that we can obtain another dynamical trajectory by applying the time-reversal operator and then running it backwards. Following this, we will say that a given segment of a dynamical trajectory of the coarse-grained dynamics is time-reversible if the corresponding statement holds true. That is, if $\rho(t)$ is a segment of a dynamical trajectory (for $t \in [t_1, t_2]$) then it is reversible iff $T\rho(-t)$ is a segment of a dynamical trajectory (for $t \in [-t_2, -t_1]$).[7]

Although the microdynamics is time-reversible, in general the coarse graining process is not, and this tends to prevent the existence of time-reversible coarse-grained trajectories. It is, in fact, possible to define a function S_G—the *Gibbs entropy*—on distributions, such that S_G is preserved under microdynamical evolution and under time reversal, but such that for any distribution ρ, $S_G(\mathcal{C}\rho) \geq S_G(\rho)$, with equality only if $\mathcal{C}\rho = \rho$. (And so, since the forward dynamics consists of alternating microdynamical evolution and coarse-graining, S_G is non-decreasing on any dynamical trajectory of the forward dynamics.) In the classical case, we take

---

[7] Note that I assume, tacitly, that the dynamics is time-translation-invariant, as is in fact the case in both classical and quantum systems in the absence of explicitly time-dependent external forces.

$$S_G(\rho) = - \int \rho \ln \rho \qquad (9.19)$$

and in the quantum case we use

$$S_G(\rho) = -\text{Tr}(\rho \ln \rho). \qquad (9.20)$$

(At the risk of repetitiveness: I am assuming *absolutely nothing* about the connection or otherwise between this function and thermodynamic entropy; I use the term 'entropy' purely to conform to standard usage.) Of the coarse-graining methods described above, the facts that correlation-discard, measurement, and smearing increase Gibbs entropy are well-known results of (classical or quantum) information theory; the exemplar rule will increase Gibbs entropy provided that the exemplars are chosen to be maximal-entropy states, which we will require.

The existence of a Gibbs entropy function for $\mathcal{C}$ is not itself enough to entail the irreversibility of the $\mathcal{C}+$ dynamics. Some coarse-grained distributions might actually be carried by the microdynamics to other coarse-grained distributions, so that no further coarse-graining is actually required.

I will call a distribution *Boring* (over a given time period) if evolving its coarse-graining forward under the microdynamics for arbitrary times within that time period leads only to other coarse-grained distributions, and *Interesting* otherwise. The most well-known Boring distributions are stationary distributions—distributions whose forward time evolution under the microdynamics is themselves—such as the (classical or quantum) canonical and microcanonical distributions; any distribution whose coarse-graining is stationary is also Boring. On reasonably short timescales, generic states of many other systems—planetary motion, for instance—can be treated as Boring or nearly so.[8] However, if the ergodic hypothesis is true for a given system (an assumption which otherwise will play no part in this chapter), then on sufficiently long timescales the only Boring distributions for that system are those whose coarse-grainings are uniform on each energy hypersurface.

If a segment of a dynamical trajectory of the $\mathcal{C}+$ dynamics contains any distributions that are Interesting on timescales short compared to the segment's length, that segment is irreversible. For in that case, nontrivial coarse-graining occurs at some point along the trajectory, and so the

---

[8] More precisely, in general a system's evolution will be Boring on timescales short relative to its Lyapunov timescale; cf. Ch. 3.

final Gibbs entropy is strictly greater than the initial Gibbs entropy. Time reversal leaves the Gibbs entropy invariant, so it follows that for the time-reversed trajectory, the initial Gibbs entropy is higher than the final Gibbs entropy. But we have seen that Gibbs entropy is nondecreasing along any dynamical trajectory of the forward dynamics, so the time-reversed trajectory cannot be dynamically allowed by those dynamics.

So: the coarse-graining process C takes a dynamical system (classical or quantum mechanics) which is time reversal invariant, and generates a new dynamical system ($C+$, the forward dynamics induced by C) which is irreversible. Where did the irreversibility come from? The answer is hopefully obvious: it was put in by hand. We could equally well have defined a *backward* dynamics induced by C ($C-$ for short) by running the process in reverse: starting with a distribution, coarse-graining it, evolving it backwards in time by some time interval, and iterating. And of course, the time reversal of any dynamical trajectory of $C+$ will be a dynamical trajectory of $C-$, and vice versa.

It follows that the forward and backwards dynamics in general make contradictory claims. If we start with a distribution at time t_i, evolve it forwards in time to t_f using the $C+$ dynamics, and then evolve it backwards in time using the $C-$ dynamics, in general we do *not* get back to where we started.

This concludes the purely mathematical account of irreversibility. One more physical observation is needed, though: the forward dynamics induced by coarse-graining classical or quantum mechanics has been massively empirically successful. Pretty much all of our quantitative theories of macroscopic dynamics rely on it, and those theories are in general *very* well confirmed by experiment. With a great deal of generality—and never mind the conceptual explanation as to *why* it works—if we want to work out quantitatively what a large physical system is going to do in the future, we do so by constructing a coarse-graining-induced forward dynamics.

On the other hand (of course), the *backwards* dynamics induced by basically any coarse-graining process is not empirically successful at all: in general it wildly contradicts our actual records of the past. And this is inevitable given the empirical success of the forward dynamics: on the assumption that the forward dynamics are not only predictively accurate now but also were in the past (a claim supported by very extensive amounts of evidence) then—since they are in conflict with the backwards dynamics—it cannot be the case that the backwards dynamics provides accurate ways of retrodicting the past. Rather, if we want to retrodict we do so via the usual methods of scientific inference: we make tentative guesses about the past, and test those guesses by evolving them forward via the forward dynamics and comparing them with observation. (The best-known and best-developed account of this practice is the Bayesian one discussed in Chapter 6: we place a credence function on possible past states, deduce how likely a given present state is conditional on each given past state, and then use this information to update the past-state credence function via Bayes' Theorem.)

9.5 Microdynamical underpinnings of the coarse-grained dynamics

In this section and the next, I turn my attention from the practice of physics to the justification of that practice. That is: given that (we assume) it is really the macrodynamics induced by the microdynamics—and not the coarse-grained dynamics—that describe the actual world, under what circumstances do those two processes give rise to the *same* macrodynamics?

There is a straightforward technical requirement which will ensure this: we need to require that for every history α,

$$\mathcal{C}H(\alpha)\rho = H^{C+}(\alpha)\rho. \tag{9.21}$$

That is, the result of alternately evolving ρ forward under the fine-grained dynamics and restricting it to a given term in a sequence of macroproperties must be the same, up to coarse-graining, as the result of doing the same with the coarse-grained dynamics. If ρ and $\mathcal{C}$ jointly satisfy this condition (for a given history space), we say that ρ is *forward predictable* by $\mathcal{C}$ on that history space. (Mention of a history space will often be left tacit.) Note that in the quantum case, if ρ is forward predictable by $\mathcal{C}$, it follows that the macrohistories are decoherent with respect to ρ.

I say 'Forward' because we are using the coarse-grained *forward* dynamics. Pretty clearly, we can construct an equivalent notion of backwards predictability, using the backward coarse-grained dynamics and the anti-histories mentioned in section 9.2. And equally clearly, ρ is forward predictable by $\mathcal{C}$ iff its time reverse is backwards predictable by $\mathcal{C}$.

Forward predictability is closely related to the (slightly weaker) notion of *forward compatibility*. A distribution ρ is *forward compatible* with a given coarse-graining map $\mathcal{C}$ if evolving ρ forward under the microdynamics and then coarse-graining at the end gives the same result as evolving ρ forward (for the same length of time) under the coarse-grained dynamics. (Note that forward compatibility, unlike forward predictability, is not defined relative to any given history space.) Forward predictability implies forward compatibility (just consider the trivial history, where the macrostate at each time is the whole state space), and the converse is true in systems that are macrodeterministic. More generally, if $H(\alpha)\rho$ is forward compatible with $\mathcal{C}$ for all histories α in some history space, then ρ is forward predictable by $\mathcal{C}$ on that history space.

Prima facie, one way in which forward compatibility could hold is if the coarse-graining rule is actually physically implemented by the micro-dynamics: if, for instance, a distribution ρ is taken by the micrograined dynamics to the distribution $\mathcal{C}\rho$ on timescales short compared to those on which the macroproperties evolve, then all distributions will be forward compatible with $\mathcal{C}$. And indeed, if we want to explain how one coarse-grained dynamics can be compatible with another even coarser-grained dynamics, this is very promising. We can plausibly explain the coarse-graining rule for local equilibrium thermodynamics, for instance, if we start from the Boltzmann equation and deduce that systems satisfying that equation really do evolve quickly into distributions which are locally canonical. (Indeed, this is the usual defence given of local thermal equi-librium models in textbooks.)

But clearly, this cannot be the explanation of forward compatibility of the *fine-grained* dynamics with any coarse-graining rule. For by construc-tion, the coarse-graining rules invariably increase Gibbs entropy, whereas the fine-grained dynamics leave it static. One very simple response, of course, would be just to postulate an explicit modification to the dynamics which enacts the coarse-graining. In classical mechanics, Ilya Prigogine has tried to introduce such modifications (see e.g. Prigogine 1984 and references therein); in quantum mechanics, the introduction of an explicit, dynamical rule for the collapse of the wavefunction could be thought of as a coarse-graining, and Albert (2000: ch. 7) has speculated that the version of this rule introduced in the GRW dynamical-collapse variant of QM[9] carries out just the sort of coarse-graining required to induce macrodynamical irreversibility.

---

[9] First proposed in Ghirardi et al. (1986); see Bassi and Ghirardi (2003) for a review.

However, at present there remains no direct empirical evidence for any such dynamical coarse-graining. For this reason, and since the whole point of this book is to investigate the status of unmodified, unitary, quantum mechanics, I will continue to assume that the unmodified microdynamics (classical or quantum) should be taken as exact.

Nonetheless, it would not be surprising to find that distributions are, in general, forward compatible with coarse-graining. Putting aside exemplar rules for coarse-graining, there are strong heuristic reasons to expect a given distribution generally to be forward compatible with the other three kinds of rules:

- A distribution will be forward compatible with a smearing coarse-graining rule whenever the microscopic details of the distribution do not affect the evolution of its overall spread across phase space. Whilst one can imagine distributions where the microscopic structure is very carefully chosen to evolve in some particular way contrary to the coarse-grained prediction, it seems heuristically reasonable to suppose that generically this will not be the case, and that distributions (especially reasonably widespread distributions) which differ only on very small lengthscales at one time will tend to differ only on very small lengthscales at later times. (However, I should note that I find this heuristic only *somewhat* plausible, and in light of the dearth of practical physics examples which use this rule, would be relaxed if readers are unpersuaded!)

- A distribution will be forward compatible with a correlation-discard coarse-graining rule whenever the details of the correlation do not affect the evolution of the macroscopic variables. Since macroscopic properties are typical local, and correlative information tends to be highly delocalized, heuristically one would expect that generally the details of the correlations are mostly irrelevant to the macroscopic properties—only in very special cases will they be arranged in just such a way as to lead to longer-term effects on the macroproperties.

- A distribution will be forward compatible with a measurement coarse-graining rule (which, recall, is nontrivial only for quantum theory) whenever interference between components of the distribution with different macroproperties does not affect the evolution of those macroproperties. This is to be expected whenever the macroproperties of the system at a given time leave a trace in the microproperties at that time which is not erased at subsequent times: when this is the case, constructive or destructive interference between branches of the wavefunction cannot occur. Decoherence theory tells us that this will very generically occur for macroscopic systems: particles interacting with the cosmic microwave background radiation or with the atmosphere leave a trace in either; the microscopic degrees of freedom of a non-harmonic vibrating solid record a trace of the macroscopic vibrations, and so forth. These traces generally become extremely delocalized, and are therefore not erasable by local physical processes. In principle one can imagine that eventually they relocalize and become erased—indeed, this will certainly happen (on absurdly long timescales) for spatially finite systems—but it seems heuristically reasonable to expect that on any realistic timescale (and for spatially infinite systems, perhaps on any timescale at all) the traces persist.

At least in the deterministic case, forward compatibility implies forward predictability; even in probabilistic cases, these kind of heuristics suggest—again, only heuristically—that forward predictability is generic.

In any case, my purpose here is not to prove detailed dynamical hypotheses but to identify those hypotheses that we need. So, given the above heuristic arguments, we could try postulating a

Bold Dynamical Conjecture. For any system of interest to studies of irreversibility, all distributions are forward predictable by the appropriate coarse-grainings of that system on the appropriate history space for that system.

It is clear that, were the Bold Dynamical Conjecture correct, it would go a long way towards explaining why coarse-graining methods work.

But the line between boldness and stupidity is thin, and—alas—the Bold Dynamical Conjecture strides Boldly across it. For suppose $X = \mathcal{C}\rho$ is the initial state of some Interesting segment of a dynamical trajectory of the forward coarse-grained dynamics (Interesting so as to guarantee that Gibbs entropy increases on this trajectory) and that X' is the final state of that trajectory (say, after time t). Then by the Bold Dynamical Conjecture, X' can be obtained by evolving ρ forward for time t under the fine-grained dynamics (to some state ρ', say) and then coarse-graining.

Now suppose we take the time-reversal TX' of X' and evolve it forward for t seconds under the coarse-grained forward dynamics. By the Bold

Dynamical Conjecture, the resultant state could be obtained by evolving $T\rho'$ forward for t seconds under the fine-grained dynamics and then coarse-graining. Since the fine-grained dynamics are time-reversible, this means that the resultant state is the coarse-graining of $T\rho$. And since coarse-graining and time reversal commute, this means it is just the time reverse TX of X.

But this yields a contradiction. For Gibbs entropy is invariant under time reversal, so $S_G(TX) = S_G(X)$ and $S_G(TX') = S_G(X')$. It is non-decreasing on any trajectory, so $S_G(TX) \geq S_G(TX')$. And it is increasing (since the trajectory is Interesting) between X and X', so $S_G(X') > S_G(X)$. So the Bold Dynamical Conjecture is false; and, more generally, we have shown that if $C\rho$ is any coarse-grained distribution on a trajectory of the forward coarse-grained dynamics which has higher Gibbs entropy than the initial distribution on that trajectory, then $T\rho$ is *not* forward compatible with C.

So much for the Bold Dynamical Conjecture. But just because not *all* distributions are forward compatible with C, it does not follow that none are; it does not even follow that most aren't. Indeed, the (admittedly heuristic) arguments above certainly seem to suggest that distributions that are in some sense 'generic' or 'typical' or 'non-conspiratorial' or some such term will be forward compatible with the coarse-grainings. In general, the only known way to construct *non*-forward compatible distributions is to evolve a distribution forward under the fine-grained dynamics and then time-reverse it.

This suggests a more modest proposal:

Simple Dynamical Conjecture (for a given system with coarse-graining C). Any distribution whose structure is at all simple is forward predictable by C; any distribution *not* so predictable is highly complicated and as such is not specifiable in any simple way *except* by stipulating that it is generated via evolving some other distribution in time (e.g. by starting with a simple distribution, evolving it forwards in time, and then time-reversing it).

Of course, the notion of 'simplicity' is hard to pin down precisely, and I will make no attempt to do so here. (If desired, the Simple Dynamical Conjecture can be taken as a family of conjectures, one for each reasonable precisification of 'simple'.) But for instance, any distribution specifiable in closed functional form (such as the microcanonical or canonical distributions, or any distribution uniform over a given, reasonably simply specified macroproperty) would count as 'specifiable in a simple way'.

In fact, it will be helpful to define a *Simple* distribution as any distribution specifiable in a closed form in a simple way, without specifying it via the time evolution of some other distribution. Then the Simple Dynamical Conjecture is just the conjecture that all Simple distributions are forward predictable by the coarse-graining. Fairly clearly, for any precisification of the notion of Simple, a distribution will be Simple iff its time reverse is.

Are individual states (i.e. classical single-system states or quantum pure states) Simple? It depends on the state in question. Most classical or quantum states are not Simple at all: they require a great deal of information to specify. But there are exceptions: some product states in quantum mechanics will be easily specifiable, for instance; so would states of a classical gas where all the particles are at rest at the points of a lattice. This in turn suggests that the Simple Dynamical Conjecture may well fail in certain classical systems (specifically, those whose macrodynamics is in general indeterministic): Simple classical systems will generally have highly unusual symmetry properties and so may behave anomalously. For example, a generic self-gravitating gas will evolve complex and highly asymmetric structure because small density fluctuations get magnified over time, but a gas with no density fluctuations whatever has symmetries which cannot be broken by the dynamics, and so will remain smooth at all times.

This appears to be an artefact of classical mechanics, however, which disappears when quantum effects are allowed for. A quantum system with a similar dynamics will evolve into a superposition of the various asymmetric structures; in general, the classical analogue of a localized quantum wavefunction is a narrow Gaussian distribution, not a phase-space point. So I will continue to assume that the Simple Dynamical Conjecture holds of those systems of physical interest to us.

9.6 Microdynamical origins of irreversibility: the classical case

It is high time to begin addressing the question of what all this has to do with the real world. I begin with the classical case, although of course the quantum case is ultimately more important. The question at hand is: on the assumption that classical microphysics is true for some given system, what additional assumptions need to be made about that system in order to ensure that its macroscopic behaviour is correctly predicted by the irreversible dynamics generated by coarse-graining?

The most tempting answer, of course, would be 'none'. It would be nice to find that absolutely any system has macroscopic behaviour well described by the coarse-grained dynamics. But we know that this cannot be the case: the coarse-grained dynamics is irreversible, whereas the microdynamics is time-reversal-invariant, so it cannot be true that all microstates of a system evolve in accordance with the coarse-grained dynamics. (A worry of a rather different kind is that the coarse-grained dynamics is in general probabilistic, whereas the classical microdynamics are deterministic.)

This suggests that we need to supplement the microdynamics with some restrictions on the actual microstate of the system. At least for the moment, I will assume that such restrictions have a probabilistic character; I remain neutral for now as to how these probabilities should be understood.

A superficially tempting move is just to stipulate that the correct probability distribution over microstates of the system is at all times forward predictable by the coarse-graining. This would be sufficient to ensure the accuracy of the irreversible dynamics, but it is all but empty: to be forward predictable by the coarse-graining *is* to evolve, up to coarse-graining, in accordance with the irreversible dynamics.

Given the Simple Dynamical Conjecture, an obvious alternative presents itself: stipulate that the correct probability distribution over microstates is at all times Simple. This condition has the advantage of being non-empty, but it suffers from two problems: it is excessive, and it is impossible. It is excessive because the probability distribution at one time suffices to fix the probability distribution at all other times, so there is no need to independently impose it at more than one time. And it is impossible because, as we have seen, in general the forward time evolution of a Simple distribution is not Simple. So if we're going to impose Simplicity as a condition, we'd better do it once at most.

That being the case, it's pretty clear when we have to impose it: at the beginning of the period of evolution in which we're interested. Imposing Simplicity at time t guarantees the accuracy of the forward coarse-grained dynamics at times later than t; but by time reversibility (since the time-reverse of a Simple distribution is Simple) it also guarantees the accuracy of the *backwards* coarse-grained dynamics at times earlier than *t*, which we need to avoid. So we have a classical recipe for the applicability of coarse-grained methods to classical systems: they will apply, over a given period, only if at the beginning of that period the probability of the system having a given microstate is specified by a Simple probability function.

So, exactly when should we impose the Simplicity criterion? There are basically two proposals in the literature:

1. We should impose it, on an ad hoc basis, at the beginning of any given process that we feel inclined to study.
2. We should impose it, once and for all, at the beginning of time.

The first proposal is primarily associated with the objective Bayesian approach pioneered by Jaynes (see e.g. Jaynes 1957a; 1957b; 1968) and briefly reviewed in Chapter 4—and I have to admit to finding it incomprehensible. In no particular order:

• We seem to be reasonably confident that irreversible thermodynamic processes take place even when we're not interested in them.
• Even if we are uninterested in the fact that our theories predict anti-thermodynamic behaviour of systems before some given time, they still do (i.e. the problem that our theories predict anti-thermodynamic behaviour doesn't go away just because they make those predictions before the point at which we are "inclined to study" the system in question).
• The direction of time is put in by hand, via an a priori assumption that we impose our probability measure at the beginning, rather than the end, of the period of interest to us. This seems to rule out any prospect of understanding (for instance) humans themselves as irreversible physical systems.

Perhaps the most charitable way to read the first proposal is as a form of strong operationalism, akin to the sort of operationalism proposed in the foundations of quantum mechanics by e.g. Fuchs and Peres (2000a), which I discussed and criticized in sections 1.5 and 1.6. In this book, though, I presuppose a more realist approach to science, and from that perspective the second proposal is the only one that seems viable: we must impose Simplicity at the beginning of time. The time asymmetry in irreversible processes is due to the asymmetry involved in imposing the condition at one end of time rather than the other.

(Incidentally, one can imagine a cosmology—classical or quantum—according to which there is no well-defined initial state—for instance, because the state can be specified at arbitrarily short times after the initial singularity but not at the singularity itself, or because the notion of spacetime itself breaks down as one goes further into the past. If this is the case, some somewhat more complicated formulation would presumably be needed, but it seems unlikely that the basic principles would be unchanged. For simplicity and definiteness, I will continue to refer to 'the initial state'.)

At this point, a technical issue should be noted. My definition of the Simple Dynamical Conjecture was relative to a choice of system and coarse-graining; what is the appropriate system if we want to impose Simplicity at the beginning of time? The answer, presumably, is that the system is the Universe as a whole, and the coarse-graining rule is just the union of all the coarse-graining rules we wish to use for the various subsystems that develop at various times. Presumably there ought to exist a (probably imprecisely defined) maximally fine-grained choice of coarse-graining rule such that the Simple Dynamical Conjecture holds for that rule; looking ahead to the quantum-mechanical context, this seems to be what Gell-Mann and Hartle (2007) mean when they talk about a maximal quasi-classical domain.

So: if the probabilities we assign to possible initial states of the Universe are given by a Simple probability distribution, and if we accept classical mechanics as correct, we would predict that the coarse-grained forward dynamics are approximately correct predictors of the probability of the later Universe having a given state. We are now in a position to state an assumption which suffices to ground the accuracy of the coarse-grained dynamics.

Simple Past Hypothesis (classical version). There is some Simple distribution ρ over the phase space of the Universe such that for any point x, $\rho(x)\delta V$ is the objective probability of the initial state of the Universe being in some small region δV around x.

(By 'objective probability' I mean that the probabilities are not mere expressions of our ignorance, but are in some sense objectively correct; cf. the discussion in Chapter 4.)

To sum up: if (a) the world is classical; (b) the Simple Dynamical Conjecture is true of its dynamics (for given coarse-graining C); (c) the

Simple Past Hypothesis is true, then the initial state of the world is forward predictable by the $\mathcal{C}+$ dynamics: the macrodynamics defined by the $\mathcal{C}+$ dynamics is the same as the macrodynamics induced by the microdynamics.

9.7 Microdynamical origins of irreversibility: the quantum case

Rather little of the reasoning above actually made use of features peculiar to classical physics. So the obvious strategy to take in the case of quantum mechanics is just to formulate a quantum-mechanical version of the Simple Past Hypothesis involving objective chances of different pure states, determined by some Simple probability distribution.

There are, however, two problems with this: one conceptual, one technical. The technical objection is that quantum distributions are density operators, and the relation between density operators and probability distributions over pure states is one-to-many. The conceptual objection is that, as argued *in extenso* in Part II, quantum mechanics already incorporates objective chances, and it is inelegant, to say the least, to introduce additional such.

However, it may be that no such additional objective chances are in fact necessary, for two reasons.

1. There may be many pure states that are Simple and which are reasonable candidates for the state of the very early Universe.
2. It is not obvious that pure, rather than mixed, states are the correct way to represent the states of individual quantum systems.

To begin with the first: as I noted previously (p. 349) there is no problem in quantum mechanics in regarding certain pure states as Simple, and the (as always, heuristic) motivations for the Simple Dynamical Conjecture are no less true for these states. As for the second, mathematically speaking, mixed states do not seem obviously more alien than pure states as representations of quantum reality. Indeed, we have already seen (Chapter 8) that if we wish to speak at all of the states of individual systems in the presence of entanglement, the only option available is to represent them by mixed states. And since the Universe appears to be open, and the vacuum state of the Universe appears to be entangled on all lengthscales (cf. Redhead

1995 and references therein), even the entire observable Universe cannot be regarded as in a pure state. (I return to this issue in Chapter 10.)

This being the case, I tentatively formulate the quantum version of the Simple Past Hypothesis as follows.

Simple Past Hypothesis (quantum version). The initial quantum state of the Universe is Simple.

What is the status of the Simple Past Hypothesis? One way to think of it is as a hypothesis about whatever law of physics (fundamental or derived) specifies the state of the very early Universe: that that law requires a Simple initial state. Indeed, if one assumes that probabilistic physical laws must be simple (which seems to be part of any reasonable concept of 'law'), and that simplicity entails Simplicity, all the Simple Past Hypothesis amounts to is the

Past Law Hypothesis. The initial quantum state of the Universe is determined by some law of physics.

Alternatively, we might think of the Simple Past Hypothesis as a (not very specific) conjecture about the contingent facts about the initial state of the Universe, unmediated by law. Indeed, it is not clear that there is any very important difference between these two readings of the Hypothesis. In either case, the route by which we come to accept the Hypothesis is the same: because of its power to explain the present-day observed phenomena, and in particular the success of irreversible macrodynamical laws. And on at least some understandings of 'law' (in particular, on a Humean account like that of Lewis 1986c, where laws supervene on the actual history of the Universe), there is not much metaphysical gap between (i) the claim that the initial state of the Universe has particular Simple form X and this cannot be further explained, and (ii) the claim that it is a law that the initial state of the Universe is X.

9.8 Avoiding a low-entropy postulate

The suggestion that the origin of irreversibility lies in constraints on the state of the early Universe is hardly new: it dates back to Boltzmann,[10]

---

[10] For historical details, see Sklar (1993: ch. 8) and references therein.

and has been espoused in recent work by, among others, Penrose (1989; 2004), Goldstein (2001), Albert (2000), and Price (1996). But their Past Hypotheses differ from mine in an interesting way. Mine is essentially a constraint on the *microstate* of the early Universe which is essentially silent on its macrostate (on the assumption that for any given macroscopic state of the Universe, there is a Simple probability distribution with support on that macrostate). But the normal hypothesis about the past is instead a constraint on the macrostate of the early Universe:

Low-Entropy Past Hypothesis. The initial macrostate of the Universe has very low thermodynamic entropy.

Is such a hypothesis needed in addition to the Simple Past Hypothesis? I think not. For if the Simple Dynamical Conjecture is correct, then it follows from the Simple Past Hypothesis and our best theories of microdynamics that the kind of irreversible dynamical theories we are interested in—in particular, those irreversible theories which entail that thermodynamic entropy reliably increases—are empirically reliable. These theories in turn entail that the entropy of the early Universe was at most no higher than that of the present Universe, and was therefore 'low' by comparison to the range of entropies of possible states (since there are a great many states with thermodynamic entropy far higher than that of the present-day Universe). So the Low-Entropy Past 'Hypothesis' is not a Hypothesis at all, but a straightforward prediction of our best macrophysics alone—and thus, indirectly, of our best microphysics combined with the Simple Past Hypothesis.

It will be helpful to expand on this a bit. On the assumption that the relevant irreversible dynamics (in this case, non-equilibrium thermodynamics) is predictively accurate, predictions about the future can be made just by taking the current state of the Universe and evolving it forward under those dynamics. Since the dynamics do not allow retrodiction, our route to obtain information about the past must (as noted earlier) be more indirect: we need to form hypotheses about past states and test those hypotheses by evolving them forward and comparing them with the present state. In particular, the hypothesis that the early Universe was in a certain sharply specified way very hot, very dense, very uniform, and very much smaller than the current Universe—and therefore much lower in entropy than the current Universe[11]—does very well under this method: conditional on that hypothesis, we would expect the current universe to be pretty much the way it in fact is. On the other hand, other hypotheses—notably the hypothesis that the early universe was much higher in entropy than the present-day universe—entail that the present-day Universe is fantastically unlikely, and so very conventional scientific reasoning tells us that these hypotheses should be rejected.

In turn, we can derive the assumption that our irreversible dynamical theories are predictively accurate by assuming (i) that our microdynamical theories are predictively accurate, and (ii) that the Simple Past Hypothesis and the Simple Dynamical Conjecture are true. So these hypotheses jointly give us good reason to infer that the early Universe had the character we believe it to have had. On the other hand, (i) alone does not give us reason to accept (ii). Rather, we believe (ii) because combined with (i), it explains a great deal of empirical data—specifically, the success of irreversible dynamical theories.

The difference between the Simple Past Hypothesis and the Low-Entropy Past Hypothesis, then, does not lie in the general nature of our reasons for believing them: both are epistemically justified as inferences by virtue of their explanatory power. The difference is that the Simple Past Hypothesis, but not the Low-Entropy Past Hypothesis, is justified by its ability to explain the success of thermodynamics (and other irreversible processes) *in general*. The Low-Entropy Past Hypothesis, by contrast, is justified by its ability to explain *specific features* of our current world (although the hypothesis that does this is better understood as a specific cosmological hypothesis about the state of the early Universe, rather than the very general hypothesis that its entropy was low).

Albert (2000: 96) gives a particularly clear statement of his framework for inducing the (Low-Entropy) Past Hypothesis, which makes an interesting contrast to my own. He makes three assumptions:

1. that our best theory of microdynamics (which for simplicity[12] he pretends is classical mechanics) is correct;
2. that the Low-Entropy Past Hypothesis is correct;

---

[11] It is widely held that (i) such a Universe ought to be much *higher* in entropy than the *present-day* Universe, but (ii) this supposed paradox is solved when gravity is taken into account. This is very confused; I attempt to dispel the confusion in Wallace (2010c).

[12] This oversimplifies somewhat: Albert is also concerned to contrast the classical case with the GRW version of quantum mechanics.

3. that the correct probability distribution to use over current microstates is the uniform one, conditionalized on whatever information we know (notably, the Low-Entropy Past Hypothesis).

He also makes a tacit mathematical conjecture, which is a special case of the Simple Dynamical Conjecture: in my terminology, he assumes that those distributions which are uniform over some given macrostate and zero elsewhere are forward compatible with coarse-graining.

Now, (2) and (3) together entail that the correct distribution to use over initial states (and Albert is fairly explicit that 'correct' means something like 'objective-chance-giving') is the uniform distribution over whatever particular low-entropy macrostate is picked out by the Low-Entropy Past Hypothesis. Since these distributions are Simple, Albert's two assumptions entail the Simple Past Hypothesis. But the converse is not true: there are many Simple distributions which are not of the form Albert requires, but which (given the Simple Dynamical Conjecture) are just as capable of grounding the observed accuracy of irreversible macrodynamics.

Put another way: let us make the following abbreviations.

SPH   Simple Past Hypothesis
LEPH   Low-Entropy Past Hypothesis
UPH   Uniform Past Hypothesis: the hypothesis that the initial distribution of the Universe was a uniform distribution over some macrostate
SDC   Simple Dynamical Conjecture
PAµ   Predictive Accuracy of Microphysics (i.e. our current best theory of microphysics is predictively accurate)
PAM   Predictive Accuracy of Macrophysics (i.e. the macrodynamics derived from microphysics by coarse-graining is predictively accurate)

My argument is that

$$SPH + SDC + PA\mu \longrightarrow PAM. \tag{9.22}$$

Albert's (on my reading) is that

$$LEPH + UPH + SDC + PA\mu \longrightarrow PAM. \tag{9.23}$$

But in fact

$$UPH \rightarrow SPH \qquad (9.24)$$

so actually *LEPH* plays no important role in Albert's argument. All that really matters is that the initial distribution was uniform over some macrostate; the fact that this macrostate was lower entropy than the present macrostate is then a straightforward inference from *PAM* and the present-day data.

9.9 Summary

There are extremely good reasons to think that, in general and over timescales relevant to the actual Universe, the process of evolving a distribution forward under the microdynamics of the Universe commutes with various processes of coarse-graining, in which the distribution is replaced by one in which certain fine structures—most notably the small-scale correlations and entanglements between spatially distant subsystems—are erased. The process of alternately coarse-graining in this manner and evolving a distribution forwards leads to dynamical processes which are irreversible: for instance, when probabilistic, they will have a branching structure; where a local thermodynamic entropy is definable, that entropy will increase. Since coarse-graining, in general, commutes with the microdynamics, in general we have good grounds to expect distributions to evolve under the microdynamics in a way which gives rise to irreversible macrodynamics, at least over realistic timescales.

Given that the microdynamics is invariant under time reversal, if this claim is true then so is its time reverse, so we have good reason to expect that, in general, the evolution of a distribution both forward and backwards in time leads to irreversible macrodynamics on realistic timescales. It follows that the claim can be true only 'in general' and not for *all* distributions, since—for instance—the time evolution of a distribution which does behave this way cannot in general behave this way. However, we have no reason to expect this anomalous behaviour except for distributions with extremely carefully chosen fine-scale structure (notably those generated from other distributions by evolving them forwards in time). I take this to be a more accurate expression of Goldstein's idea of 'not being ridiculously special': it is not that systems are guaranteed *to achieve equilibrium* unless they or their dynamics are 'ridiculously special'; it is that only in 'ridiculously special' cases will the micro-evolution of a distribution not commute with coarse-graining. Whether, and how fast, a system approaches thermal equilibrium is then something that can be determined via these coarse-grained dynamics.

In particular, it seems reasonable to make the Simple Dynamical Conjecture that reasonably simple distributions do not show anomalous behaviour. If the correct distribution for the Universe at some time t is simple in this way, we would expect that macrophysical processes after t are well described by the macrodynamics generated by coarse-graining (and so exhibit increases in thermodynamic entropy, dispersal of quantum coherence, etc.), in accord with the abundant empirical evidence that these macrodynamics are correct. But we would also expect that macrophysical processes *before* t are not at all described by these macrodynamics—are described, in fact, by the time reversal of these macrodynamics—in wild conflict with the empirical evidence. But if t is the first instant of time (or at least, is very early in time), then no such conflict will arise.

It follows that any stipulation of the boundary conditions of the Universe according to which the initial distribution of the Universe is reasonably simple will (together with our microphysics) entail the correctness of our macrophysics. Since any law of physics specifying the initial distribution will (essentially by the nature of a law) require that initial distribution to be reasonably simple, it follows that any law which specifies the initial distribution suffices to ground irreversible macrodynamics.

It is virtually tautologous that if microscopic physics has no time asymmetry but the emergent macroscopic dynamics does have a time asymmetry, that time asymmetry must be due to an asymmetry in the initial conditions of the Universe. The most common proposal for this asymmetry is the proposal that the initial distribution is the uniform distribution over a low-entropy macrostate. From the point of view of explaining irreversibility, all the work in this proposal is being done by the 'uniform distribution' part: the low-entropy part alone is neither necessary nor sufficient to establish the correctness of the irreversible macrodynamics, though of course if the initial macrostate is a maximum-entropy state then its macro-evolution will be very dull and contradicted by our observations.

And in fact, the only special thing about the uniformity requirement is that we have good (if heuristic) grounds to expect the microdynamical evolution of uniform distributions to be compatible with coarse-grainings.

But we have equally good (if equally heuristic) grounds to expect this of any simply specified distribution. So really, the asymmetry of the Universe's macroscopic dynamics is not a product of the particular form of the physical principle which specifies the initial conditions of the Universe: it is simply a product of some such principle being imposed at one end of the Universe and not at the other.

CHAPTER 9. The time asymmetry of branching has the same cause as the other time asymmetries of macrophysics, such as the thermodynamic ones: it arises from the fact that only very special systems do not display this behaviour and the initial state of the Universe was not 'very special' in this sense. Since the early Universe was also very far from being maximally branched (and, indeed, from being maximally entropic), if there is a final state of the Universe, it will be very special, simply by virtue of being time-evolved from a not-maximally-branched state. The direction of time defined by branching and the direction defined by other irreversible processes must (and, of course, do) point in the same direction.

CHAPTER 10. What else does Everettian quantum mechanics have to tell us about the Universe?

10

A Cornucopia of Everettian Consequences

There is something fascinating about science. One gets such wholesale returns of conjecture out of such a trifling investment of fact.

Mark Twain[1]

This last chapter is something of a mixed bag: here I want to develop
a number of consequences of Everettian quantum mechanics which
have relatively little in common with one another but which—though
they seem worth discussing—did not merit their own chapter. I begin
philosophically: first I reconsider quantum chaos and ask just how predictable the future is according to the Everett intepretation; then I explore
Everettian probabilities in some rather exotic circumstances, including the
(in)famous 'quantum suicide' thought experiment.

After that, I move on to the suggestion (most famously made by David
Deutsch) that parallel universes are after all directly observable, through
interference experiments and through quantum computations. I then
reconsider the nature of density operators (a.k.a. mixed states) in the
light of the Everett interpretation; I finish the chapter on a speculative
note by exploring the idea (also due to Deutsch) that Everettian quantum
mechanics has major consequences for the theory of time travel.

The six sections below all essentially stand alone, and they can (with
negligible loss) be read in any order. Both for reasons of space and because
some of the topics are by their nature rather speculative, this chapter's arguments are a bit more sketchy, and their conclusions a bit more tentative,
than has (I hope!) been true elsewhere.

---

[1] Twain (1996)

10.1 Indeterminism, chaos, and the predictability of the future

Can we predict what the future will be like? The popular-science answer looks something like this:

- Where quantum mechanics is relevant, the future is unpredictable *in principle*. All we can do is say that certain outcomes will occur with certain probabilities, but we cannot say which outcomes will occur.
- Where classical mechanics is relevant, the future is unpredictable *in practice*. The underlying theory is perfectly deterministic, so that if we had infinite computing power, and infinitely accurate knowledge of the current state of the system in question, we could predict with certainty what it will do in the future. But in most realistic cases, the dynamics, though deterministic, is chaotic. This means that the future is extremely sensitive to small details of the present: to use the standard example, whether or not a hurricane occurs in a month's time may depend on whether or not a butterfly on the other side of the world is currently flapping its wings. So we cannot, in practice, predict the future in classical mechanics either.

Actually, though, quantum mechanics is always relevant, and Everettian quantum mechanics offers a unified account of the two stories. To begin with the quantum case: of course, from a God's-eye view the quantum dynamics is deterministic. We can predict with certainty what will happen in some quantum decay process: with certainty, the Universe (or a local bit of it, in any case) will branch into parallel versions, in some of which the decay occurs at a certain time, in some of which it occurs at another time, and in others of which it does not occur at all. Furthermore, the relative weights of these branches can be determined with very great accuracy. However, from the point of view of an individual observer, the theory is indeterministic: an observer cannot, in principle, predict what branch he will end up in; and as we saw in Chapter 7, even making sense of the phrase 'what branch he will end up in' is less than trivial. All that the observer can know is the probabilities of each outcome; these, though, he can know with high precision.

And the case where classical mechanics is relevant . . . is exactly the same. As we saw in Chapter 3, chaos (i.e. evolution under Hamiltonian whose classical cousin is chaotic) rapidly leads to quantum branching, because wavepackets initially very localized in phase space will rapidly get smeared out and then decohered. This means that the reason I cannot predict the weather in a year's time is not that I cannot in practice determine the current state of the atmosphere accurately, nor that I cannot in practice do the appropriate calculations. The reason, instead, is that there is no fact of the matter as to what the weather is in a year's time.[2] From a God's-eye perspective, the weather is perfectly predictable, but the prediction is that it varies widely from branch to branch. All that the branch-bound observer can do, though, is predict with what probability each outcome will occur—this, though, he can do without an exponentially expensive requirement for computational power. (He would require a rather better theory of quantitative meteorology than we currently have, and probably better data collection too, but this is a practical constraint based on the current state of climate science, not something more fundamental.)

Furthermore, it follows from this that even in the presence of chaos, there is not really any strong sensitivity of the future on the present.[3] Small shifts in the current state of a chaotic quantum system lead only to small shifts in its (very indeterminate) future state. (For *sufficiently* small shifts, ultimately this is a corollary of the fact that Hilbert-space distance is conserved under quantum evolution, whereas there is in general no equivalent conserved distance in classical mechanics: hence, states which initially are close together in Hilbert-space terms remain close together.) The ability of butterflies to affect the weather has been greatly exaggerated.

This has two philosophical consequences, of widely differing nature. The first concerns how we think about chaos. Classically, it is normal[4] to define a system as chaotic if phase-space trajectories diverge exponentially from one another, which is essentially another way of saying that chaotic systems are those whose future dynamics are exponentially sensitive to their initial conditions. A corollary of this is that phase-space

---

[2] Readers of Ch. 7 will recognize that on the 'Lewisian' account presented in that chapter, there is a fact of the matter but it is in principle unknowable; they will also recall that the choice of whether to adopt that account is a choice about how to describe the underlying physics, not a choice as to what that underlying physics actually is.

[3] To the best of my knowledge, this observation was first made in print in Deutsch (1997: 201–3).

[4] See e.g. Cvitanović et al. (2009), and references therein.

blobs in chaotic systems are exponentially stretched in some directions and squeezed in others.

Given that quantum mechanics is true and classical mechanics is false, though, we should understand this the other way round. Classical chaos is a real phenomenon even in the actual (quantum) world, but it does not actually involve high degrees of sensitivity to initial conditions or exponential divergence of trajectories, because the former is false and the latter does not correspond to anything that actually occurs. Rather, the key feature of a classically chaotic system is its tendency to exponentially contract and stretch phase-space blobs: it is these blobs, not the individual phase-space points, that correspond (approximately) to some physically real feature of the world, and so it is their evolution, and not the fictional evolution of the phase-space points, which is physically salient.

The other philosophical consequence—bizarrely—is in theoretical ethics. It is prima facie tempting to adopt what Parfit (1984: 351–80, 393–6) calls the *person-affecting principle*: the principle that when an act is bad, it is bad because it is bad for some person or persons. Deliberately or negligently causing a spillage of nuclear waste, for instance, is bad because it causes harm to many people, by exposing them to an increased risk of cancer.

The problem with this principle, Parfit argues, is that when we consider acts whose consequences occur many years in the future, any such act seems certain also to perturb the highly sensitive processes which determine which human beings exist in the first place. In particular, the processes by which a given set of chromosomes ends up in a given sperm or egg cell, and by which a given sperm cell fertilizes an egg cell, are so sensitive to external conditions that it seems very likely that any significant act I perform will entirely change the particular people who exist in, say, two centuries' time. If so, and if the person-affecting principle is correct, then my negligence in, say, causing nuclear waste to be spilled two centuries from now is morally irrelevant: the people who get cancer from the spillage have not been harmed by my causing that spillage, because if I had not done so, they would not even have existed!

For this reason, Parfit advocates the rejection of the person-affecting principle. He (and others) have explored attempts to replace the principle with others that rank the morality of our actions in terms of the quantity and quality of the lives in those situations brought about by our actions, but it has proved difficult to find generally acceptable principles of this kind.

Given Everettian quantum mechanics, though, Parfit's objection to the person-affecting principle seems to lose its force. The human population two hundred years from now does not have the extreme sensitivity to my actions that Parfit suggests: rather, whether or not I cause the nuclear spillage, the future consists of vastly many branches all containing different persons. In general, my action will not change those persons who exist, nor will it materially affect the weight of a branch in which a given person exists. What it will do is cause many people, in many branches, to get cancer when they would not otherwise have done so. (This is not, of course, a positive argument in favour of the person-affecting principle; there may be other reasons for rejecting it.)

10.2 Exotic cases of quantum probability

I have been assuming for the last hundred or so pages that Everettian quantum mechanics has no problem with probability—at least in relatively straightforward contexts like experimental verifications of quantum predictions. In this section, I want to consider probability in some more exotic contexts; I will raise more questions than I answer.

10.2.1 Cosmological probabilities and anthropic reasoning

To start with, suppose that our best theory of quantum cosmology tells us that some phenomenon—something about the inhomogeneities in the microwave background radiation or in the distribution of galaxies, perhaps—is such that in branches of collective weight 0.999 a certain phenomenon occurs, and in branches of collective weight 0.001 the phenomenon does not occur. If we look up into the night sky and fail to find the phenomenon, does that count against our best theory?

(This is not a purely hypothetical example: the microwave background really does appear to have fluctuations which can be well modelled by stochastic assumptions about primordial density inhomogeneities;[5] in an Everettian context those stochastic assumptions become assumptions about branch weight.)

The answer, I think, has to be yes: but that answer presupposes that branch weight = probability even in cosmological contexts. This isn't

---

[5] See e.g. Weinberg (2008: 101–48) and references therein.

something which the decision-theoretic arguments of Chapter 5 can really address, for the simple reason that they would require the agent to have the power to rearrange the cosmos on a vast scale. Put another way, the symmetries which ultimately motivated Chapter 5's argument remain present in the cosmological case, but the operational argument that forced (rather than merely motivated) agents to conform their personal probabilities to those symmetries is absent.

However, I don't think this is profoundly important. Any attempt to operationalize probability is always going to have to make idealizations; sometimes those idealizations are pretty large; there is nothing specifically quantum-mechanical about this. Once more, Chapter 4's moral holds true: quantum mechanics does not really create new problems with probability, it just throws old ones into sharp relief.

Now let's make things more interesting: suppose that the 'certain phenomenon' that occurs (according to our best theory) in branches of weight 0.999 is *the absence of intelligent life*. In other words, suppose that our best cosmological theories predict that it is 99.9% probable that the Universe has evolved in such a way as to rule out creatures such as ourselves.

Now, pretty clearly, if we are completely certain that our 'best theory' (call it Theory A) is correct, then our own existence becomes just one more data point on which to conditionalize. But the interesting case is where we have some rival theory: suppose the rival theory—theory B, say—says that sentient life occurs in all (or at least, most by branch weight) branches. Does this count in favour of theory B?

We could argue in two ways:

First way. Obviously, yes. Theory B says that some phenomenon is certain to occur; Theory A grants it probability 0.001. If it occurs, that's tantamount to a falsification of Theory A.

Second way. Obviously, no. Actually, both Theory A and Theory B say that the phenomenon is certain to occur. They just disagree as to which branches it occurs in (all vs some). So the occurrence of the phenomenon is no evidence at all against Theory A.

This second way might remind us uncomfortably of the problems of falsifying the Everett interpretation: since the interpretation predicts that all possible outcomes of an experiment occur on some branches, how can any result count against the Everett interpretation? We saw in Chapter 6 that actually, consideration of the strategy which a rational experimenter should adopt shows that the experimenter will indeed reject the Everett interpretation if he detects results which occur in sufficiently low-weight branches.

But those kinds of arguments do not apply here. They rely on the agent's existence (conditional on the Everett interpretation being true) in both the high-weight and low-weight branches. If the high-weight branches are empty of sentient life, the arguments fail—or put another way, it's not coherent to suppose that an experimenter can *do an experiment* to determine whether or not sentient life exists in his region of the Universe, for his very ability to carry out the experiment presupposes the answer. This being the case, I lean towards the second way of running the argument, according to which Theory A is not disconfirmed by the low branch weight it assigns to intelligent life.

However, that isn't the end of the story. Suppose (fancifully) that actually Theory A and Theory B both give large branch weights to sentient life, but theory A distributes most of that branch weight among non-humanoid life forms. (Maybe, according to Theory A the high-weight branches are hostile to carbon-based life, friendly to intelligent excitations of the Solar magnetic field.) Does *this* count against Theory A? Again, the argument could be run both ways: is Theory A disconfirmed because of the low weight it assigns to an observed phenomenon, or does the fact that the experimenter *is* a carbon-based life-form mean that only the branches in which such life forms exist are relevant?

This kind of defence of Theory A had better stop somewhere. Suppose (even more fancifully) that Theory A and Theory B both predict carbon-based sentient creatures rather like us with high weight, but Theory A gives a very low weight to their using DNA as their genetic basis whereas Theory B gives a very high weight to this assumption. Does the fact that we do actually rely on DNA count against Theory A, and would this conclusion have changed if physics had got far enough ahead of biology that we were considering the A-vs-B choice long before we actually knew how human genetics work? If we are not careful, we rule out any data at all as disconfirmatory of Theory A, and find ourselves back in the Everettian confirmation paradox.

The more one contemplates these sorts of puzzles the more one feels the need for a systematic theory of anthropic reasoning: that is, reasoning which takes into account the existence of, and perhaps more detailed information about, the scientific community itself.[6] And in fact, these puzzles are no longer just philosophers' games: it is quite common now in physics—especially in inflationary cosmology and in string theory—to invoke multiverses, and anthropic reasoning, to explain otherwise unexplained facts about the constants of nature and the laws of physics. (For an introduction to this subject, see Carr 2007.)

In fact, the Everettian multiverse is rather well behaved and well understood compared to some which have been postulated. The *mere* postulation of a vast multiplicity of Universes, and a probability distribution across them, with no justification other than the desire to explain certain features of the dynamical laws, feels (rightly or wrongly) rather tenuous as a piece of scientific reasoning, and rather painfully removed from empirical confirmation. By contrast, in Everettian quantum mechanics we have a rather precisely constrained theory of a multiverse, generating detailed—and thoroughly confirmed—predictions, and arguably allowing us to directly detect multiple Universes in certain cases (see section 10.3). For instance, it is not infrequently suggested that the value of the charge on the electron[7]—which appears (Barrow and Tipler 1986: 288–366) to be fairly tightly constrained by the existence of life—can be understood by postulating a multiverse in which it takes all conceivable values, and noting that most of them are barren of life.[8] To make such a story work using an *Everettian* multiverse, we would require some underlying quantum theory in which the charge of the electron was dynamically determined in such a way as to take different values in different branches; such a theory would, presumably, make other concrete and testable predictions. (Note that our current best theory of electromagnetism, the Standard Model, does not have this feature: the electron charge is dynamically determined (via spontaneous symmetry breaking) but its value is branch-independent.[9] All this suggests that—*if* anthropic reasoning can be made sense of at all in a

---

[6] So far as I am aware, Bostrom (2002) is the only sustained attempt at developing such a systematic theory.

[7] To be strictly accurate, one should talk about the dimensionless fine structure constant, not the unit-dependent electron charge *per se*.

[8] The issue actually has some cultural and political relevance: the smarter proponents of the argument that God designed the Universe (e.g. Stannard 1993; Polkinghorne 1986) appeal not to biology but to the fine-tuning of the constants of nature so as to make their point. (See Leslie 1989 for a (somewhat sympathetic) philosophical analysis.)

[9] See e.g. Cheng and Li (1984: 336–63).

coherent way—the Everett interpretation offers space to raise these kinds of anthropic arguments to a more scientifically rigorous level.

10.2.2 Quantum Russian roulette

A very different probability puzzle also concerns branches where the observer does not exist, but in this case, his nonexistence is a matter of his own actions. I refer to the *quantum Russian Roulette* or *quantum suicide* problem, much discussed in informal conversations about the Everett interpretation and presented formally in e.g. Squires (1986), Tegmark (1998), and David Lewis (2004). In the paradigm form of this problem, some agent who accepts Everettian quantum mechanics places a bet, at steep odds, on the outcome of some quantum process, and then arranges for himself to be swiftly and painlessly killed in the event he loses the bet. From the agent's point of view (goes the argument) he should expect to win the bet with subjective certainty, and so will be willing to take it in the expectation of certainly becoming rich. Call this the *sensational* interpretation of quantum suicide, as opposed to the *deflationary* interpretation according to which Russian roulette is as foolish in the quantum as in the classical case.

The arguments for sensationalism, I think, do not really withstand close inspection. First, it is argued that since it does not make sense to expect to experience nothing, we have no choice but to renormalize our probabilities over only those future branches in which we do have experiences. Here, for instance, is David Lewis, describing the Schrödinger cat experiment from the cat's perspective:

There's nothing it's like to be dead. Death is oblivion. (Real death, I mean. Afterlife is life, not death.) The experience of being dead should never be expected to any degree at all, because there is no such experience. So it seems that the [expectation = branch weight] rule does not work for the life-and-death branching that the cat undergoes. . . . When we have life-and-death branching, the [branch weight] rule as so far stated does not apply. We must correct it: first discard all the death branches, because there are no minds and no experiences associated with death branches. Only then divide expectations of experience between the remaining branches in proportion to their [branch weights]. (Lewis 2004: 17)

But if Lewis is right (and it sounds pretty plausible) that 'the experience of being dead should never be expected to any degree at all, because there is no such experience', then this is as true in classical as in quantum mechanics, and so as true in ordinary Russian roulette as in the quantum version. What do I *expect to experience next* as I contemplate some action which may well lead to my death? If it doesn't make sense to expect nothing, then presumably I expect one of the possible outcomes in which I survive—but nothing about probability follows from this.[10]

An alternative defence of sensationalism, offered by Peter Lewis (2000) and Tappenden (2004), considers not my *expectations* of death but why I should care about death. Tappenden, for instance, writes:

> It would seem highly plausible that, as materialists, we conventionally think of death as being egocentrically very, very bad because it will be the end of our personal experience.

And indeed, if all that matters to me is that I continue to have certain experiences somewhere in the multiverse, sensationalism looks attractive. But *pace* Tappenden, it is not at all clear that this analysis of the badness of death is correct. Indeed, the question of why we should fear death is famously tangled: as far back as the Roman Republic, Lucretius wrote:[11]

> Look back again to see how the past ages of everlasting time, before we are born, have been as naught to us. These then nature holds up to us as a mirror of the time that is to come, when we are dead and gone. Is there aught that looks terrible in this, aught that seems gloomy? Is it not a calmer rest than any sleep?

Perhaps part of the badness of death is the failure of continued consciousness *somewhere*; perhaps part of it is the failure of continued consciousness *in this branch*;[12] perhaps part of it is the ethical significance of our death to other agents; perhaps part of it is the failure of our desires to be fulfilled, and our intentions realized, in which case their being realized in only some branches may be only partial comfort.[13]

Furthermore, it can be questioned why it is appropriate in this case to adopt such an egocentric perspective. The decision-theoretic analysis of probability advanced in Chapter 5 makes no such egocentric assumption:

---

[10] Although it does follow that expecting X *to happen* with nonzero (personal) probability is not the same as expecting *to experience X*.

[11] Lucretius, *De Rerum Natura* III 972–7; this translation by Cyril Bailey (Lucretius 1947).

[12] For this to be a legitimate response to sensationalism, something like the Lewisian account of branch identity in Ch. 7 is probably needed.

[13] For further discussion on the philosophy of death, see Luper (2009) and references therein.

it considers an agent's preferences between ways in which a branch might turn out but does not rely on reducing these to preferences between the agent's own state within each branch. Papineau (2003) argues strongly that no such reduction is justified; if so, personal death does not interfere with the arguments of Chapter 5, and so an agent who prefers certain life to certain death is rationally compelled to prefer life in high-weight branches and death in low-weight branches to the opposite.

It is probably fair to say, though, that precisely because death is philosophically complicated, my objections fall short of being a knock-down refutation of sensationalism. Nor are they intended to be; the interested reader is referred to the papers of David Lewis, Peter Lewis, David Papineau, and Paul Tappenden cited above, and also to Cirkovic (2006). I should stress, though, that the question—however interesting—does not bear on the epistemic status of the Everett interpretation, for the fairly obvious reason that the experiments which provide the evidential basis for quantum mechanics do not generally involve the death of the experimenter, far less of third parties such as the writer!

Peter Lewis, however, disagrees (Lewis 2004). He compares (i) a situation in which an agent dies with (quantum) probability 0.5, with (ii) a situation in which an agent experiences a survivable punishment (heinous torture, say), which he regards as equally bad to death, with probability 1/6. Lewis claims that

1. Anyone with personal probability $> 2/3$ in the Everett interpretation will maximise their expected utility by making choice (i).
2. No one would actually make choice (i). So
3. No one can actually have personal probability $> 2/3$ in the Everett interpretation.

(The 2/3 is just illustrative, of course: Lewis's claim is that no one can really believe the Everett interpretation given quantum suicide.)

Premise (1), of course, presupposes sensationalism. I have no idea what Lewis's basis for premise (2) is: if he has surveyed people sympathetic to the Everett interpretation, he does not give details. (My own anecdotal experience, for what little it's worth, is that Everettians tend either to be outright deflationists, or at least to regard the effects of their death on third parties to be a dominant aspect of the badness of their death; either way, they reject premise (1).)

But suppose, just for the sake of argument, that Lewis is correct in premise (2) and that (at least in some circumstances) sensationalism is also correct. Still, all that would follow is that people—even advocates of the Everett interpretation—do not always allow rational considerations to override their strong instinctive preferences. And after all, that we have such strong instinctive preferences is scarcely a surprise. Whatever the answer to the question of why we *should* fear the possibility of death, evolutionary theory makes it pretty obvious why we *do* fear it.[14]

10.3 Observing the multiverse

This book has presented the case for believing in a multiverse—a reality of many non-interacting quasi-classical worlds—by analysis of our extant physical theories. The argument goes: we have excellent evidence for quantum mechanics, therefore we have good reason to believe quantum mechanics is approximately true; quantum mechanics entails that there are many worlds; so there are.

But there is an alternative route to the multiverse which has been argued for by David Deutsch, most notably in Deutsch (1997): namely, independent of the mathematical formalism of quantum theory, observational evidence *directly* supports the thesis that there are multiple worlds. For Deutsch, both interference phenomena and the (albeit not yet empirically realized) behaviour of quantum computers can only be made sense of by the existence of parallel universes, independent of our detailed quantum-mechanical analysis of how they work.

In this section, I want to see how far this line of argument can be pushed; my conclusion is that while Deutsch overreaches a bit, there is quite a lot of truth in his argument. I will consider three case studies: neutron interferometry,[15] quantum computation, and interaction-free measurement phenomena.

---

[14] Before moving on, I feel obliged to note that we ought to be rather careful just how we discuss quantum suicide in *popular* accounts of many-worlds quantum mechanics. Theoretical physicists and philosophers of physics (unlike, say, biologists or medical ethicists) rarely need to worry about the harm that can come from likely misreadings of their work by the public, but this may be an exception: there are, unfortunately, plenty of people who are both scientifically credulous and sufficiently desperate to do stupid things.

[15] Here, my argument is close to that of Vaidman (1998).

10.3.1 Classical particles, and a criterion for reality

Start by supposing that we are ignorant of quantum theory, and possess some object labelled 'neutron source', which we can take to be some sort of black box with a hole in it and a button on top, and also a collection of objects labelled as 'neutron detectors'.

Experimenting, we find that if we place one of the detectors directly in front of the hole and press the button on the black box, the detector will click; careful measurements show that the time between pressing the button and the click is directly proportional to the distance between the box and the detector. If we place a detector anywhere else, it will not click; regardless of where we place a detector, if we don't press the button then we don't get a click. If we place two detectors in a straight line, both in front of the hole, then the result will depend on the sort of detector. Some of them (we could call them type one detectors), if placed in front of another detector, don't prevent the second one from clicking; others (type two detectors) do.

Let us introduce further pieces of apparatus. Certain blocks, placed between source and detector, delay the clicks; others stop them altogether. Blocks labelled 'reflectors', when placed in front of the source, require us to put the detectors elsewhere to get the clicks; the position to which we have to move the detectors depends on the orientation as well as the position of the reflectors.

Of course, this is a terribly cumbersome way of speaking. Clearly, everything is proceeding *as if* something

- emerges from the hole every time the button is pressed;
- travels directly away from the hole at some constant speed;
- is detected but not stopped by the type one detectors;
- is detected and halted by the type two detectors;
- is slowed down by passage through some of the blocks, halted by others, deflected onto a new trajectory by yet others.

Does this justify our assuming that something *is* emitted from the hole and behaves in this way? Deutsch certainly thinks so (and I concur):

We must adopt a methodological rule that if something behaves as if it existed, by kicking back, then one regards that as evidence that it does exist. (Deutsch 1997: 88)

We might formalize this as something like

> **Deutsch's criterion**[16] **(first variant).** Something is real if it interacts in a complex, autonomous way with other real things.[17]

Deutsch's criterion has a status somewhere between a metaphysical principle and a definition of terms. In one sense it seems to capture something about the way the world has to be; in another, it simply seems to say what we mean by 'real'. To see this, consider how you would react to the following claim:

> The paper you are reading does not exist; the region of space you think it occupies is empty. But photons bounce off that region, or disappear *as if* they were reflected by the white space of the paper, or absorbed by the type. Thermal radiation appears and disappears at the boundary of the region, *as if* it were at room temperature. Other physical objects interact with the region *as if* there were a paper in it.

This is, I take it, an absurd thing to think; but *why* it is absurd is more complicated (and takes us back to the debates about realism sketched in section 8.8). However, it will not matter for my purposes whether we should regard the claim as being perfectly coherent but obviously false, or as being internally contradictory (in the sense that if the laws of nature are such as to cause the rest of reality to interact with the region in the way described, then it doesn't make sense to deny that there is something real there with those properties).

To clarify what 'complex and autonomous' means: by this phrase I mean that the entity in question is acted upon by the rest of reality, and acts back on it, in such a way that our explanation of how the rest of reality behaves cannot be given without reference to the entity. This suggests an alternative formulation:

> **Deutsch's criterion (second variant).** Something is real if it plays an explanatorily indispensible role in accounting for the behaviour of other real things.

---

[16] Deutsch in fact calls it 'Dr Johnson's criterion', following Samuel Johnson's famous refutation of idealism by kicking a stone; cf. Deutsch (1997: ch. 4).

[17] Note that this principle has a somewhat circular character, or better, a recursive one, in that it gives a criterion for reality in terms of other real things. This is to be expected: once we have a few things which we are already confident are real (e.g. our own existence), we can work out the rest. Presumably, some such starting point in an epistemic principle is unavoidable.

As a historical example, consider our changing attitude to the electromagnetic field. When only electrostatic forces were understood, the electric field could be seen as a convenient *façon de parler*: we could regard the force between charges as being due to a field, but there would be no need, as the field itself could be simply defined in terms of those charges. But once it was recognized that the field was a dynamical player in its own right—and, in particular, that radiation was just a manifestation of those dynamics—this changed. The field could no longer be seen as reducible to the charges: it carried degrees of freedom of its own, and interacted with matter in a complex autonomous way. As such it was seen as being real in the same sense as the charges themselves were real.

There have, of course, been various attempts (Wheeler and Feynman 1945; Hoyle and Narlikar 1995) to remove the field from our ontology, but they always involve a return to the days when we could define the field, and explain the phenomena associated with the field, in terms of the charge distribution. The authors of these proposals have appreciated that they need to modify the structure of electromagnetism in this way: it would not be good enough to carry on using electromagnetism in its standard form and then just declare that the field isn't real.

Deutsch's criterion is rather similar to Ian Hacking's famous slogan (formulated with particles in mind), 'If you can spray it, it's real' (Hacking 1983: 23).[18]

It also bears more than a passing similarity (especially in the second variant) to Dennett's criterion for the reality of higher-order ontology, discussed in Chapter 2. But there is an important difference: Dennett's criterion is a metaphysical principle concerned with the status of structures within a given already-known (or already-hypothesized) underlying theory; Deutsch's is an epistemic principle telling us when the phenomena entail the reality of some entity, and is neutral as to whether or not that entity is fundamental or emergent.

10.3.2 Interferometry

Returning to the behaviour of the neutron sources: if our best theory of their behaviour is essentially a theory of things behaving 'as if' some kind of particle was emitted by the sources and interacts with the detectors and

---

[18] This slogan characterises *entity realism*, a weaker form of scientific realism which accepts the reality of the objects of science but declines to accept the laws that govern them. For defences of this position, see Cartwright (1983) and Hacking (1983).

blocks, then Deutsch's criterion tells us that we should indeed believe that there are particles being admitted. Let's call these particles *c-neutrons* ('c' for 'classical', since their behaviour so far has been entirely classical).

Now we move out of the realm of classical physics and introduce a new device: a 'splitter', a box with two input and two output slots. We find that if we send a c-neutron through either input slot and put a detector in front of one of the output slots, it clicks exactly half the time. Furthermore, if between the output slot and the detector we put an obstacle course of absorbers, reflectors and so forth, we find that:

- if a c-neutron could have travelled from the output slot to the detector in some time t, then the detector will click—exactly half the time;
- if on the other hand a c-neutron could not get from slot to detector then the detector never clicks.

We can put one detector in front of each slit; if we do this we find that each one clicks half the time, but for each input neutron we get exactly one click.

Thus it is tempting to conclude that an input c-neutron which enters the splitter is emitted through one of the output slots, with a 50% probability of each.

If we concatenate two splitters as shown in Fig. 10.1, we find that the results bear out this idea: half of the time neither detector clicks, one

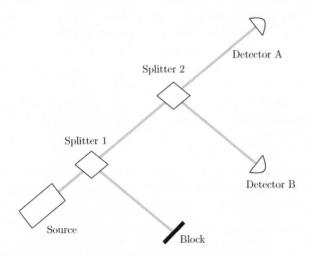

Figure 10.1. Concatenating two neutron-beam splitters

quarter of the time the first clicks, and the remaining quarter of the time the second clicks. Furthermore, when a detector does click it does so after exactly the period of time that a c-neutron would have needed to reach it from the source.

Our tentative conclusion is, of course, spoiled as soon as we consider the alternative setup shown in Fig. 10.2.[19] Here the two paths are set up so that a c-neutron following either path from the first splitter would take the same amount of time to reach the second splitter. Our theory would predict that each detector triggers exactly half the time, but in general we will not find this. Minuscule adjustments to either path length will alter the fraction of clicks at each detector, allowing us to reach the extreme situation where the first detector *always* clicks, but the second one *never* does; we call this 'interference'. Altering the path length enough to make an appreciable[20] difference in travel time destroys this strange effect, and

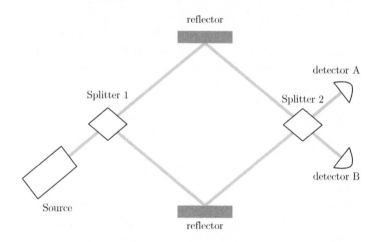

Figure 10.2. Interferometry with neutron-beam splitters

[19] The 'reflectors' shown in the diagram are somewhat unrealistic: in practical interferometry, bending neutron beams is not feasible and instead the beam is split again and one half is discarded. Since this would have complicated the discussion without adding anything of conceptual significance (for our purposes) I have abstracted from this detail. See Kaloyerou and Brown (1992), Brown (1996a), and Brown et al. (1995), however, for foundational discussions of neutron interferometry from a more experimentally careful perspective.

[20] i.e. appreciable compared to the tiny variations which we actually find in travel time, if we measure it *very* carefully. Obviously, what we require in terms of the quantum formalism is that appreciable adjustments in length are large compared to the spatial extent of the neutron wavepacket.

restores what we previously expected: a 50% chance for each detector to click. However we adjust the length, though, we can never arrange for both detectors to click simultaneously.

If we try putting blocks, reflectors etc. in one of the paths, we find that:

- anything placed in the path which would stop a c-neutron destroys the effect: if this is done, half the time no clicks will be observed and the other half of the time one or other detector clicks, with equal probability of each;
- if we put anything in the path which does not interact with c-neutrons at all, the effect is preserved;
- if we put anything in the path which would change the travel time for a c-neutron, then the changed path is used to predict whether we get the effect or not: if the path is changed so that the two path lengths are equal, then we get the effect; if they are unequal, we don't;
- if we put in the path any object which would deflect a c-neutron away from the second splitter, we get the same effect as for a blocked path: half the time no clicks at all, the other half one click, in one detector or the other at random.

This is as cumbersome a description as of the c-neutron's behaviour, and can be simplified in exactly the same way: everything occurs *as if*:

- one particle travels down *each* of the possible paths;
- each particle is affected by matter in precisely the same way as a c-neutron;
- if both particles arrive at the second splitter at the same time then they interact, and this interaction causes the results of the clicks to be dependent upon both particles, and hence on both paths;
- if the two particles arrive at the second splitter at different times then they cannot interact, and so the clicks observed are the sum of those we would observe if each path were opened separately.

Applying Deutsch's criterion again: just as with the c-neutrons, there is no way of explaining the phenomena without use of these particles, or of things that behave just like these particles, other than using them and then, at the end of the explanation, claiming that they do not exist. So their existence is as robust as that of the c-neutron, and warranted on the same grounds. I will call these particles s-neutrons[21] ('s' is for 'shadow', following Deutsch 1997).

Why not call them 'c-neutrons'? Because we know they're not c-neutrons: if two c-neutrons were emitted by the splitter then we could get two simultaneous clicks when one detector was placed at each output slot, and we never do.[22]

The difference, then, lies in the way the s-neutrons act back on matter: in particular, on the detectors. We now turn our attention to this. As was mentioned before, if a detector is placed in the path of an s-neutron it clicks precisely half of the time; if one detector is placed in the path of each s-neutron then exactly one of them clicks. We also find that the interference effect is destroyed whenever a detector is placed in the path of an s-neutron. This applies whether the detector is type two, and stops c-neutrons, or is only type one, and detects them.

We can now state the general principles by which s-neutrons and matter interact. In most cases, they follow directly from the properties of c-neutrons.

- Any object which would change state as a consequence of interacting with a c-neutron will destroy the interference effect.
- Any object which would deflect a c-neutron from its path, or delay it appreciably, will destroy the interference effect and halve the number of clicks.
- However, any object will itself change state in such a situation only half of the time.
- Furthermore, if two such objects are placed in the two s-neutron paths, exactly one of them will change state. That is: exactly one of the s-neutrons causes the detector we observe to change state; the other is acted on by the detector but does not act back.

So, half the time our s-neutron does not interact with the objects in its path, since they are totally unaffected by it. However, we know that it interacts with something, and furthermore that it interacts with something

---

[21] Mathematically, of course, they are actually (represented by) the two components of a standard two-peak wavepacket; the point is, we don't need that mathematical structure to establish their existence.

[22] Also—though we haven't discussed it—the interaction of two c-neutrons when they collide is different from that observed when the s-neutrons collide.

which affects c-neutrons in precisely the same way as the actual objects in the path, and is in precisely the same places. One final application of Deutsch's criterion tells us that, in those situations where the detector would have changed state upon being hit by a c-neutron but is instead hit by an s-neutron, after the collision there is what we might call an 's-detector', structurally isomorphic to the c-detector except for being undetectable to us. Or, equivalently: s-neutrons cause detectors to become s-detectors, and just as we only see the interactions of one of each pair of s-neutrons with detectors, so we only see the interactions of one of each pair of s-detectors with our eyes.

Couldn't the detector just act asymmetrically on the s-neutrons, so that half of the time they affect it but all of the time it affects them? But the detector is not a magic box. We know—we can show empirically—that it stops the s-neutrons *because* it detects c-neutrons. This is the only relevant property to determine whether or not the s-neutrons are stopped: detectors which fail to interact with c-neutrons, never interact with s-neutrons. Furthermore, on the 50% of occasions when the s-neutrons are detected, we could track the detection process, and see (in principle) that the dynamics inside the detector are precisely those expected in the detection of c-neutrons. We have a full dynamical story that could be told about the way the s-neutrons are detected.

The other 50% of s-neutrons react to their arrival at the detector *as if* a similar process were occurring, and so by Deutsch's criterion, we should accept the reality of that process. Or, to be more precise,[23] they react at least as if the first stages of such a process were occurring. I have not presented empirical evidence to show that the process must occur all the way to a macroscopic detection result; this matter will be discussed later.

Put another way, the detector stops all s-neutrons because it *can* detect c-neutrons. This might be called a modal property of the detector: it is possible for it to detect c-neutrons, therefore it does stop s-neutrons. But the merely possible cannot influence the actual: if a possible event has a complex and autonomous effect on actual events, then Deutsch's criterion tells us that the possible event is real.

We can make one further observation about these copies (called s-detectors, above). It is not sufficient for every object to split into copies separately. The copies must be correlated with one another, so that if one

---

[23] This is one point where I think Deutsch overreaches slightly.

element of a given set of copies can interact with a given s-neutron, then all elements of that set can. If not, then (for instance) we would never find—as we unfailingly *do* find—that if we place multiple (type one) detectors in an s-neutron's path, either all interact with it or none do. We can summarize our deductions as follows:

> When a neutron passes through a splitter, reality (at least in the vicinity of the neutron) is split into what we might call 'layers'. Each layer is initially identical to reality prior to the split, and the layers interact strongly within themselves but only very weakly (via interference effects) with one another.

A few observations:

1. We have not deduced anything about the details of the process of splitting. It may occur to the whole Universe at once, at the time that the c-neutron enters the splitter; it may occur piecemeal, as the s-neutrons reach each object (this would require the s-neutrons to carry the information about correlations with them). In the latter case, a given object would only be split if interaction with a c-neutron would change it. A reflector, for instance, reflects a c-neutron with negligible internal changes, and thus might not be split.
2. We do not know whether the layers persist. It is possible that in certain circumstances one or more layers simply disappears. This is an empirical claim which could in principle be tested, and which will be discussed later.
3. We do not know whether the layers are written into the low-level structure of the world (like fundamental fields) or whether they are higher-level elements of reality (like tigers). However, their existence as real things has been established independently of this fact.

Reconsidering neutron interferometry from the perspective of the quantum formalism: the c-neutron is a fairly localized wavepacket of the neutron wavefunction, following a fairly classical trajectory, and the various blocks, reflectors, etc. simply modify this classical trajectory. The splitter causes a wavepacket to split in two, each with equal amplitude, and the two packets are our two s-neutrons; they can interfere with one another, but only when in the same spatial region. Furthermore, when one of the wavepackets hits a detector it entangles the state of the detector with the state of the neutron. As this entanglement spreads out through decoherence, it gives us the layering mentioned in the summary above.

But we did not need the formalism to reach this conclusion: we just needed the experimental data. Of course, to actually *set up* the experiment would in practice require considerable knowledge of the underlying theory—neutron interferometry is not easy to do—but once it has been set up, the conclusions do not depend on that theory. There is an analogy to experimental violations of Bell's Inequality: in practice, experimental tests of the inequality require considerable knowledge of quantum physics to set up, but nonetheless the violation is independent of quantum mechanics, and is just an observed fact about the world. (See the introduction to Maudlin 2002 for a very clear discussion of this point.)

It is important to note that the thought experiment, or variants of it, do not tell us that *all* quantum superpositions can be interpreted in terms of parallelism. It is crucial to the argument (at both an empirical and a theoretical level) that we have an experimental setup something like a Michelson interferometer, with two distinct approximately classical s-neutrons following spatially separated paths. In situations where this does not occur, such as the two-slit experiment, there is no parallel-Universe description[24] to offer of the interference process: there is just a quantum system in a very nonclassical state.[25]

10.3.3 Digression: what about other interpretations?

How is all this compatible with the existence of multiple 'interpretations of quantum mechanics'? Well, some interpretations—those of an operationalist bent—just reject the idea that science tells us about reality, and so would not accept Deutsch's criterion for reality; for the reasons I sketch in sections 1.5 and 1.6, I ignore this possibility here.

Other 'interpretations', however, are really alternative physical *theories*, inasmuch as they accept the principles of scientific realism but apply them to a modified version of quantum mechanics, and—given the nonrelativistic nature of neutron interferometry—it is possible to see how the arguments above apply to them. (This is probably a good

---

[24] It is possible to use the *heuristic* of parallel worlds to think about the quantum state even in such cases, but the notion of 'world' that thus emerges is rather bloodless, and rather different in character from the way I use 'world' in this book; for more on this, see Wallace (2002b).

[25] Deutsch (1997) used the two-slit experiment—erroneously, I think—to argue for the existence of parallel worlds; however, in subsequent lectures he has shifted to something more like the Michelson setup I give here, and has noted (in conversation) that his point is better made in that context.

place to remind the reader (cf. section 1.7) that, notwithstanding some interesting recent developments, no satisfactory relativistic version of these modifications is currently accepted as reproducing the predictions of quantum field theory.)

Dynamical collapse theories, recall, postulate that at some scale beyond anything we have currently probed, the Schrödinger equation breaks down, so that all but one term in the superposition decays away. These theories accept the existence of 'many worlds' on the small scale, but change the dynamics to eliminate them on the large scale.

This is a coherent alternative to (Everettian) quantum mechanics, and is in principle testable: any successful test of the superposition principle is also a constraint on dynamical collapse theories. Since it is always healthy to explore the limits of our theories, it is to the credit of at least the more developed collapse theories that they give a concrete alternative set of predictions to test.

On the other hand, I find it hard to see a good motivation for finding such a theory to be at all likely. We don't have any empirical evidence at all for the disappearance of other branches—*ex hypothesi* we couldn't have, since their virtually nonexistent interactions with our branch are what makes the collapse theories so far empirically equivalent to no-collapse theories. An argument often given in favour of such interpretations over many-worlds interpretations—that the latter are ontologically extravagant—doesn't seem to apply well here: a theory which avoids the need to postulate extra ontology is one thing, a theory which causes it to wither away dynamically is another.[26]

Hidden variable theories are in a way complementary to collapse theories, in that the dynamics is left alone but the ontology is modified. The wavefunction is taken to evolve unitarily, but in addition some extra objects—the hidden variables—are added so as to define a determinate reality. In the most popular hidden-variable theory, the de Broglie–Bohm or pilot-wave theory,[27] the hidden variables are point particles (sometimes called 'corpuscles'[28]), which are intended to represent the positions of particles, and which follow determinate trajectories in a potential deter-

---

[26] One possible motivation to take such theories seriously, of course, is the belief (refuted, I hope, in Part II) that the problem of probability is fatal to the Everett interpretation.

[27] Bohm (1952); see Cushing et al. (1996) and references therein for further discussion.

[28] I borrow this terminology from Brown (1996b).

mined by the wavefunction. Probability enters the theory via ignorance: we do not know the precise positions of the corpuscles so we have to assign a probability distribution.

The problem with this approach can be seen from reconsidering our interferometry experiment. In that experiment the c-neutron wavepacket contains a corpuscle, and when it splits into two s-neutrons the corpuscle moves with one of them. When the s-neutrons recombine, they create a complicated potential which causes the corpuscle to follow one or other of the paths to the detectors.

What work is the corpuscle doing here? The wavefunction continues to play an essential role in the dynamics, but the corpuscle is, it seems, just along for the ride. The theory predicts that the wavepackets will enjoy a robust existence, and in any case our empirical argument for this conclusion is still in force. But the corpuscle plays no explanatory role, and indeed has no dynamical effect whatsoever—so why include it?

The usual answer to this question is that it is the corpuscle that is detected by the detectors, so its role in the explanation is to tell us what is finally observed. Furthermore, in doing so it explains in a straight-forward way (it is claimed) how probabilistic concepts enter the theory. This would be fine if neutron detectors could reasonably be postulated to detect corpuscles directly. However, the detectors follow the same physical laws as does anything else (according to the pilot-wave theory itself, if it isn't obvious enough a priori); and the theory predicts that the detector wavefunction will entangle with the neutron wavefunction to create a superposition of different detection results. Each component of the detector wavefunction will perform perfectly well as a detector of neutron wavepackets within its own layer, but only one of them will be accompanied by corpuscles. Again there doesn't seem to be any particular reason to concentrate our attention on the detector with corpuscles: according to Deutsch's criterion (applied either to the formalism of the pilot-wave theory, or—as was done earlier—directly to the detectors in the interferometry experiment), all the components of the detector wavepacket count as real. If anything, it is the corpuscles whose existence looks suspect, since they play no dynamical role in the evolution of the wavefunction.[29]

---

[29] Brown et al. (1995) observe, along similar lines, that attributing properties such as mass and charge solely or even partially to the corpuscle is problematic given that, dynamically speaking, those properties are more relevantly associated with the wave function.

This chain of reasoning can be applied almost indefinitely, with meta-detectors observing the detectors and so forth. As reality splits into layers, Deutsch's criterion continues to declare each layer equally real, with the corpuscles serving only to point at one particular layer.

There is really only one place to stop this regress: as has been observed by Brown (1996b), the only way for the corpuscles to play an explanatory role is for them to be the entities responsible for conscious thought.[30] Then although all branches would be equally real—and contain physical entities behaving exactly the same as people—only one of them would contain *conscious* observers. Of course, this commits us to a radical form of substance dualism—of the philosophical doctrine that the mind is made of a fundamentally different substance from the body, and to the abandonment of any hope of understanding the mind in physical terms; this, I think, renders it beyond the pale.

But absent this move, the corpuscle's role is minimal indeed: it is in danger of being relegated to the role of a mere epiphenomenal 'pointer', irrelevantly picking out one of the many branches defined by decoherence, while the real story—dynamically and ontologically—is being told by the unfolding evolution of those branches. The 'empty waves' which the corpuscle do not 'point at' are none the worse for its absence: they still contain cells, dust motes, cats, people, wars, and the like.[31] The point has been made acerbically by Deutsch (1996: 225): 'pilot-wave theories are parallel-universe theories in a state of chronic denial.'

10.3.4 Quantum computation

The case of quantum computation has the advantage of showing in more detail the autonomous behaviour of quantum layers which justify calling them 'worlds', and the sheer number of them which can exist. Its disadvantage is that—unlike the case of neutron interferometry—it has not yet been carried out in the laboratory.

Schematically, a certain subset of quantum algorithms—including Shor's famous (1994) algorithm for polynomial-time factorization—have the following form:

---

[30] I should emphasize that Brown himself does not advocate this approach, but simply (and correctly, in my view) notes its necessity for advocates of the pilot-wave theory.

[31] For a more detailed version of this argument, see Brown and Wallace (2005); for responses, see Valentini (2010) (though see also Brown's 2010 reply) and Lewis (2007a). Lewis's response is essentially to reject the account of macro-ontology advocated here and in Ch. 2.

1. The computer is assumed to consist of subsystems ('registers') each of which have some basis of 'computational' states to represent natural numbers; in this basis the state can be readily measured. We denote elements of this basis by $|n\rangle$, where n ranges from 0 to $N-1$ ($N-1$ is taken to be the maximum number representable by the subsystem). We begin with the computer in the state $|0\rangle \otimes |0\rangle$, where the two kets $|0\rangle$ each belong to a different subregister, and then rotate the first component of this product into an equally weighted superposition of all N computational-basis states:

$$|0\rangle \otimes |0\rangle \longrightarrow \sum_{n=0}^{N-1} |n\rangle \otimes |0\rangle.$$

(Here and subsequently I omit normalization factors.)

2. We now perform some classical computational task on each element in the superposition; without loss of generality we can regard the task as giving some number as output. The result could be different for each n, so we denote it by $f(n)$ and store it in the second register:

$$\sum_{n=0}^{N-1} |n\rangle \otimes |0\rangle \longrightarrow \sum_{n=0}^{N-1} |n\rangle \otimes |f(n)\rangle.$$

This task can be performed provided that we are able to perform the operation $|n\rangle \otimes |0\rangle \longrightarrow |n\rangle \otimes |f(n)\rangle$ for all n.

3. The next step is the most subtle: we need to perform some further quantum operation on the combined state which will extract some global property of the function $f(n)$. Whatever the specifics of the transformation (which will not concern us), it will have to leave the computer in a superposition of relatively few computational basis states.

4. In the last step, we measure the computer in the computational basis. If the computation is to be useful, we must have a pretty good chance of obtaining the information we need on measurement, hence the requirement of the previous step that most of the basis states will have destructively interfered.

Note that stages 1 and 3 are necessarily quantum, in the sense that they do not preserve the computational basis. Stage 2, on the other hand, is a purely classical computation (more accurately, it is a superposition of N

classical computations) in the sense that it takes computational basis states to other computational basis states. (Stage 4, being a quantum *measurement*, is difficult to classify as either.)

The Everettian explanation for the power of such algorithms is that they share the computational task between an enormous number of universes. However, as Steane (2003) has pointed out, this claim cannot be established just by notation: many manifestly one-Universe processes may formally be written using a superposition notation. To establish the sense in which the Everettian claim is true, we need to look at the process more closely and to invoke Deutsch's criterion.

In practice, building a quantum computer is enormously difficult owing to decoherence. However, for the purpose of our thought experiment we assume decoherence to be absent, in the following sense: we assume that we can build *classical* computers such that the computer's internal state after the calculation (excluding the outputted bits), and the state of the environment, is the same as before. In this case, we can use such a classical computer to do the second stage of the algorithm, provided that computer is also reversible. Specifically, we suppose that the computer has some input slot for a number, and then performs a classical (and reversible) computation on that number and outputs both it and the result of the computation, writing the result on a previously blank bit:

$$(n, 0) \longrightarrow (n, f(n)).$$

Then the linearity of quantum mechanics[32] entails that inputting the quantum state $\sum |n\rangle \otimes |0\rangle$ will lead to $\sum |n\rangle \otimes |f(n)\rangle$.

In the spirit of section 10.3.2, we can now consider this setup without further reference to the structure of quantum theory. We can imagine making various changes to the innards of the computer, and ask what effect these have on the computation. It turns out that any change which would impede the computer's ability to compute $f(n)$ for a significant[33] number of values of n, or which would record the internal state of the

---

[32] In *this* sense, of course, the computer is *not* classical, because it makes sense to give it a superposition as input. That much follows from the assumption that quantum theory is universally applicable.

[33] Quantum algorithms would in practice be tolerant to a small number of miscalculations. This would not appear to be of foundational significance: the same could be true of a classical algorithm which determined some global property of a function by computing all values of it.

computer, will cause the algorithm to fail, and that no other changes will do so.

We have very wide latitude in the changes here: we can delay the electrons[34] in the computer's circuits, send them on diversions and bring them back again, even arrange, for some values of n, that the calculation of f is performed using a different sub-algorithm or sent to a completely different computer. Furthermore, whenever we look into the computer we find that it is performing some computation of $f(n)$ for a specific n.

The situation is just interferometry on a grand scale: everything proceeds *as if* the computer is performing all N classical calculations and then combining them together; hence by Deutsch's criterion all the calculations are to be regarded as real—as are all the *s-bits* (as we might call them) which are required to perform the calculations, and which must have been created from the original bit.

All of this can be seen a little more directly if we allow ourselves access to the quantum formalism. The wavefunction splits into layers, and each layer carries out its own autonomous computation, obeying some effectively classical dynamics and evolving independently from the other layers. Once we start to perform the final interference experiment (stage 3), the layers cease to evolve either classically or independently, and are no longer conceptually useful.

(Note that all three interesting structural properties of quantum theory have a role in this process: without superposition we would not have the s-bits which are inputted into the classical computer; without entanglement we would not have had the development of layers; without interference we would not have had the interaction of layers which gives us evidence for their existence—and more pragmatically would not have been able to extract useful information from the computation.)

It is the independent, autonomous evolution of the computation which grants them reality. If we could only perform some highly restricted set of operations on the layers, then their status as real things would be questionable.

---

[34] To stress again: this is a thought experiment. Ordinary silicon-chip computers are overwhelmingly too prone to decoherence—not least due to the computers' own irreversibility—for this story to be practically possible. But if real quantum computers are ever built, the only differences will be in the technological specifics.

I now return briefly to non-Everettian approaches, to see what explanation they give for the efficacy of quantum computation. There are basically two options. Those interpretations which deny reality to the wavefunction have no explanation at all to offer: they predict the results, of course, but give no account as to why. For instance, in some versions of the de Broglie–Bohm pilot-wave theory, the wavefunction is treated as not representing anything physical (it represents, at most, some physical *law*), so that the corpuscles (the possessors of the hidden variables) constitute the entirety of the system under study. In these versions of the pilot-wave theory, the corpuscles enter the classical computer, perform, at random, exactly one evaluation of $f(n)$ and then leap, as if by magic, into the output state which gives the result of the quantum algorithm. (Note, incidentally, that interpreting the wavefunction as describing an ensemble of systems is obviously not viable here: in an ideal quantum algorithm, the correct result is outputted *every time*.)

In those interpretations which accept the wavefunction as real, on the other hand, the explanation is exactly the same as for Everettian quantum theory. In collapse theories, provided that the collapse doesn't destroy the computation, the wavefunction behaves exactly as described above, and of course has exactly the same claim to real status. The same is true in those versions of pilot-wave interpretations which do accept the wavefunction as real, with the irrelevant addition of corpuscles which randomly trace one of the computations but contribute nothing to the final result.

In a sense, of course, this discussion of quantum computation tells us nothing philosophical that we didn't already know from neutron interferometry: the really crucial step is from one layer of reality to more than one, and the further considerations here are just quantitative changes. Nonetheless there is something rather striking about the idea of empirical proof that such a huge number of different realities must exist. Deutsch puts it this way:

When Shor's algorithm has factorized a number, using 10^{500} or so times the computational resources that can be seen to be present, where was the number factorized? There are only about 10^{80} atoms in the entire visible universe, an utterly minuscule number compared with 10^{500}. So if the visible universe were the extent of physical reality, physical reality would not even remotely contain the resources required to factorize such a large number. Who did factorize it, then? How, and where, was the computation performed? (Deutsch 1997: 217)

Now, if by this Deutsch means that the very fact of the calculation entails multiple universes, he has overstated the case. It is unproven that there are no classical algorithms for efficient factorization. And it is not *logically* impossible that the calculation just happens 'by magic', as it were, without any detailed account at all: there is no logical contradication, although it goes against everything we have learned of science.[35] in the suppostion that the laws of physics contain primitive factorization-implementing processes.

But to object thus is to miss the point: which is not that there *could be* no other explanation for the factorization, but that we actually have a very good, in principle thoroughly testable explanation. Namely, it involves simple, well-understood algorithms operating in a massively parallel way, within a single computer. It presumes that each computation happens independently, the empirical prediction is that everything will happen *as if* each computation is occurring independently, and there is no way of explaining the *actual* computational process taking place which does not assume the computations are happening independently. By Deutsch's criterion, then, there is no way of so explaining the algorithm which does not accept the reality of all of the independent computations.[36] At least within the quantum computer, there would be many worlds.

Finally, we can link this to the discussion in Chapter 8. There, recall, I pointed out that *generically* branching propagates due to the ubiquity of decoherence. In a quantum computer, the name of the game is to keep the decoherence under control and the branching localized. If Fig. 8.1 depicted decoherence-induced branching, the sort of branching required for (this kind of) quantum computation looks, schematically, more like Fig. 10.3.

10.3.5 The quantum Zeno 'paradox' and the bomb problem

The quantum Zeno *effect* is not really paradoxical: it just consists of the observation that if a system, whose state would otherwise evolve from out

---

[35] Cf. Deutsch's own discussion in Deutsch (1997), or Dennett (1995: Ch. 3).

[36] Hidden variable theories are a nice illustration of this fact. In (say) a Bohmian account of Shor's algorithm, no less than an Everettian account, it is essential to the computation that each classical process is performed by some component of the wavefunction. If any significant fraction of those computations are in any way prevented or disrupted, the algorithm will fail, and this is so even if the single classical computation performed by the actual hidden variables is not in any way prevented.

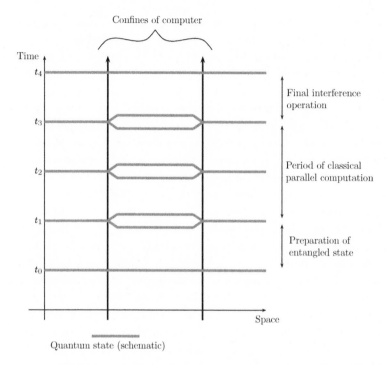

Figure 10.3. Schematic view of parallel quantum computation

of some subspace $\mathcal{S}$ of its Hilbert space $\mathcal{H}$, is subject to some measurement process which (a) determines whether or not it is still in that subspace and (b) occurs on timescales fast compared to the evolution, then the system's evolution out of the subspace will be dramatically suppressed. If, for instance, we model the evolution of the system over some time t by

$$|\psi(t)\rangle = \cos(\omega t)|0\rangle + \sin(\omega t)|1\rangle, \tag{10.1}$$

where $|0\rangle \in \mathcal{S}$ and $|1\rangle$ is orthogonal to $\mathcal{S}$, and if at intervals of $1/N\omega$ we make a projective measurement of the system to determine whether or not its state is in $\mathcal{S}$, then if $N \gg 1$, its density operator after time t will be approximately

$$\left(1 - \frac{(\omega t)^2}{N}\right)|0\rangle\langle 0| + \frac{(\omega t)^2}{N}|1\rangle\langle 1|. \tag{10.2}$$

As N tends to infinity, the evolution of the system is entirely halted: as the slogan goes, 'a watched pot never boils'. The effect is ubiquitous, insensitive to most of the details of this particular example, and has been empirically verified; see Home and Whitaker (1997) for a very clear review.

Where things appear to get paradoxical (and here I again follow Home and Whitaker) is when we use a so-called 'negative-result' measurement: a detector which only fires—i.e. only interacts with the system—if the system's state is found by the measurement not to be in $\mathcal{S}$. This requirement in no way undermines the argument for the Zeno effect, so a sufficiently sensitive negative-result detector will (with arbitrarily high probability) entirely prevent the transition occurring—*even though* if that transition does not occur, the detector remains entirely inert.

Essentially the same phenomenon occurs in the so-called 'Bomb problem' (Elitzur and Vaidman 1993; see Penrose 1994b: 5.2, 5.9 for a clear discussion).[37] Here, we are to imagine that we have some supply of bombs whose triggers—which are taken to be perfect reflectors of light—are so sensitive that *a single photon* bouncing off them will set them off. The catch is, some of the bombs are duds. The only difference between dud bombs and real bombs is that the latter explode if a photon hits their trigger—and the puzzle is to separate out the dud bombs from the real bombs without destroying the real bombs.

This sounds close to a formal impossibility; but it can be done. An elementary method starts with an Michelson interferometer like the one in Fig. 10.2, aligned so that all photons arrive at detector A, and then replaces one of the mirrors with the bomb trigger. If the interferometer is aligned so that photons are sent through one at a time, if the bomb is real then we will get a detection at A one quarter of the time, a detection at B one quarter of the time, and an explosion half of the time; if the bomb is a dud, we will get a detection at A with certainty. Any detection at B therefore indicates a real bomb without that bomb blowing up. By sending photons through until we get either an explosion, a detection at B, or a few hundred consecutive detections at A (making it statistically almost certain that the bomb is dud), we can distinguish real from dud bombs with 1/3 probability—and in fact (Kwiat et al. 1995) we can improve this to

---

[37] The general term for processes of this kind is 'interaction-free measurement'; the term is due to Elitzur and Vaidman.

arbitrarily high (though never perfect) accuracy (essentially by exploiting the quantum Zeno effect, in fact.)

In both these cases—unless one accepts the many-worlds concept —whether or not something in the actual world (a transition occuring; a detection happening) actually occurs depends not on some object's actual behaviour, but on how it *would* have behaved if things had been otherwise. If the possible cannot influence the actual, this entails that some other version of the object (another bomb, another detector)—or at least something that behaves just like the object—must exist, elsewhere in reality. According to the Everett interpretation, this is exactly what happens.

The interpretive situation in the quantum Zeno 'paradox' is just as we we saw in sections 10.3.2–10.3.4. In summary: the firing of the detector, or the explosion of the bomb, in another branch destroys the interference effect between that branch and this one, and this has empirically detectable consequences. According to dynamical collapse theories, exactly the same is true—and the rapid, dynamical withering away of one of the branches is of questionable explanatory relevance. According to hidden-variable theories which take the reality of the wavefunction seriously, exactly the same is true—and the fact that one but not the other branch has some dynamically inert hidden variables associated to it is of questionable explanatory relevance. According to 'interpretations' of quantum mechanics which do not take the reality of the wavefunction seriously, the whole thing is an inexplicable miracle.

10.4 The status of mixed states

Traditionally in the foundations of quantum mechanics, a sharp distinction is drawn between two kinds of mixed state (a.k.a. density operator). Timpson and Brown (2005: 1) state the distinction as follows:

[I]t has proved essential to distinguish between density operators which can be given an 'ignorance' interpretation; and those which cannot. In the former case, the system whose state the density operator represents is in some definite quantum state from a specified set—say a pure state $|\psi_i\rangle$—but we don't know which. Our ignorance of the actual state can be represented by a probability distribution $\{p_i\}$ over the different (not necessarily orthogonal) possibilities $\{|\psi_i\rangle\}$, and the density operator may be written as

$$\rho = \sum_j p_j |\psi_j\rangle\langle\psi_j|.$$

...Ensembles that can be given an ignorance interpretation are called *proper* mixtures; the terminology is due to [d'Espagnat (1976)]. ...By contrast, density operators which cannot be given an ignorance interpretation are said to represent *improper* mixtures. Here the density operator arises from tracing out irrelevant, or unavailable, degrees of freedom. In this case, individual systems cannot be thought to be in some definite state of which we are ignorant; rather, the (reduced) density operator is the only description that they can have.

Timpson and Brown are here presenting the orthodox position, without endorsing it. Indeed, as they go on to point out—and as I, largely following their example, demonstrate below—in Everettian quantum mechanics, the two types of density operator are elided considerably.

If the density operator really does come in these two very different variants, it is rather extraordinary in non-Everettian versions of quantum theory that

(i) it is impossible in principle to distinguish a proper from an improper mixture, if they are represented by the same density operator;

(ii) it is likewise impossible to distinguish two distinct proper mixtures which are represented by the same density operator. (In the case of spatially separated, entangled systems, it is this extraordinary fact which prevents the faster-than-light interactions entailed by Bell's Theorem from allowing macrolevel violations of Lorentz covariance.)

To put the oddness of this in another way: if I say that a system is in a proper mixture, then on the orthodox picture I am making an epistemic statement about my limited knowledge of the system. If I say that it is in an improper mixture, I am saying something about the system itself.

In the Everett interpretation, though, the 'extraordinary' fact becomes less remarkable, as we can see by considering how an Everettian agent might prepare each type of mixture.

Suppose, then, that the agent prepares a pair of spin-half particles in the singlet state

$$\frac{1}{\sqrt{2}}(|\uparrow\rangle \otimes |\downarrow\rangle - |\downarrow\rangle \otimes |\uparrow\rangle) \tag{10.3}$$

and then removes the first particle to some distant location. The second particle is now in the mixed state

$$\frac{1}{2}(|\uparrow\rangle\langle\uparrow| + |\downarrow\rangle\langle\downarrow|) \tag{10.4}$$

and this is an improper mixture, since it results from tracing over the degrees of freedom of the other particle.

Now, on the standard account, as described above, when a system is in an improper mixture, then 'individual systems cannot be thought to be in some definite state of which we are ignorant'. In Everettian quantum mechanics,[38] this is a little misleading: the particle is indeed in a definite state, the mixed state given by equation (10.4). But since it is entangled with the second particle, even knowing the definite states of both particles separately does not suffice to determine the joint state of the two particles.

Suppose the agent now measures the spin of the particle in some nondisturbing way, but does not look at the result. Writing $|\text{'ready'}\rangle$, $|\text{'up'}\rangle$ and $|\text{'down'}\rangle$ for states of observer-plus-surroundings corresponding repectively to no measurement, to a measurement which gives spin-up, and to a measurement which gives spin-down, the particle-plus-agent-plus-surroundings system undergoes an interaction something like

$$\frac{1}{2}(|\uparrow\rangle\langle\uparrow| + |\downarrow\rangle\langle\downarrow|) \otimes |\text{'ready'}\rangle\langle\text{'ready'}| \longrightarrow$$

$$\frac{1}{2}(|\uparrow\rangle\langle\uparrow| \otimes |\text{'up'}\rangle\langle\text{'up'}| + |\downarrow\rangle\langle\downarrow| \otimes |\text{'down'}\rangle\langle\text{'down'}|). \tag{10.5}$$

In this evolution, the state of the *particle* doesn't change at all: it remains the state given by (10.4). What changes is that the particle becomes entangled with the agent and with its surroundings: before, but not after, the measurement, the joint state of particle-plus-agent-plus-surroundings was just the product of the individual states of the particle and the agent-plus-surroundings.

So, from a God's-eye viewpoint, there really is no difference between the two kinds of mixtures, at least not locally: the difference resides in the form of their entanglement with their surroundings. One way to see this is to relativize to a branch: prior to the measurement, the state of the particle *relative to the agent's branch* is given by (10.4), whereas after the measurement, the relative state is either $|\uparrow\rangle\langle\uparrow|$ or $|\downarrow\rangle\langle\downarrow|$ but the agent

[38] Here and after, where relevant I assume the 'spacetime state realism' of Ch. 8 when discussing the quantum state.

doesn't know which. So *relative to the agent* there is a difference between the two mixtures. Nonetheless, an agent who says 'That system is in proper mixture ρ' can still be understood as saying something about the system itself (albeit something about a 'system' as spread across multiple branches), and not just something about his epistemic state. It might seem, then, that all Everettian mixtures are improper.

There is an important assumption in all this, though, which is that the probabilities in proper mixtures have a quantum-mechanical origin. If it were possible to prepare a system in state $|\uparrow\rangle$ with probability p and $|\downarrow\rangle$ with probability $1 - p$ *where these probabilities are not to be understood quantum-mechanically*, the account above would fail. It is interesting to ask, then, if this is possible.

One way in which it *could* be possible is if there were another source of objective chances in physics, other than those in quantum mechanics. But there seems to be no evidence for such objective chances, and—as I argued in Part II—some reason to think that objective chance does not even make sense outside the quantum-mechanical context.

Similarly, classical random processes like the spinning of roulette wheels and the rolling of dice ultimately rest on certain assumptions about phase-space probability distributions, which in turn—as we saw in Chapter 9 and again in section 10.1—are ultimately quantum-mechanical in nature, being caused by quantum branching in the evolution of the Universe away from its Simple initial state.

However, there are two remaining sources of non-quantum probabilities which should be taken seriously:

Pseudo-random processes. Though it is not philosophically clear just how to make sense of them, the fact is that various fully deterministic processes—notably, those used in the 'random' number generators in computers, which in reality are deterministic but very complex—do seem to behave for many purposes as if they really were random. At least prima facie, if I am told that my assistant has used a pseudo-random source of ones and zeroes to decide whether to prepare a system in spin-up or spin-down, it seems to make sense to represent the system (and my ignorance about it) using a proper mixture of spin-up and spin-down.

Personal probability. As we saw in Chapters 4 and 6, many philosophers and decision theorists take it as coherent to model agents' actions by assuming that they assign probabilities not only to objectively random statements, but also to claims like 'Bloggs will win the Presidential election' or 'There is no extra-terrestrial intelligent life'. The claim that either of these has (say) 80% likelihood does not seem to be a claim about objective probabilities, but rather some statement of confidence, quantified by the claimant's willingness to accept certain bets.

Not everyone accepts that it does make sense to assign probabilities to claims like this; if it does, though, it would be possible to model those agents' beliefs via mixed states which would, again, have to be treated as proper mixtures.

In both of these cases (both of which, to be sure, are controversial), an agent's use of a density operator does seem to be epistemic: there is no reason to suppose that the claim that Bloggs is 80% likely to be re-elected entails that in branches of relative weight 0.8 he is re-elected and in branches of relative weight 0.2 he is not. In this case, it really does seem as if we have two fundamentally different concepts described by the same mathematical entity. In general, though, and in particular in those cases which actually occur in physics, proper and improper mixtures alike are physically real things, and there is no intrinsic difference between them.

10.5 Must the Universe as a whole be in a pure state?

As I argued in Chapter 8, there is nothing contradictory about the notion that the quantum state of a quantum system such as a spatial region really does fully describe the state of that region, even if the quantum state is mixed. Assuming the Universe as a whole is in a pure state, however, mixed states of subsystems always tell us that the subsystems are entangled with one another, so that there are some nonlocal properties not specified by the individual subsystem states.

But should we assume that the Universe as a whole is in a pure state? This is the question I wish to address in this section.

One 'argument' for purity of the Universe can be dismissed right away. Sometimes, mixed states are taken to be in some sense incomplete descriptions of a physical system—incomplete because they must be entangled with other systems, so that some information is contained in the entanglement rather than in the states of the system and its environment separately.

If so (so might the argument go) then the Universe as whole, which by definition cannot be incomplete, must be describable by a pure state.

This argument, though, is question-begging. The only reason why we assume that a mixed state indicates entanglement, and thus incompleteness, is because we assume that the state of the Universe as a whole is pure. But if this is not the case, a state can be mixed without being entangled with anything.

Indeed, once we have accepted that subsystems can be in mixed states, it can seem a little odd and abstract to require purity of the Universe as a whole. This oddness is, I think, sharpened once we remember that the quantum vacuum appears to be entangled on all lengthscales, so that *any* subsystem of the Universe is entangled with its surroundings. This being the case, even if the Universe as a whole is in a pure state, the *observable* Universe's state is mixed.[39]

A rather better argument against taking the Universal state to be mixed is given by Deutsch (1991):

Suppose that the contents of a chronology-respecting[40] spacetime are described by a density operator $\rho(t)$. The evolution of $\rho(t)$ will be unitary

$$\rho(t) = \widehat{U}(t)\rho(0)\widehat{U}^{\dagger}(t).$$

This implies that $\rho(t)$ has the same spectrum, say $\{p_i\}$, as $\rho(0)$, and has eigenstates $\{|\psi_i(t)\rangle\}$ that are related to those of $\rho(0)$ by

$$|\psi_i(t)\rangle = \widehat{U}(t)|\psi_i(0)\rangle.$$

$\rho(t)$ therefore describes a collection of 'universes' (each one consisting of multiple universes under the Everett interpretation), one for each nonzero p_i. Each evolves precisely as if the others were absent and it had a pure state $|\psi_i(t)\rangle\langle\psi_i(t)|$. This is quite unlike the Everett multivaluedness caused by the linear superposition of components of a state vector, which is detectable through the phenomenon of interference. Thus the cosmology described by $\rho(t)$ contains a multiplicity of mutually disconnected and un-observable entities and is vulnerable to the 'Occam's razor' argument that is sometimes erroneously levelled against the Everett interpretation.

---

[39] A certain amount of caution is needed here, for the observable universe is restricted both by spatial distance and by the fact that we can only observe our own branch. But in practice, it seems clear that the branch-relative state of a quantum field possesses entanglement on all lengthscales just as does its absolute state.

[40] Deutsch adds this qualifier due to his view—to be examined in section 10.6—that spacetimes which permit time travel do require mixed Universal states.

This is powerful, but I think not completely convincing. Whatever our route to knowledge of the initial state of the Universe, it is bound to be pretty theoretical, given how little of the Universe we can actually observe (both because of its spatial size and because of quantum-mechanical branching). It is not difficult to imagine that such theoretical considerations lead us to a mixed rather than a pure state.

However, suppose that these 'theoretical considerations' led us to suppose that the initial state was the mixed state

$$\rho = \sum_i p_i |\psi_i\rangle\langle\psi_i|. \tag{10.6}$$

Arguably, it would always be open to us to say that the initial state was actually probabilistically determined: that it was certainly one of the $|\psi_i\rangle$, and that p_i gives the probability of the initial state being $|\psi_i\rangle$. If this strategy is available, then Deutsch's appeal to Ockham's razor seems to have force again.

I see two problems with the strategy, however. First, recall that any density operator admits a multiplicity of decompositions into sums of pure states. Knowing that the initial density operator is ρ is insufficient to tell us the actual set of pure states available to the early Universe (though of course considerations of simplicity might well point us to the spectral (orthogonal) decomposition, which is unique for generic ρ[41]).

Secondly, and more interestingly, we have already seen (in Part II) that probability is a philosophically shaky concept, and that a major *philosophical* advantage of the Everett interpretation is that it allows us to make sense of probability. If probability cannot be understood except in Everettian terms, then the strategy of treating ρ as merely probabilistic is just a redescription of the strategy of taking ρ as the physical state.

Recall, however, that the theorems of Part II all assumed a pure universal state. I do not know whether they can be generalized to hold in situations where the dynamics remain unitary but the universal state is allowed to be mixed. Even if they can, we would be accepting a major expansion of our ontology, from admitting only pure states, to admitting also mixed states, *entirely* on the grounds that this expansion solves a purely philosophical problem. This might or might not be justified, but it is very different

---

[41] That is to say: it is unique except for density operators some of whose eigenvalues are not distinct, and the set of such density operators is in various senses 'small' (contains no open sets, has measure zero in the standard measure on the space of density operators, etc.).

from the situation with the Everett interpretation itself, whose ontological commitments are forced on us for reasons which have nothing to do with problems in the philosophy of probability and for which solving those problems is a bonus.

A very different reason for taking mixed universal states seriously comes from the physics of black holes. Suppose, for instance, that I prepare a pair of spin-half particles in a singlet state and then toss one of them into a black hole. The other particle is now in a mixed state, but the Universe as a hole remains pure.

However, we have good theoretical reasons to suppose that the black hole will evaporate by Hawking radiation (Hawking 1976), which is thermal and carries no information about the interior of the black hole. When the hole has evaporated entirely, the particle in the black hole will be *gone* and will have had no opportunity to transfer its half of the entanglement to another system. The particle outside the hole remains in its mixed state; but there is no longer a pure state such that tracing over part of it returns the mixed state. Accordingly, the Universe as a whole now seems to be in a mixed state. (This is the famous *information-loss paradox* of quantum gravity; see Wald 1994 and references therein for an introduction.)

If this last view really is the correct account of black hole decay, mixed universal states appear compulsory, for the probabilistic account of mixed states is not available: that is, we cannot suppose that the pure-to-mixed-state transition that occurs in black hole evaporation is just an indeterministic pure-to-pure-state transition. To see this, suppose that we take the other half of the singlet (whose mixed state is as in (10.6), say) many light years from the black hole. At the point of the black hole's decay, then—according to the probabilistic reading of density operators—the particle instantaneously and randomly collapses into some pure state in the set $\{|\psi_i\rangle\}$, with probability p_i of entering state $|\psi_i\rangle$.

Now, for one thing, the choice of the set of pure states is not fixed by ρ. But this is not a new problem: we saw it when considering mixed initial states, and it does not appear fatal. What is fatal, I think, is that when the particle is far from the black hole, relativity tells us that there is no fact of the matter as to when the transition occurs. On pain of violating Lorentz covariance,[42] then, if pure-to-mixed states occur in black hole evaporation then they must be understood as deterministic transitions to

---

[42] Of course, the violation will not be empirically detectable. For a realist, though, this makes things worse, not better: an *empirically detectable* violation of Lorentz covariance is

a genuinely mixed universal state, not as shorthand for some underlying stochastic law.

Of course, none of this means anything if black hole evaporation is genuinely unitary. This remains intensely controversial, with the substantial majority position (at least in string theory) being that it is (see Susskind 2008 for an accessible account of the controversy by one of the participants). But the controversy, I think, is really (at least for our purposes) about whether the quantum theory of gravity allows for the possibility of quantum systems *ceasing to exist* rather than just changing state.[43] If it does allow this possibility, pure-to-mixed transitions, and so the physical possibility of universal mixed states, seem unavoidable. In the next section, when I consider the quantum mechanics of time travel, we will see that the same phenomenon occurs there too.

10.6 The quantum mechanics of time travel

Other than their shared place in science fiction, what do time travel and Everettian quantum theory have in common that justifies a discussion of the former in a book about the latter? The answer goes back to yet another argument of David Deutsch (1991), who has argued that in regions of spacetime which permit time travel, classical mechanics fails even at the macroscopic level and direct empirical evidence for parallel Universes can be acquired. My purpose in this section is to present and critique this argument, and to comment on some of its consequences.

Any consideration of quantum mechanics in the vicinity of closed timelike curves is necessarily extremely speculative, in the absence of a satisfactory quantum theory of gravity. My conclusions in this section of the book, therefore, should be taken as, at most, tentative.

10.6.1 A computational model of time travel

Following Deutsch, I abstract away nearly all of the detailed physics of the time-travel situation, and consider cases where a finite number of (classical or quantum) bits—i.e. physical systems with two distinguishable internal

---

an exciting experimental prospect (albeit a remote one when we're considering black hole evaporation), but a conspiratorially hidden violation is just an embarassment.

[43] Recall in this context that while quantum field theory on flat spacetime does permit the creation and annihilation of *particles*, this does not correspond physically to the creation or annihilation of a *system* in the quantum-mechanical sense, but only to a transition from one state in the system to another with different particle numbers.

states—travel along fixed, timelike classical trajectories. I model the time travel itself by requiring the background spacetime to contain closed timelike curves: paths which could be followed by a bit moving slower than light but which nonetheless intersect themselves. Probably the simplest such spacetime is the so-called 'Deutsch–Politzer' spacetime (Deutsch 1991; Politzer 1992) shown in Fig. 10.4, in which (in this example) four bits approach the causality-violating region, two of them travel backwards through time, and then all four leave again. This spacetime essentially simulates an instantaneous, though spatially extended, wormhole.

A slightly more realistic spacetime is shown in Fig. 10.5, which is a very stylized representation of some wormhole which remains open for a finite period; in practice, the Deutsch–Politzer spacetime is easier to work with than that in Fig. 10.5, and nothing crucial seems to turn on the choice.

The model is somewhat more general than it may seem, since we can take the state of a bit to represent not just the internal state of a definitely present system, but also whether or not the system is present at all. Furthermore, Deutsch points out that in quantum mechanics at least, we have good reason (Deutsch 1985b) to suppose that (in the absence of time travel) any physical system can be simulated by some set of quantum bits, so it is at least reasonable to explore the possibility that this is also true in the presence of time travel.

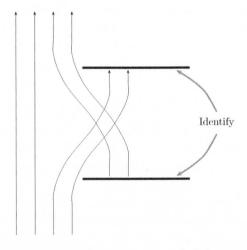

Figure 10.4. Four particles traversing Deutsch–Politzer spacetime

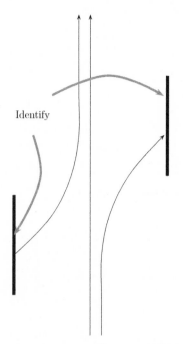

Figure 10.5. Two particles traversing spacetime with temporally extended wormhole

Identify

Nonetheless, the model is extremely stylized and simplified. There has been considerable work done on more realistic models (see Earman and Wuthrich 2008, and references therein, for more details). But concrete models rapidly become intractably difficult, and general theorems are both hard to come by and of questionable significance given that we do not possess the full physics of this regime. The computational model, by contrast, is extremely tractable and seems worth exploring for this reason alone; but I caution once more that conclusions drawn from it are only tentative.

That said, it will be convenient to consider the model in the following form (shown in Fig. 10.6). N bits approach the causality-violating region of the spacetime. A further M bits emerge from the past end of the wormhole, interact with the N bits via some $N + M$-bit interaction U (unitary in the quantum case, reversible in the classical case), then travel through time once more. The N bits continue on their way. Because the interaction can include swap operations where two bits exchange states, this is actually a very general form of the model, and includes

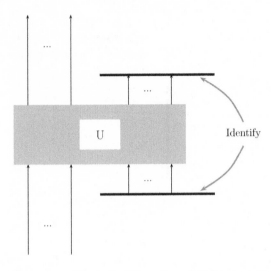

Figure 10.6. The general Deutsch model of time travel

circumstances where some bits arrive from before the wormhole formed, plunge into it, and then emerge and continue their journey.

The fundamental ('consistency') condition which we will impose—following Deutsch—is just that we will enforce the 'Identify' instruction in Figs. 10.4–10.6. That is, the state of a bit, or set of bits, coming out of the wormhole must be the same as it was going in.

If this is not obvious, consider that for the particles traversing the wormhole, there is no magic dividing line across the entrance: from the particle's point of view, everything proceeds smoothly. However, in the quantum case, the obviousness does depend on the idea that the quantum state represents something physically real just as the classical state does. An alternative model of time travel (the so-called 'post-selected closed timelike curve model', or 'P-CTC model' for short) has been studied recently (cf. Lloyd et al. 2010 and references therein); it does not appear compatible with this view of the quantum state, but it may be better suited to nonrealist interpretations of quantum mechanics, and possibly also to path-integral approaches to quantum theory. In this section, however, I will adopt a straightforwardly realist account of the quantum state; this does seem to force Deutsch's consistency condition.

It will also be convenient to adopt quantum-computational notation. Each bit can be in the states $|0\rangle$ and $|1\rangle$. Depending on the particular physics we want to consider, we will also need to allow for superpositions, incoherent mixtures, and entangled multi-bit states.

10.6.2 The paradoxes of classical time travel

In the classical case, there are no superpositions and no entanglements. The consistency condition then just requires that if a bit enters the wormhole in state $|0\rangle$ then it also emerges in state $|0\rangle$, and similarly for $|1\rangle$. I will generally assume for simplicity that all the N bits arriving at the wormhole from before it came into existence are in state $|0\rangle$, and will specify U only for these states. For uniformity of notation, I write e.g. the state of a pair of bits as $|i\rangle \otimes |j\rangle$, but of course $\otimes$ denotes Cartesian product, not tensor product.

We can now identify three sorts of paradoxes, each of which has been discussed in the literature. My discussion here will be brief; interested readers should consult e.g. Deutsch (1991), Arntzenius (2006), and Lewis (1976).

10.6.2.1 Contradiction paradoxes
Let one bit arrive at the wormhole from outside, before its formation (in state $|0\rangle$), and one from inside, i.e. from the 'past' mouth of the wormhole. Let the interaction U be

$$U|0\rangle \otimes |0\rangle \rightarrow |0\rangle \otimes |1\rangle$$
$$U|0\rangle \otimes |1\rangle \rightarrow |1\rangle \otimes |0\rangle$$
$$U|1\rangle \otimes |0\rangle \rightarrow |1\rangle \otimes |0\rangle$$
$$U|1\rangle \otimes |1\rangle \rightarrow |1\rangle \otimes |1\rangle \tag{10.7}$$

Ignore the last two rows for now, since we assume that the first bit arrives in state $|0\rangle$. In this situation, what is happening is that the non-time-travelling bit records the state of the time-travelling bit, and then flips it, putting '1' into the wormhole if '0' emerged and vice versa. If we interpret $|0\rangle$ and $|1\rangle$ as recording the presence and absence, respectively, of a particle on the trajectory, we can think of this interaction as modelling a particle which enters the wormhole if and only if it does not see itself leaving.

It should be obvious that neither state of the time-travelling bit can satisfy the consistency condition. This is, in fact, a stylized version of the infamous *Grandfather paradox*: what happens if you travel backwards through time and kill your grandfather, thus preventing your birth, thus preventing your travelling through time, thus preventing the killing itself...?

The general consensus in discussions of time travel is that this sort of thing should be understood as a constraint on the possible states of the universe prior to the time travel: since contradictions are impossible, conditions must be arranged such that events that would lead to a contradiction do not in fact take place. In the simple model above, for instance, the consistency condition can be satisfied if the first bit arrives in state $|1\rangle$ but not if it arrives in state $|0\rangle$; hence, circumstances must be such that it arrives in state $|1\rangle$. In the 'grandfather' case, it must be that even if you try to kill your grandfather, somehow events are arranged such that your attempts are frustrated—maybe your gun jams; maybe you suffer a heart attack; maybe you have pangs of remorse.

Call restrictions like this *prior constraints*; the philosophical question is then whether prior constraints are acceptable, and here opinions vary widely. Earman and Wuthrich (2008), for instance, are very relaxed about them, even comparing them to constraint equations in physics, like $\nabla \cdot E = 4\pi\rho$. Hawking and Ellis (1973: 189) regard them as all but unacceptable due to their conflict with our free will (on which subject see also Lewis 1976, Sider 2002, Vihvelin 1996, and Arntzenius 2006: 611–14). Arntzenius (2006) and Arntzenius and Maudlin (2010) take an intermediate line.

I should flag one concern with the use of a *computational* model to discuss the contradiction paradoxes: it has been argued (Wheeler and Feynman 1945; Maudlin 1990) that continuous systems will not have this sort of anomalous behaviour. If so, then the use of a discrete computational basis may be inappropriate for analysis of *classical* time travel. See Arntzenius and Maudlin (2010: section 3) for more on this point; see also footnote 46 below.

10.6.2.2 Knowledge-generating paradoxes Suppose f is a function (which, for nonessential convenience, I suppose to be one-to-one) from the first 2^K numbers to themselves and that for one and only one number n_0 in this range, $f(n_0) = n_0$. Using binary notation, K bits suffice to encode any number between 1 and 2^K, and I write $|n\rangle$ for the combined state of those K bits when they encode the number n.

Suppose, then, that K bits arrive at the wormhole from outside (in state $|0\rangle$) and K bits arrive from inside, and let the interaction be

$$U|0\rangle \otimes |n\rangle = |n\rangle \otimes |f(n)\rangle. \qquad (10.8)$$

It is easy to see that the only consistent possibility is that the bits emerge from the wormhole in state $|n_0\rangle$. As such, the state of the non-time-travel-ling qubits after the time travel has occurred is also guaranteed to be n_0: in other words, the time machine has infallibly solved the problem of finding the fixed point of f. This can be generalized straightforwardly to a wide class of computational problems: given a wormhole with a large enough bit capacity, any problem whose solution is checkable is easily solved by this method.

Generally speaking, the knowledge-generating paradoxes have been seen as less paradoxical than the contradiction paradoxes. An exception is Deutsch, who objects:

It is a fundamental principle of the philosophy of science ... that the solutions of problems do not spring fully formed into the Universe, i.e. as initial data, but emerge ... only through evolutionary or rational processes. (Deutsch 1991: 3201)

This objection has not been widely explored, to my knowledge, in the time travel literature.

10.6.2.3 Indeterminism-generating paradoxes Let one bit arrive at the worm-hole from outside (in state $|0\rangle$), and one from inside. Let the interaction U be

$$U|0\rangle \otimes |0\rangle \rightarrow |0\rangle \otimes |0\rangle \qquad (10.9)$$

$$U|0\rangle \otimes |1\rangle \rightarrow |1\rangle \otimes |1\rangle \qquad (10.10)$$

In this interaction, the state of the time-travelling bit is copied onto the state of the external bit, and the time-travelling bit then drops back into the time machine. It is equivalent to a situation where one bit approaches the wormhole, records the state of the bit emerging, and changes its own state to match the state of that bit. Clearly, this completely underdeter-mines the state of the time-travelling bit, and therefore of the emerging bit: either $|1\rangle$ or $|0\rangle$ are legitimate states for the time-travelling bit. In the presence of closed timelike curves, then, classical systems have the potential to become indeterministic. In fact, investigations of more dynamically detailed classical models of time travel (see Thorne 1994 and references therein) suggest that while contradiction paradoxes are quite hard to set up classically, indeterminism paradoxes are ubiquitous. They are also common in the better class of time travel fiction: consider, for instance, scenarios where the older version of the inventor travels backwards through time to explain to his younger self how to build a time machine. (See McCaffrey 1968, Egan 1995, Rowling 1999, and Hamilton 2002 for examples of the genre.)

My own view (for which I claim no originality) is that indeterminism paradoxes just point us in the direction of new physics. If the total state of the Universe prior to the time-travel-permitting region fails to determine (even probabilistically) the subsequent state of the Universe according to our best physics, this just means that we need to improve on our best physics. From this perspective, an indeterministic time machine, like a naked singularity, represents a breakdown of our current physics.

It is not uncommon, though, for writers simply to deny that any such improvement is required: Arntzenius (2006), for instance, writes:[44]

Jimmy's father taught him how to build a time-travel machine. So Jimmy asked him: 'How do you know how to build one?' 'Oh,' said Jimmy's father, 'I got it from you. You are going to time travel back to my youth just to tell me how to do it.'

Some people are perturbed by circular explanation. But what do you expect? Circular time, circular causation, circular explanation. Get used to it.

I worry, though, that this instead tells us that the connection between causation and explanation, whatever its general virtues (as extolled by e.g. Lewis 1986a), breaks down in the presence of time travel. Suppose, for instance, that when we find a wormhole throat, a copy of the King James version of the Bible (or, if preferred, of H. G. Wells' *The Time Machine*) invariably emerges. There might be no difficulty in tracking the (perhaps circular) causal history of each copy, but nonetheless this regularity seems in urgent need of *explanation*. (Put it this way: the atheist who claims that the invariable appearance of the Bible out of the time machine provides no evidence in favour of God probably has to say more than just 'Circular causation, get used to it'!)

---

[44] Arntzenius also acknowledges, though, that the existence of this kind of occurence in a theory might be prima facie evidence against that theory.

Be this as it may, my real interest here is the *quantum* theory of time travel, and we will see that things are interestingly different once quantum phenomena are taken into account.

10.6.3 Quantum mechanics and time travel

Generalizing the computational model from classical to quantum physics seems fairly straightforward: as noted above, we replace the bits by qubits (i.e. by quantum systems with a two-dimensional internal Hilbert space) and we take U to be a unitary and not merely reversible operation.

If we assume that the collective quantum state of the qubits remains pure at all times, this does not interestingly change the situation. Consider the contradiction paradox above, for instance, in which one external bit (initially in state $|0\rangle$) and one time-travelling bit interact via

$$U|0\rangle \otimes |0\rangle \rightarrow |0\rangle \otimes |1\rangle$$

$$U|0\rangle \otimes |1\rangle \rightarrow |1\rangle \otimes |0\rangle.$$

If the overall state just after the time-travelling bit emerges is pure, then (since the external bit is in pure state $|0\rangle$ *ex hypothesi*) the time-travelling bit must be in some pure state $\alpha|0\rangle + \beta|1\rangle$. The overall effect of the U operator is then

$$\alpha|0\rangle \otimes |0\rangle + \beta|0\rangle \otimes |1\rangle \longrightarrow \alpha|0\rangle \otimes |1\rangle + \beta|1\rangle \otimes |0\rangle \qquad (10.11)$$

and since this entangles the two qubits, the post-U state of the time-travelling qubit will no longer be pure. So this scenario leads to a contradiction paradox in quantum mechanics as surely as in classical mechanics.

Deutsch, however, points out (1991) that if we allow the time-travelling qubit to be in a *mixed* state, things are different. Suppose in particular that it emerges in the state

$$\rho_{2,i} = p|0\rangle\langle 0| + (1-p)|1\rangle\langle 1|. \qquad (10.12)$$

Since the external qubit is in a pure state, the time-travelling qubit cannot be entangled with it, and the overall system's state just after the qubit emerges from the wormhole is also mixed. The overall effect of U is now

$$p|0\rangle\langle 0| \otimes |0\rangle\langle 0| + (1-p)|0\rangle\langle 0| \otimes |1\rangle\langle 1| \rightarrow p|0\rangle\langle 0| \otimes |1\rangle\langle 1|$$

$$+ (1-p)|1\rangle\langle 1| \otimes |1\rangle\langle 1| \qquad (10.13)$$

in which case the final state of the time-travelling qubit is

$$\rho_{2,f} = p|1\rangle\langle 1| + (1 - p)|0\rangle\langle 0| \qquad (10.14)$$

and so satisfies the consistency condition provided that $p = 1/2$. In this situation, the external qubit also finishes up in the state $1/2(|0\rangle\langle 0| + |1\rangle\langle 1|)$.

Consistency is only possible here because the mixed state effectively represents multiple parallel worlds. In one world, the time-travelling bit exits the wormhole in state $|0\rangle$ and is flipped into state $|1\rangle$ before re-entering. It travels backwards through time and 'sideways' into a different branch, emerges in state $|1\rangle$ and is flipped into state $|0\rangle$, and then goes backwards through time and 'sideways' back into the first branch.[45]

More generally, for any finite number of qubits (internal and external) and for any initial state of the external qubits, there is guaranteed to be at least one solution to the consistency conditions. (This is a special case of the Schauder fixed point theorem (Schauder 1930), the infinite-dimensional generalization of the Brouwer fixed point theorem, since the transformation induced by U on the time-travelling qubits is a continuous map of the mixed states to themselves, and the mixed states form a compact, convex subset of a Hilbert space; in the finite-dimensional case Deutsch (1991, 3023) explicitly gives a construction for solutions.[46]) There are therefore (it appears) no contradiction paradoxes in quantum mechanics.

This is too quick, though: the story so far actually has little or nothing to do with quantum mechanics *per se*.[47] For suppose we wish to avoid contradiction paradoxes in classical mechanics by appeal to parallel universes: we can do so simply by permitting probability distributions over classical bits and interpreting those distributions realistically, as ensembles of classical worlds. This amounts to permitting only those mixed states which are diagonal in the qubit basis; this set, too, is compact and convex and is mapped to itself by any classical U, so by the Brouwer fixed point theorem, it too always contains at least one point satisfying any given consistency condition.

---

[45] This kind of parallel-universe account of time travel is also familiar in fiction; see e.g. Hogan (1986) or Baxter (2004) for well-worked-out realizations.

[46] From this perspective we can see why classical systems are not always proof against contradiction paradoxes even when we abandon the computational idealization: the phase spaces of classical systems are not always compact and convex, and do not always have fixed points.

[47] I'm grateful to David Albert (in conversation) for pressing this point.

Conversely, the multiplicity present in a mixed state is, as we have seen previously, of a rather different kind from conventional Everettian multiplicity. The parallel universes of the Everett interpretation are present even in pure states, are emergent phenomena whose concrete form is given by decoherence theory, and affect one another dynamically via interference; the multiple 'Universes' (actually, multiple entities which are themselves multiplicities of quasi-classical branches) in mixed states are present by interpretive fiat.

So far, then, all I think we have established (and all Deutsch establishes) is that accepting parallel universes can solve time travel's contradiction paradoxes. There is nothing quantum-mechanical about this observation, and it would be reasonable to regard the quantum and time-travel parallel universes as being of an entirely different kind.

But this too is too quick, as I will show in the next section.

10.6.4 Entanglement and time travel

Let two external qubits approach a wormhole; let one qubit emerge from it; let the overall dynamics be

$$U|k\rangle \otimes |i\rangle \otimes |j\rangle \rightarrow |k\rangle \otimes |j\rangle \otimes |i\rangle \qquad (10.15)$$

where i, j, k are each 0 or 1, and where the state of the time-travelling qubit is given by the rightmost ket. In this process, the second external qubit swaps with the internal qubit and the remaining qubit is just along for the ride; it is equivalent to a process where one of two qubits travels backwards through time, emerges, and then continues to travel freely without interacting with either itself or the other qubit. Classically this leads to no paradox of any kind: the internal bit just emerges in the same state as the second external bit, they swap, and the overall state of the external bits is unchanged.

But now let the two external qubits be in the entangled state

$$\frac{1}{\sqrt{2}}(|0\rangle \otimes |0\rangle + |1\rangle \otimes |1\rangle), \qquad (10.16)$$

so that each qubit individually is in the mixed state $1/2(|0\rangle\langle 0| + |1\rangle\langle 1|)$. In this situation, the *only* state that can satisfy the consistency conditions is therefore a mixed state. And in this case, the overall state of the system just after it emerges must likewise be mixed, since the external qubits are jointly in a pure state.

There is nothing particularly special about my example: essentially any situation in which part but not all of an entangled system undergoes time travel will require mixed states to satisfy the consistency condition. So either (i) in the presence of closed timelike curves the Universe can be in a genuinely mixed state, or (ii) entangled states cannot travel on closed timelike curves. (That is, if genuinely mixed states are not allowed, then the initial conditions of the Universe must satisfy constraints which conspire to ensure that genuinely mixed states are not needed.)

But genuinely unentangled states are hard to come by, in a world where even the vacuum is entangled on all lengthscales. Not very entangled, but the argument above is all-or-nothing: *any entanglement at all* suffices to require mixed states. I conclude, then, that a bar on entangled states undergoing time travel is tantamount to a ban on time travel full stop. Or equivalently: if time travel is possible in a quantum universe, so are genuinely mixed universal states. And once we recognize that this is the case, as a bonus Deutsch's argument tells us that there are no consistency paradoxes, and so no time-travel-induced constraints on initial data.

10.6.5 Features of quantum time travel

In this section, I want to touch briefly on some of the interesting features of a world in which both (Everettian) quantum mechanics is true and time travel is possible. In some cases (not all) the issue has been more broadly explored in the literature, in which case I give references.

10.6.5.1 Time travel does not generically appear to give rise to indeterminism paradoxes It is easy to construct specific examples of quantum processes which lead to indeterminism when combined with time travel: the classical example above also works in quantum mechanics, for instance. Deutsch, recognizing this, suggests as an explicit rule that the supplementary data should maximize von Neumann entropy. However, there are good reasons to expect that generically this does not occur. The most general effect of the interaction on the time-travelling qubits is given by the Kraus representation theorem[48] as

$$\rho \longrightarrow \sum_k \widehat{E}_k \rho \widehat{E}_k^\dagger \tag{10.17}$$

[48] See e.g. Benenti, Casati, and Strini (2004: 282–3).

for some set of operators $\{\widehat{E}_k\}$ satisfying $\sum_k \widehat{E}_k^\dagger \widehat{E}_k = \widehat{1}$, and the consistency condition requires that ρ is a fixed point of this mapping. In the special case where the identity of this map is a fixed point, it is known (Blume-Kohout et al. 2008) that fixed points are precisely those density operators which commute with all the $\widehat{E}_k$; generically, this will mean that only the (normalized) identity is a fixed point. In the more general case, Bruzda et al. (2009) have studied the spectrum of randomly selected operators of this kind. Clearly the spectrum has to contain one eigenvalue of value 1 (since the map must have at least one fixed point); their arguments entail that only a measure-zero subset of such operators have other eigenstates with eigenvalue 1 (i.e. other fixed points). Since (if only because quantum theory is at most an approximation to the truth) we should expect that the physics is not arbitrarily sensitive to arbitrarily small perturbations in the design of the time-travel interaction, this suggests (though does not prove) that indeterminism paradoxes in quantum time travel are artefacts of the formalism, and can be disregarded.

10.6.5.2 Time travel breaks entanglement Classical intuition would suggest that if a particle is transported backwards through time for a few seconds and then allowed to continue on its trajectory, nothing is actually changed. Quantum-mechanically, though, this appears not to be the case. As we have seen, if the particle which is transported through time is actually part of an entangled pair, its intrinsic (mixed) state is unchanged by the time travel but the entanglement is destroyed.

Some more insight can be gained into this by considering not the Deutsch–Politzer spacetime (with its instantaneous wormhole) but the 'more realistic' spacetime of Fig. 10.5. Consider the following scenario (Fig. 10.7), where the dotted line indicates entanglement, Particle A flies past the wormhole mouth, but particle B enters it. Because there is nothing locally special about the wormhole boundary—from the point of view of particle B, the wormhole mouth is just an ordinary region of spacetime—the entanglement can 'thread' the wormhole, and we would expect it to be empirically detectable by bringing B back through the wormhole, or A through the wormhole, or transferring the entanglement from A or B to a third particle and sending that particle through the wormhole. But we would not expect to find entanglement between A and B if we compare them via experiments which do not thread the

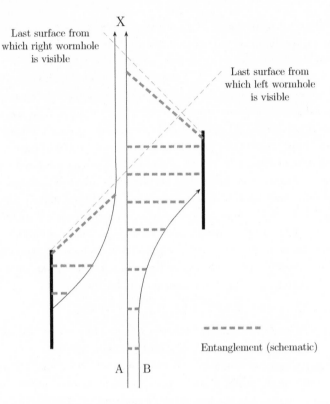

Figure 10.7. Removal of entanglement by time travel

wormhole; and in particular, after the wormhole is closed and the particles come back together at X, they will not be entangled. This suggests that if time travel is possible, it will require some kind of formalism which allows us to distinguish between entanglement of A and B with respect to different paths linking them.

10.6.5.3 Time travel leads to nonlinear dynamics Suppose that a qubit approaches a wormhole from which its future self emerges. A measurement is made of the first qubit and the result recorded on its future version; then the first qubit enters the wormhole. The gate implementing this is

$$U|0\rangle \otimes |0\rangle \rightarrow |0\rangle \otimes |0\rangle$$
$$U|0\rangle \otimes |1\rangle \rightarrow |1\rangle \otimes |0\rangle$$

$$U|1\rangle \otimes |0\rangle \to |1\rangle \otimes |1\rangle$$

$$U|1\rangle \otimes |1\rangle \to |0\rangle \otimes |1\rangle \qquad (10.18)$$

If the input qubit is in (mixed) state

$$\rho_{1,i} = \cos^2\theta|0\rangle\langle0| + \sin^2\theta|1\rangle\langle1| + \alpha|0\rangle\langle1| + \alpha^*|1\rangle\langle0| \qquad (10.19)$$

(the most general state it can be in) then the measurement process may affect the off-diagonal terms but will not change the on-diagonal terms. It follows that the qubit that enters the wormhole—and hence the qubit that exits the wormhole—has the form

$$\rho_{2,i} = \cos^2\theta|0\rangle\langle0| + \sin^2\theta|1\rangle\langle1| + \beta|0\rangle\langle1| + \beta^*|1\rangle\langle0|. \qquad (10.20)$$

A quick calculation then shows that consistency requires $\beta = 0$; in other words, the particle that emerges from the past mouth of the wormhole has state $\cos^2\theta|0\rangle\langle0| + \sin^2\theta|1\rangle\langle1|$. It follows that the particle that eventually leaves the wormhole region has state

$$\rho_f = \left(\frac{1}{2} + \frac{1}{2}\cos^2(2\theta)\right)|0\rangle\langle0| + \left(\frac{1}{2}\sin^2(2\theta)\right)|1\rangle\langle1|. \qquad (10.21)$$

So if we input $|0\rangle$ ($\theta = 0$) or $|1\rangle$ ($\theta = 1$) then in both cases, the output bit will be in state $|0\rangle$. But if we input the superposition $(1/\sqrt{2})(|0\rangle + |1\rangle)$ or the mixed state $(1/2)(|0\rangle\langle0| + |1\rangle\langle1|$, then the output state will be $(1/2)(|0\rangle\langle0| + |1\rangle\langle1|$, i.e. it will sometimes be observed to be $|1\rangle$. It follows that dynamical processes in the vicinity of closed timelike loops can be nonlinear, even though locally the dynamics remains unitary.

It follows from this that closed timelike loops can be used to distinguish proper from improper mixtures. Suppose that I know that some bit is either (i) definitely in state $|0\rangle$ or $|1\rangle$, but I don't know which, or (ii) in some mixed state that is a mixture of the two. If it is the former, putting it into the time-travel process above will always give result $|0\rangle$; if it is the latter, it will sometimes give result $|1\rangle$.

This seems to indicate a breakdown of the probability interpretation of density operators in the presence of time travel. (Needless to say, this is no argument against Everettian quantum mechanics, since the empirical evidence for the probability–weight link rests entirely on data from experiments not carried out in the presence of closed timelike loops.) I conjecture, however, that this breaks down above some macroscopic threshold: at some point, we would expect a quantum theory of gravity to entangle spacetime structure with the matter fields, so that some situations would be better described not as 'a mixed state interacting with a time machine' but rather as 'a superposition of two time machines with different inputs'.

10.6.5.4 Time travel is computationally extremely powerful It has been shown by a variety of methods (see Bacon 2004 and Aaronson and Watrous 2008, and references therein) that given closed timelike curves, we can efficiently compute a very wide class of functions: any problem for which the answer can be *checked* efficiently can be calculated efficiently (more precisely, with access to closed timelike curves we can compute efficiently any problem in the class P_{SPACE}). There is nothing especially quantum-mechanical about the result (the same speedup is available for classical time machines) but it appears that quantum mechanics at least does not prevent this kind of speedup. (Bacon provides strong arguments, for instance, that the noise which would develop in realistic time-travel-based quantum computations can be corrected for by the standard methods of quantum error correction.)

As such, knowledge-generating paradoxes are alive and well in quantum time travel. If Deutsch is correct to think that this violates 'a fundamental principle of the philosophy of science' then, presumably, so much the worse for time travel (or for that fundamental principle). Scott Aaronson makes a similar suggestion, arguing that certain constraints on computational power—specifically, the 'NP hardness assumption' that problems in the NP-complete complexity class ought to be unsolvable in polynomial time—ought to be respected in any reasonable physical theory:

> Even many computer scientists do not seem to appreciate how different the world would be if we could solve NP-complete problems efficiently.... If such a procedure existed, then we could quickly find the smallest Boolean circuits that output (say) a table of historical stock market data, or the human genome, or the complete works of Shakespeare. It seems entirely conceivable that, by analyzing these circuits, we could make an easy fortune on Wall Street, or retrace evolution, or even generate Shakespeare's 38th play. For broadly speaking, that which we can compress we can understand, and that which we can understand we can predict.... The NP Hardness Assumption is the belief that such power will be forever beyond our reach. (Aaronson 2005: 19–20)

10.6.5.5 Time-travelling systems end up in maximally decohered states Suppose that some reasonably large number of qubits emerge from the past mouth of a wormhole, evolve forward to the future mouth, and re-enter the wormhole. In general, as we have seen in Chapter 3, the qubits will become entangled with their environment. This will select some preferred basis and suppress off-diagonal elements with respect to that basis. But because the qubits must collectively be in the same state upon entering the wormhole as they were upon exiting it, it follows that they must have emerged from the wormhole in an already maximally decohered state.

Indeed, since in general there is no basis completely immune from decoherence, we would expect that the quantum state of matter emerging from a macroscopically large wormhole would be completely random: up to normalization, they should be in the state $\widehat{1}$ (or perhaps some restriction of that state to a subset specified by conserved quantities). This suggests that macroscopic time travel generally does not allow the transmission of information, and that doing something useful with a closed timelike curve requires the same kind of decoherence-suppressing tricks already familiar from quantum computation theory. (One imagines that by the time we are anywhere near being able to build even a microscopic time machine, we'll have those tricks down to a fine art!)

10.6.5.6 Time travel explicitly breaks time reversal invariance In general, qubits in pure states which pass through regions containing closed timelike curves evolve into mixed states, because they become entangled with systems that then—from the point of view of a distant observer—cease to exist. A qubit initially in state $|0\rangle$ interacting via (10.7), for instance, encounters a qubit in the mixed state $(1/2)(|0\rangle\langle 0| + |1\rangle\langle 1|)$ and gets entangled with it, forming the mixed state $(1/2)(|01\rangle\langle 01| + |10\rangle\langle 10|)$. When the second qubit disappears back down the wormhole, the first qubit is left in the mixed state $(1/2)(|0\rangle\langle 0| + |1\rangle\langle 1|)$.

If processes like these are run backwards, they describe a mixed state which turns out to be mixed partly because it's entangled with another state which emerges from the 'future' end of the wormhole. It disentangles itself from that other state and heads off in a pure state.

Is this time-reversed description dynamically allowed? So far I have said nothing about what happens when initially mixed states arrive at a region containing closed timelike curves. In Deutsch's model, it is assumed that if the arriving state is ρ_1 and the state of the qubits emerging from the past mouth of the wormhole is ρ_2, then the overall state of the system just after the second set of qubits emerge is $\rho_1 \otimes \rho_2$, i.e. entanglement is ruled out by fiat and the time-reversed process is not dynamically allowed. In the model considered above, for instance, if the incoming qubit is in state $(1/2)(|0\rangle\langle 0| + |1\rangle\langle 1|)$ and the time-travelling qubit is in state ρ_2 then the overall state must be

$$(1/2)(|0\rangle\langle 0| + |1\rangle\langle 1|) \otimes \rho_2 \qquad (10.22)$$

and the consistency condition then tells us that $\rho_2 = (1/2)(|0\rangle\langle 0| + |1\rangle\langle 1|)$ and that the outgoing qubit is likewise in state $(1/2)(|0\rangle\langle 0| + |1\rangle\langle 1|)$, and not in a pure state.

However, Politzer (1992) has pointed out that the consistency condition allows for the overall state to be any ρ such that the partial trace over the time-travelling qubits yields ρ_1 and the partial trace over the non-time-travelling qubits yields ρ_2. This suggests that the time reversed description is allowed after all.

The suggestion should be resisted, though. For suppose that the reason that the arriving qubits are in state ρ is because they are entangled with some other, distant group of qubits and the total system of qubits is in a pure state. For instance, maybe the incoming qubit is in state $(1/2)(|0\rangle\langle 0| + |1\rangle\langle 1|)$ because it is one of two qubits jointly in the state

$$\frac{1}{\sqrt{2}}(|0\rangle\otimes|0\rangle + |1\rangle\otimes|1\rangle), \qquad (10.23)$$

with the other qubit very distant. Then the only possibility is that the overall state of the qubits after the time-travelling bit emerges is a product state (systems in pure states cannot be entangled), and so once we trace over the distant qubits, the joint state of the remaining qubits is again a product state.

Of course, all this assumes that systems in 'genuinely' mixed states behave the same way as systems which are in mixed states because they are entangled with other systems. But this assumption is unavoidable, because the one can be transformed into the other by causing the distant qubits to traverse closed timelike paths: as we have seen, this destroys entanglement. If 'genuinely' and 'ersatz' mixed states had dynamically different behaviour, this would allow for distant signalling, in violation of Lorentz covariance (and indeed, if there really were a distinction between 'genuinely' and 'ersatz' mixed states, the process of transforming one into the other could not be defined covariantly even if it were not empirically detectable via their dynamics).

I conclude (assuming that nothing is wrong with the arguments that support Deutsch's version of quantum time travel) that time travel is an explicitly time-reversal-non-invariant process. Ultimately, any process which destroys or creates quantum systems (black hole evaporation as well as time travel) has to decide what to do about the nonlocal information which can be created or destroyed in the process, and the case of time travel suggests that the rule for this defines an arrow of time.

This may seem to sit a little uneasily with the discussion of emergent irreversibility in Chapter 9, but every problem is an opportunity. Were this a less careful book I would enthusiastically endorse the genuine time-reversal-non-invariance of processes involving creation and destruction of quantum systems as the ultimate explanationn of the Simple Past Hypothesis. As it is, though, I will settle for the duller but safer expedient of noting that definitive results here will have to await a quantum theory of gravity.

Conclusion

True, when quantum mechanics was new, some physicists thought
that it put humans back into the picture, because the principles
of quantum mechanics tell us how to calculate the probabilities of
various results that might be found by a human observer. But, starting
with the work of Hugh Everett forty years ago, the tendency of
physicists who think deeply about these things has been to reformu-
late quantum mechanics in an entirely objective way, with observers
treated just like everything else. I don't know if this program has been
completely successful yet, but I think it will be.

Steven Weinberg[1]

Although it began with dinosaurs and ended with time machines, this has
been a very conservative book. It has been concerned, from beginning
to end, with *quantum mechanics*: with deterministic, Lorentz-covariant
quantum theory, interpreted in a straightforwardly realist way.

And when we interpret it that way, we find (I have argued):

- That applying the same analysis of emergence and high-level structure
  we use all over the special sciences tells us that the unitarily evolving
  state has an emergent branching structure, with each branch giving
  approximately definite positions and momenta to macroscopic objects
  and with those objects evolving in an approximately classical way
  within their branch. (Chapters 2 and 3)
- That the emergent branching structure has a measure on it, defined
  in terms of the underlying physics, which satisfies the probability
  calculus and which, if we allow ourselves no more license than is
  anyway needed in classical physics to interpret a physical magnitude
  as probability, can indeed be so interpreted. (Chapter 4)
- That even if we don't allow ourselves this license, we can *still* justify why that physical magnitude is probability, because probability
  actually makes more sense in quantum than in classical physics.
  (Chapters 5 and 6)

---

[1] Weinberg (2002: 233).

- That although the theoretical basis for claims about our uncertainty about the future, and about personal identity more generally, may be different in Everettian quantum mechanics as compared to classical mechanics, nonetheless the role that these concepts actually play in our lives will be unchanged. (Chapter 7)

- That relativistic versions of quantum theory are fully Lorentz-covariant, and comply with both the letter and the spirit of relativity theory. Quantum systems can have nonlocal states, but only in the same sense that gauge theories can have nonlocal states; they do not undergo nonlocal interactions. (Chapter 8)

- That the decoherence process defines an arrow of time whose explanation is essentially the same as the explanation for the arrows of time defined by other in-practice-irreversible processes in physics. (Chapter 9)

I conclude that there is no measurement problem in quantum mechanics. As I stated in the Preface, the apparent measurement problem is just caused by an insufficiently thorough understanding of the structure of quantum physics. In actuality, there is no clash between unitary, realist-interpreted quantum mechanics and our observations or our experimental data. The Everett interpretation does not *solve* the measurement problem: it *dissolves* it.

To be sure, one thing we learn when we really look at the structure of quantum mechanics is that the classical world which we inhabit is one of a vast number of such worlds, most of which are in practice dynamically isolated from one another. That is surprising—perhaps even shocking—and highly counterintuitive, but none of these observations constitutes any kind of argument against the theory. Furthermore, we have seen (section 10.3) that the existence of at least some parallel Universes is an unavoidable feature of any theory adequate to explain our empirical data. We can modify the theory to get rid of many of those Universes dynamically if we want; we can even add additional hidden variables to accompany those universes if we want, and try to make a case for why *our Universe* is instantiated by those hidden variables. But in neither case does the underlying multiplicity really go away.

This is not to say that no one should try to construct alternatives to quantum theory. Nor is it even to deny them the right to be motivated in doing so by a distaste for parallel Universes. Testable alternatives to our best theories, and empirical proposals to test those theories, are always welcome[2] (in-principle-untestable ones, not so much). But it *is* to say that there is no scientific defect in quantum theory which gives us any reason to think that some such theory must be true. Given the inelegance of most of them, the ferocious difficulties in constructing relativistically covariant versions, and the phenomenal success of unitary quantum mechanics over an extraordinary range of energy (and other) scales, I doubt very much that any such theory is correct.

In one sense, this is actually a bit of a disappointment. After all, we know that quantum theory (or at least, our best current quantum theory) is going to have to fail somewhere, because of the clash with general relativity; but the experimental regime associated with *that* failure is horrifically hard to probe empirically. How convenient it would have been if we could have been led to an improved version of quantum mechanics by the measurement problem, and then found that improved version to be naturally compatible with general relativity! Alas, it was not to be.[3]

In another sense, it is a huge relief. The last eighty years of quantum mechanics have, frankly, been a bit embarrassing for physicists and philosophers of science alike: physicists because they had no fully coherent story to tell about the physical world (but nonetheless felt obliged to talk about it—to refer to quarks, atoms, fields and the like, and to lapse into operationalism only when pushed); philosophers because despite this, the physicists made astonishing progress. According to the Everett interpretation, the reason this all worked out is because all along the theory the

---

[2] Some good recent examples of such proposals are Leggett (2002), Valentini (2004), and Penrose (2004: 816–68).

[3] Let me for completeness note one argument for this strategy which survives the Everettian move. Penrose (1989) has argued that to assume the superposition principle applies to geometries is to inappropriately prioritize quantum-mechanical over general-relativistic insights. If so, we might expect that the superposition principle will fail at the point when otherwise a superposition of geometries would be created—i.e. at the point at which we try to create a superposition of spatially separated masses above some threshold of mass and separation. If so, this would in effect be a dynamical collapse theory. Penrose is only playing a hunch, of course (and is motivated in part by the quantum measurement problem), but it's a perfectly reasonable hunch to try to play, and in principle it's empirically testable. My own hunch is that the correct quantum theory of gravity will fully preserve all the interesting features of both general relativity and quantum mechanics, and so it will incorporate spacetime geometries (general-relativistic feature) which can exist in superpositions (quantum-mechanical feature); both string theory and loop quantum gravity appear to conform to this hunch, but only time will tell if it is correct.

physicists were using was correct, and straightforwardly realist—but it took a long time to understand it sufficiently well that we really grasped the connection between the theory and our own corner of empirical reality, and in the interim we got by with a series of ad hoc rules that basically worked but that fostered much confusion and forced much obfuscation. But we learned more, and refined those rules (the correspondence principle was replaced by phase-space representations; PVMs were replaced by POVMs, 'observation event' was replaced by 'macroscopic superposition', which was in turn replaced by 'decohered superposition'). Now we basically get the big picture, and the obfuscation is disappearing.

And it really *is* disappearing. It's customary to end discussions of the quantum measurement problem by bemoaning the unwillingness of the philistine physicists to take the interpretation question seriously, but I'm much more optimistic. Right across those bits of modern physics where the issue is significant, it has become all but universal to treat quantum mechanics, at any scale, as unitary whenever it matters that it is. In quantum computation, workers are relaxed about invoking the 'Church of the Larger Hilbert Space' and treating any non-unitary process as just the restriction to a smaller system of a larger unitary process. In quantum cosmology, 'measurement' and 'observation' play no role whatsoever, and the Everettian position is extremely widespread: Hawking (1976), for instance, casually speaks of observers being localized within one of 'Everett's branches', and Hartle (2010) calls quantum cosmology the 'killer app' for the Everett interpretation. In string theory (as I mentioned in section 10.5) an intense debate has raged as to whether black hole evaporation violates unitarity; one infers that if the issue is that contested, it cannot be widely felt in that community that ordinary measurement violates unitarity.

This is not to say that most physicists—even most physicists who accept universal unitarity—believe in the literal existence of many worlds. Plenty do, of course; plenty more don't, or are agnostic. But a pretty large fraction recognize that what the theory *means* is (among other things) that those worlds exist.[4] Beyond that, it's a free country: scientists are

---

[4] The most coherent and widely discussed alternative position in mainstream physics, I think, is the collection of attempts to connect quantum theory to various kinds of information-theoretic considerations (see e.g. Fuchs 2002). Aspects of this programme are very interesting even from an Everettian standpoint: an axiomatization of quantum theory in terms

welcome to disbelieve some of the claims of their theories (however shaky the philosophical ground on which they stand may be), provided they recognize what those claims are.

It's also customary to declare that much work still remains to be done on one's pet interpretation. Up to a point, I'm happy to follow this pattern: for sure, we don't yet have the final story about probability, or emergence, or subjective notions of identity and uncertainty. But in the main, the value of a (dis)solution to the measurement problem is not that it lets us study that (dis)solution in more and more detail, but that it actually lets us make progress in studying *the physical world*, without being fettered by a failure to understand what our best theory of that world says, or by a nagging doubt that the theory is wrong, or that the very idea of our theories telling us about the world is mistaken.

The point applies to philosophy and physics alike. In philosophy, David Lewis memorably complained:

I am not ready to take lessons in ontology from quantum physics as it now is. First I must see how it looks when it is purified of instrumentalist frivolity, and dares to say something not just about pointer readings but about the constitution of the world; and when it is purified of doublethinking deviant logic, and—most of all—when it is purified of supernatural tales about the power of the observant mind to make things jump. (Lewis 1986c: xi)

Well, the Everett interpretation delivers on these demands.[5] In fact, analytic metaphysics has been lamentably unwilling to engage with modern physics, too often preferring what Ladyman and Ross (2007: 24) caustically refer to as the 'philosophy of A-level chemistry'.[6] The reasons are varied, but one reason is no doubt that for too long, modern physics has shied away from telling any positive story about the world, and so philosophers fell back on classical ideas. It remains to be seen whether, now that this period may be coming to an end, more metaphysicians will rise to the challenge of re-engaging with real physics.

of information flow need not be any more hostile to the objectivity of the quantum state than axiomatizations of special relativity in terms of the light postulate are to the objectivity of the Minkowski metric (cf. Brown 2005). Other parts, I think, fall prey to old objections about the incoherence of operationalism and the inappropriacy of treating measurement as a primitive; see Timpson (2010), though, for a critical but sympathetic review.

[5] Lewis, unfortunately, had little opportunity to become acquainted with Everettian quantum mechanics; his first foray into the subject (2004) was written right before his untimely death, and published posthumously.

[6] Note to non-British readers: A-levels are (as of 2010) the standard post-16 UK secondary education qualification.

In physics, and in the philosophy of physics, similar morals hold (and are more widely accepted in the community, at least in mainstream physics). Whatever the substantive merits of the conclusions I reach in Chapters 8 and 9, and in section 10.6, I hope they at least succeed in showing how much *easier* it is to discuss knotty conceptual problems when one can, unapologetically, treat the quantum state as a complete description of physical reality and, equally unapologetically, treat the Schrödinger equation as universal.[7]

Actually, many of the most interesting conceptual features of quantum physics seem to occur once we move past the level at which branching-universe descriptions are valid. Those descriptions, after all, ultimately apply only in the domain where the theory can be treated as a stochastic classical process. But quantum information theory has shown us just how rich and complicated the *non*classical aspects of quantum mechanics— notably entanglement phenomena—really are. (And in fact, the 'parallel computations' paradigm of quantum computation, for all that it is straight-forwardly correct as a description of Shor's algorithm, has not been especially fruitful as a source of other algorithms; quantum cryptography, on the other hand—which has no parallel-universe reading as far I know—has reached the stage of commercial implementation.[8])

Deutsch (2010) provides a long list of features of this nature that may merit investigation. In many cases, we do not *explicitly* need to assume the existence of many worlds, but we do need, in practice, to take the quantum state as physically real, to reject the existence of additional hidden variables, and to regard the Schrödinger equation as universal. From there, Everettian quantum mechanics is basically inescapable.

Let me conclude by returning to the basic point of the book: that 'Everettian quantum mechanics' is just quantum mechanics itself. This is, I think, the basis of de Witt's famous remark (with which I began Chapter 1) that 'the mathematical formalism of the quantum theory is capable of yielding its own interpretation'. This is usually read as saying that quantum

---

[7] Of course, it is equally easy to discuss these problems if we assume some other realist 'interpretation' (at least in the nonrelativistic domain, where we have access to such 'inter-pretations'): the difficulty comes not from the particular details of a given theory but from obfuscation as to what the theory is. On the other hand, studying these questions on the assumption that (Everettian) quantum mechanics is true has the rather decisive advantage that we have positive reasons to believe that (Everettian) quantum mechanics *is* true.

[8] By e.g. Id Quantique (http://www.idquantique.com) or MagiQ Technologies (http://magiqtech.com).

theory, *unique among all physical theories*, yields its own interpretation—but the right reading, I think (whether or not de Witt himself meant it this way) is that quantum theory yields its own interpretation *like any other physical theory*. The claim that we can 'interpret' quantum mechanics by taking it literally is not some strange claim about quantum mechanics: it is as valid (or otherwise) as similar claims about general relativity, or electromagnetism, or any physical theory you like.

Nor is it unique to quantum mechanics for us to take a while to work out what the theory actually says. In the case of general relativity, it took time to recognize the event horizon as a mere coordinate singularity. In the case of Darwinian natural selection, it was not until the late twentieth century that biologists really came to appreciate the relevance of the gene's-eye view of evolution—and the debate as to exactly how to understand the units of selection still rages, and is as much a conceptual as an empirical debate. Scientific theories are passing strange, and have no respect for our intuitive idea of the world. It takes time—and, often, serious mathematics—to adjust our ways of thinking so that we see what the theory is really saying.

So it is with quantum mechanics. The third customary thing to do in a discussion of the quantum measurement problem is to remind the reader of how crazy quantum mechanics really is and how no one really understands it, but quantum mechanics is not crazy at all, and after eighty years of hard work I think we basically do understand it, or enough of it to see how it relates to the world we observe. Eddington[9] only got it half right: time and again, the Universe has turned out to be stranger than we *have* imagined, but so far—as long as we keep our nerve, think clearly, and accept theories as they are and not as we would like them to be—it has not proved stranger than we *can* imagine.

Oxford
2011

---

[9] The quotation ('the Universe is not only stranger than we imagine; it is stranger than we can imagine') is normally attributed to Eddington, but no written record appears to exist, and it may be a distorted version of Haldane's quote: 'the Universe is not only queerer than we suppose; it is queerer than we can suppose' (1928).

Oh well. If he didn't say it in this world, no doubt he said it in another one.

Epilogue

We have indicated that it is possible to have a complete, causal theory of quantum mechanics, which simultaneously displays probabilistic aspects on a subjective level, and that this theory does not involve any new postulates, but in fact results simply by taking seriously wave mechanics and assuming its general validity.

Hugh Everett III[1]

AUTHOR: Persuaded?

SCEPTIC: No. But it won't do me any good, since I'm your creation: presumably you're going to write some craven capitulation into my half of the dialogue.

AUTHOR: I promised not to do that. Instead I'll let some other fictional characters—from Galileo's *Dialogue Concerning the Two Chief World Systems*—have the last word.[2] They're not discussing quantum mechanics, of course, but the resemblance is interesting for all that.

SIMPLICIO: [W]e ought not to admit anything to be created in vain, or useless in the universe. Now we see this beautiful arrangement of the planets disposed round the earth at distances proportioned to the effects they are to produce upon us for our benefit. To what purpose, then, should such a vast vacancy be afterwards interposed between the orbit of Saturn and the starry spheres, containing not a single star, and altogether useless and unprofitable? to what end? and for whose use and advantage?

SALVIATI: Methinks we arrogate too much to ourselves, Simplicio, when we assume that the care of us alone is the adequate and sufficient work and limit beyond which the Divine wisdom and power do nothing and dispose of nothing. I feel confident that nothing is omitted by God's providence which concerns the government of human affairs; but that there may not be other things in the universe dependent on His supreme

---

[1] Draft of Ph.D. thesis; communicated to the author by Max Tegmark.
[2] The translation is taken from Drake (1978).

power, I cannot, with what power of reasoning I possess, bring myself to believe. So that when I am told of the uselessness of an immense space interposed between the orbits of the planets and the fixed stars, I reply that there is temerity in attempting by feeble reason to judge the works of God, and in calling vain and superfluous every part of the universe which is no use to us.

SAGREDO: Say rather that we have no means of knowing what *is* of use to us. I hold it to be one of the greatest pieces of arrogance and folly that can be in this world to say, because I know not what use Jupiter and Saturn are to me, that therefore these planets are superfluous. Nay more, that there are no such bodies in existence. To understand what effect is worked upon us by this or that heavenly body (since you will have it that all their uses must have a reference to us) it would be necessary to remove it for a while, and then the effect which I find no longer produced on me, I may say depended on that body. Besides, who will dare to say that the space (called too vast and useless) between Saturn and the fixed stars is void of other bodies belonging to the universe? Then, the four Medicean planets and the companions of Saturn came into the heavens when we began to see them, and not before! And by the same rule the innumerable host of fixed stars did not exist before men saw them. The nebulae, which the telescope shows us to be constellations of bright and beautiful stars, were, till the telescope was discovered, only white flakes! Oh, presumptuous! rather, oh rash ignorance of man.

Appendix A

Proof of the Branching-Decoherence Theorem

Recall that the theorem is

Branching-Decoherence Theorem: If $\mathcal{P} = \{\widehat{P}_j^i\}$ is a history space and $|\psi\rangle$ is a quantum state, then

(i) If $\mathcal{P}$ has branching structure (relative to $|\psi\rangle$) and α is a history then $\widehat{C}_\alpha|\psi\rangle \neq 0$ iff α is realised (with respect to $|\psi\rangle$).

(ii) If the set Hist of all histories α such that $\widehat{C}_\alpha|\psi\rangle \neq 0$ has branching structure (that is, if no two histories in Hist agree on their nth index but not on all previous indices), then $\mathcal{P}$ also has branching structure (relative to $|\psi\rangle$), and the realised histories in that branching structure are just the histories in Hist.

(iii) If $\mathcal{P}$ has branching structure (relative to $|\psi\rangle$), $\mathcal{P}$ satisfies the decoherence condition.

(iv) If $\mathcal{P}$ satisfies the decoherence condition, it is a coarse-graining of a (decoherent) history space relative which has branching structure relative to $|\psi\rangle$.

To develop the proof, it will be helpful to modify Chapter 3's index notation somewhat. Recall that a history is uniquely specified by a string of indices, one for each time-index. This being the case, I identify each history with that string of indices: a boldface index $\mathbf{j}$ denotes a sequence $\langle j_0, \ldots j_n \rangle$ of indices (one for each time $t_0, \ldots t_n$) such that the projector in the history corresponding to time t_k is $\widehat{P}_{j_k}^k$. I then generalise the notion of history so that $\mathbf{j}$ can stand for any sequence of indices for times $t_0, \ldots t_m$, where $m \leq n$; $\mathcal{L}(\mathbf{j})$ will denote the length of the sequence minus one (so if the sequence is defined for times $t_0, \ldots t_m$ then $\mathcal{L}(\mathbf{j}) = m$). I write $\mathbf{i} \sim \mathbf{j}$

to indicate that one of $\mathbf{i}$ and $\mathbf{j}$ is an initial segment of the other (so if $\mathcal{L}(\mathbf{i}) = \mathcal{L}(\mathbf{j})$, then $\mathbf{i} \sim \mathbf{j}$ entails $\mathbf{i} = \mathbf{j}$).

I define history operators for our generalised indices in the obvious way: if $\mathcal{L}(\mathbf{j}) = m$, then we have

$$\widehat{C}_{\mathbf{j}} = \widehat{P}^m_{j_m} \cdots \widehat{P}^0_{j_0}. \tag{A.1}$$

Notice that

$$\widehat{C}_{\mathbf{j}} = \left(\sum_{k_l} \widehat{P}^{m+l}_{k_l} \right) \cdots \left(\sum_{k_1} \widehat{P}^{m+1}_{k_1} \right) \widehat{C}_{\mathbf{j}} = \sum_{\mathcal{L}(\mathbf{k})=l+m; \, \mathbf{k} \sim \mathbf{j}} \widehat{C}_{\mathbf{k}}. \tag{A.2}$$

It follows from (A.2) that the decoherence criterion may be rewritten for generalized sequences as

$$\langle \psi | \widehat{C}^{\dagger}_{\mathbf{i}} \widehat{C}_{\mathbf{j}} | \psi \rangle = 0 \text{ unless } \mathbf{i} \sim \mathbf{j}. \tag{A.3}$$

Finally, I say that a generalized history is realized (for a branching history space) iff it is an initial segment of a realized history.

Part (i) of the theorem may now be proved by induction on the length of histories. For histories of length 1–2, it is a trivial consequence of the definition of a branching history; for longer histories, let $\mathbf{j}$ be an arbitrary history with $\mathcal{L}(\mathbf{j}) = m$. We have

$$\widehat{C}_{\mathbf{j}} | \psi \rangle = \widehat{P}^m_{j_m} \widehat{C}_{\mathbf{j}_-} | \psi \rangle \tag{A.4}$$

where $\mathbf{j}_-$ is the initial segment of $\mathbf{j}$ with length $\mathcal{L}(\mathbf{j}_-) = m - 1$. If the theorem holds for sequences with $\mathcal{L}(\mathbf{k}) < m$, then it follows that this vanishes unless $\mathbf{j}_-$ is realized. Assuming $\mathbf{j}_-$ is indeed realized, it also follows that

$$\widehat{P}^{m-1}_{j_{m-1}} | \psi \rangle = \widehat{P}^{m-1}_{j_{m-1}} \sum_{\mathcal{L}(\mathbf{k})=m-2} \widehat{C}_{\mathbf{k}} | \psi \rangle = \widehat{P}^{m-1}_{j_{m-1}} \sum_{\mathcal{L}(\mathbf{k})=m-2, \, \mathbf{k} \text{ realised}} \widehat{C}_{\mathbf{k}} | \psi \rangle \tag{A.5}$$

and therefore, since $\mathbf{j}_-$ is the unique realized history with final element j_{m-1},

$$\widehat{P}^{m-1}_{j_{m-1}} | \psi \rangle = \widehat{C}_{\mathbf{j}_-} | \psi \rangle. \tag{A.6}$$

Hence $\widehat{C}_{\mathbf{j}} | \psi \rangle = \widehat{P}^m_{j_m} \widehat{P}^{m-1}_{j_{m-1}} | \psi \rangle$, which, by the definition of branching, vanishes unless j_{m-1} is the penultimate term on the unique history whose final term is j_m.

To prove part (ii), notice that if Hist has branching structure, we have

$$\widehat{P}_j^m |\psi\rangle = \sum_{\mathcal{L}(\mathbf{k})=m-1, k_m=j} \widehat{C}_{\mathbf{k}} |\psi\rangle = \widehat{C}_{\mathbf{j}} |\psi\rangle \tag{A.7}$$

where $\mathbf{j}$ is the unique history in Hist with $\mathcal{L}(\mathbf{j}) = m$ and $j_m = j$. (If there is no such, then $\widehat{P}_j^m |\psi\rangle = 0$). Then

$$\widehat{P}_{j'}^{m+1} \widehat{P}_j^m |\psi\rangle = \widehat{P}_{j'}^{m+1} \widehat{C}_{\mathbf{j}} |\psi\rangle = \widehat{C}_{\mathbf{j}_+} |\psi\rangle \tag{A.8}$$

where $\mathbf{j}_+$ is the history with final element j' and initial history $\mathbf{j}$. Since $\widehat{C}_{\mathbf{j}_+} |\psi\rangle$ is nonzero for at most one such $\mathbf{j}_+$, $\widehat{P}_{j_{m+1}}^{m+1} \widehat{P}_{j_m}^m |\psi\rangle$ is nonzero for at most one j.

Part (iii) is a trivial consequence of (i). To prove part (iv), suppose that our history space $\{\widehat{P}_j^i\}$ is decoherent (for given $|\psi\rangle$). I then define, for each time index m,

$$|\mathbf{j}, m\rangle = \widehat{C}_{\mathbf{j}} |\psi\rangle / \|\widehat{C}_{\mathbf{j}} |\psi\rangle\| \tag{A.9}$$

where $\mathbf{j}$ is any sequence with $\mathcal{L}(\mathbf{j}) = m$. If $\widehat{P}_j^m$ is any projector in the mth PVM in the history space, then

$$\widehat{P}_j^m |\mathbf{j}, m\rangle = |\mathbf{j}, m\rangle \ (j_m = j)$$
$$\widehat{P}_j^m |\mathbf{j}, m\rangle = 0 \ \text{(otherwise)}. \tag{A.10}$$

Hence the set of operators

$$\widehat{Q}_{\mathbf{j}}^m = |\mathbf{j}, m\rangle \langle \mathbf{j}, m|,$$
$$\widehat{O}_j^m = \widehat{P}_j^m - \sum_{\mathcal{L}(\mathbf{j})=m; \, j_m=j} |\mathbf{j}, m\rangle \langle \mathbf{j}, m| \tag{A.11}$$

is a PVM and $\{\widehat{P}_j^m\}$ is a coarse-graining of that PVM.

To see that the history space consisting of the family of these PVMs (for each m) has branching structure, it suffices to note that:

1. Since $|\psi\rangle$ can always be written as a superposition of the $|\mathbf{j}, m\rangle$, $\widehat{O}_j^m |\psi\rangle = 0$.

2.
$$\widehat{O}_j^{m+1} \widehat{Q}_{\mathbf{k}}^m |\psi\rangle = \sum_k \widehat{O}_j^{m+1} \widehat{P}_k^{m+1} \widehat{Q}_{\mathbf{k}}^m |\psi\rangle$$
$$= \sum_{\mathbf{j}: \mathbf{k} \sim \mathbf{j}} \|\widehat{C}_{\mathbf{j}} |\psi\rangle\| \, \widehat{O}_j^{m+1} |\mathbf{j}, m+1\rangle \tag{A.12}$$

so $\widehat{O}_j^{m+1} \widehat{Q}_{\mathbf{k}}^m |\psi\rangle = 0$.

3.
$$\widehat{Q}_{\mathbf{k}}^{m+1}\widehat{Q}_{\mathbf{j}}^{m} = |\mathbf{k}, m+1\rangle\langle\mathbf{k}, m+1|\mathbf{j}, m\rangle\langle\mathbf{j}, \mathbf{m}| \qquad (A.13)$$

and

$$\langle\mathbf{k}, m+1\mathbf{j}, m\rangle \propto \langle\psi|\widehat{C}_{\mathbf{k}}^{\dagger}\widehat{C}_{\mathbf{j}}|\psi\rangle \qquad (A.14)$$

so $\widehat{Q}_{\mathbf{k}}^{m+1}\widehat{Q}_{\mathbf{j}}^{m}$ vanishes unless $\mathbf{j}$ is an initial segment of $\mathbf{k}$.

Appendix B

Classical Decision Theory

> Reason is commonly associated with logic, but it is obvious, as many
> have pointed out, that the implications of what is ordinarily called
> logic are meager indeed when uncertainty is to be faced. It has
> therefore often been asked whether logic cannot be extended, by
> principles as acceptable as those of logic itself, to bear more fully on
> uncertainty.
>
> Leonard J. Savage[1]

In this appendix I aim to give a general, self-contained, and technically
rigorous introduction to classical (i.e. non-Everettian) decision theory.
The appendix is largely included as background; its main result, however (the Diachronic Representation Theorem) will play some role in
Chapter 6.

A note on sources: none of this chapter is directly based on any particular source; however, for the most part I make no claim to originality,
and in particular, the synchronic decision theory detailed in Sections B.6
and B.7 is a close cousin of the theory in Savage (1972). On the other
hand, the focus on diachronic considerations which takes centre stage in
sections B.8–B.11 is not standard; in particular, the formal framework and
results of those sections are original insofar as I am aware.[2]

To avoid disrupting the flow of the exposition, I have relegated formal
proofs of results to Appendix C unless they are particularly informative in
their own right.

---

[1] Savage (1972).

[2] For classic discussions of representation theorems in decision theory, see Ramsey (1926),
Savage (1972), and Jeffrey (1983); for more recent discussions, see Joyce (1999) and Kaplan
(1996); for a comprehensive discussion of different (synchronic) representation theorems, see
Fishburn (1981).

B.1 The general idea of decision theory

How do rational agents act?

One way to answer the question begins at the end, with the well-known principle that rational agents should act by following the rule of maximizing expected utility (MEU). That is, if $c_1, \ldots c_n$ are the possible consequences of an agent's action, he makes use of a *utility function* $\mathcal{U}$ which maps these consequences to real numbers, a *probability function* Pr which gives the probability of each consequence actually happening, and he chooses whichever action A maximizes

$$\sum_i \Pr(c_i|A)\mathcal{U}(c_i). \tag{B.1}$$

The MEU rule is often presented as if Pr and $\mathcal{U}$ are just given to us (as if the utility of a consequence is its cash value, say, and the probability of that consequence is something given to us by the laws of physics). From this perspective, agents look up the probabilities and utilities, and then (if they are rational) they choose whichever action is best according to the MEU rule. Probabilities and utilities are inputs, preferences are outputs.

From this perspective, though, it is quite mysterious *why* maximizing expected utility is the right thing to do. Why not maximize the expected value of (utility)^2, or log(utility), for instance? Equally, where do these 'utilities' and 'probabilities' come from?

In the Dutch book argument sketched in Chapter 4, we saw an alternative strategy. In a Dutch book argument, we do take utility as an input: in fact, we take the utility of a reward to be its value in dollars. But we do not take probability as an independent input: instead, we define the probability of an event E as the maximum price, in dollars, we would pay for a bet which returns one dollar if and only if E obtains. We then argue that, unless those probabilities obey the axioms of probability calculus, the agent is committed to accepting bets which cause him to lose money.

The Dutch book method for defining probability makes MEU true by definition, at least in the case of these simple bets: the expected utility of a one-dollar bet on E is the probability of E, which by definition is the cash value of that bet. And if (as the Dutch book argument tacitly assumes) the cash value of a combined bet is the sum of the cash values of its components, this can be extended to more complicated bets (ones which return different amounts on different outcomes).

It might appear, then, that the Dutch book strategy takes as input the utility function and the agent's actual preferences, and gives as output the probability function. If this were true, it would make the MEU rule fairly useless as a guide to action. But in fact it is not true. For we have seen that (according to the Dutch book argument) not any old set of preferences counts as rational: those which cause the agent to lose money are deemed irrational, and give rise to no (consistent) probability function. That is, the Dutch book argument is a constraint on rational preferences: it says that agents with certain preferences are rationally committed to having other preferences.

And this is the basic structure of decision theory. It is not in general concerned with the rationality or otherwise of any single decision by an agent: if someone wants to jump into an alligator pit, we might deem them irrational but decision theory will not. But someone who prefers jumping into alligator pits to lying on the beach, and prefers lying on the beach to jumping into snake pits, is constrained to prefer alligator pits to snake pits.

From this perspective, what decision theory aims to do is to state general, reasonable principles of rationality and use those principles to prove that any agent conforming to those principles must be behaving as if he is using MEU with respect to some probability measure and some utility function. The Dutch book argument can be understood as a rudimentary sort of decision theory, but so far it is at most vaguely formulated and its constraints on rationality are questionable (is it *per se* irrational to choose a course of action which always loses money?). We will see, in the rest of this chapter, how to do very much better.

B.2 Synchronic decision problems

To make more progress, we need to provide a formal mathematical framework for decision theory. We will eventually look at several such frameworks, but our first is fairly minimal. It consists of two parts: the possible outcomes of some action (usually called *events*), and the *rewards* which can accrue to an agent if he makes a certain bet on what the outcomes are.

For the moment, we can take the rewards to be elements of any set we like. The events have rather more structure, though: if A and B are events, so should be (A-and-B), (A-or-B), and (not-A). There is a natural mathematical way to represent something like this: recall that a *Boolean algebra* is a set equipped with a 'maximum' element 1, a 'minimum' element 0, associative and symmetric binary operations $\vee$ and $\wedge$, and a unary 'complement' operation $E \to \neg E$, such that:

1. $\vee$ and $\wedge$ are distributive over one another;
2. $E \vee (E \wedge F) = E \wedge (E \vee F) = E$;

3. $E \vee \neg E = 1$ and $E \wedge \neg E = 0$;
4. $0 \wedge E = 0$ and $0 \vee E = E$;
5. $1 \wedge E = E$ and $1 \vee E = 1$.

The paradigm case of a Boolean algebra is the set of all subsets of some set $\mathcal{S}$.

A Boolean algebra has a natural partial ordering which (again following the paradigm case) is written as $\subset$: $E \subset F$ iff there is some G such that $F = E \vee G$. If for any countable set $\mathcal{S}$ of elements in a Boolean algebra, there is a minimal element $\vee \mathcal{S}$ such that $S \subset \vee \mathcal{S}$ for any S in $\mathcal{S}$, the algebra is said to be *complete*. Complete algebras allow the action of $\vee$ to be extended to countably infinite sets:[3] note that $\vee \{E_1, \dots E_n\} = E_1 \vee \cdots \vee E_n$. With this in mind, I define an *event space* just as a complete Boolean algebra.

One natural way to construct the event space is to start with a set of 'atomic events' and take the events to be all subsets of that set. (In betting on dice, for instance, the atomic events might be the different possible showings of the dice.) I will not require this, though: in some circumstances it represents an inappropriate idealization. But in recognition of this natural approach, we shall write $\emptyset$ and $\mathcal{O}$ in place of the minimum and maximum elements 0 and 1. (Note that if $E \in \mathcal{O}$, we can regard E as a subspace of $\mathcal{E}$ and an event space in its own right: $F \in E$ iff for some G, $E = F \vee G$. As such, I normally write $E \subset \mathcal{O}$ for $E \in \mathcal{O}$.)

The simplest decision problems are those where some chance event (a horse race, say, or a measurement, or a presidential election) is going to occur, and where an agent is considering various bets he might place on the outcome. We can represent the possibly payoffs of those bets by some set $\mathcal{R}$ of rewards; fairly obviously, the next thing we need is some formal notion of what a bet actually is.

It is easiest to see how a bet is represented mathematically by supposing that there are atomic events. In this case, we can specify a bet just by saying what reward (or cost!) is acquired (or incurred!) for each atomic events. That is, a bet is just a map from the set of atomic events to the set of rewards. Specifying this in a way which does not presume that there are atomic events is only technically more complicated. We define a *partition* Π of a Boolean algebra $\mathcal{O}$ as a set of elements of $\mathcal{O}$ such that

1. For any two distinct elements F, G of Π, $F \wedge G = \emptyset$.
2. $\vee \Pi = \mathcal{E}$.

---

[3] Via de Morgan's law, $\wedge$ can also be extended to countably infinite sets, but this will not be relevant here.

A *reward-valued partition* of $\mathcal{O}$, then, is just a map f from some reward set $\mathcal{R}$ into $\mathcal{O}$ such that its image, $f(\mathcal{R})$, is a partition of $\mathcal{O}$. Or put another way, to specify a bet we partition the space of events and associate a reward to every element of the partition. At least in the simple context we are considering, reward-valued partitions are the natural way to represent bets. And if there are atomic events, of course, then a function φ from the atomic events to the rewards determines a reward-valued partition f_φ via

$$f_\varphi(r) = \{s : \varphi(s) = r\}. \tag{B.2}$$

It is perfectly reasonable to formulate decision theory with no more structure than a single event space and a set of rewards, and this is indeed the way in which many of the pioneers of the subject (notably Ramsey 1926 and Savage 1972) chose to formulate it. However, it does have the disadvantage that it excludes situations where the agent has a choice not only over which bets to take on a given chancy process, but over which such process is enacted. (The agent might choose how many dice to roll, for instance, or how many times to repeat the experiment.)

It is possible to handle this formally in a fully general way by entirely blurring the distinction between processes and bets. A decision theory of this kind was developed by another pioneer of the field, Jeffrey (1983), and is lucidly discussed and forcefully advocated by Joyce (1999). But it is cumbersome to work with from a technical viewpoint, and is if anything too general for our needs (we are concerned, ultimately, with relatively stylized decision-theoretic problems, like betting on the outcomes of quantum-mechanical experiments). So I will develop this chapter's decision theory through a somewhat unorthodox half-way house: I define a *synchronic decision problem* as specified by:

- a Boolean algebra **E**, corresponding to all the possible outcomes of any possible process;
- a set of *chance setups*, corresponding to the particular processes the agent might choose to set in motion;
- for each chance setup $\mathcal{E}$, a Boolean subalgebra $\mathcal{O}_\mathcal{E}$ of **E**, the *outcome space* of $\mathcal{E}$, corresponding to the possible outcomes of that particular process;
- a set $\mathcal{R}$ of rewards.

A bet, for a given synchronic decision problem, is an ordered pair of one of the chance setups and a reward-valued partition of that chance setup's outcome space. A bet, in other words, corresponds to performing a certain process (represented by the chance setup) and betting on its outcome (represented by the partition). It is natural to call the bet $\langle \mathcal{E}, f \rangle$ a bet *on*

$\mathcal{E}$; a bet $\langle \mathcal{E}, f \rangle$ *has outcomes in* $\mathcal{T}$ (where $\mathcal{T}$ is a subset of $\mathcal{R}$) if $f(r) = \emptyset$ whenever r is not in $\mathcal{T}$.

Two kinds of bets will be particularly useful in the rest of the chapter. If $\{E, E'\}$ is a two-element partition of $\mathcal{O}_\mathcal{E}$, a bet on $\mathcal{E}$ which returns reward r on E and s on its complement is written $C_{r,s}(E|\mathcal{E})$; a bet on $\mathcal{E}$ which returns r whatever the outcome is written $C_r(\mathcal{E})$.

B.3 A rudimentary decision theory

Were I in the reader's shoes, I would probably be getting a sinking feeling at this point: definitions all around, and no end in sight! To break the flow of formality and provide some motivation, then, I will pause to consider a particularly simple decision problem: one with only one chance setup, the event space of which is just the subsets of some finite set of atomic events.

Now, in the bets we considered when discussing the Dutch book argument, the rewards were sums of money, and we made the idealization that any real-valued sum of money could be a reward (never mind that one rarely gets charged 2π dollars for a burger). In our current framework, this amounts to the assumption that $\mathcal{R}$ is actually the real line, and this in turn allows us to add bets up: $(f_1 + f_2)(s) \equiv f_1(s) + f_2(s)$. Let us make this assumption, too, about our decision problem.

Previously, we defined the probability an agent gives to an event as the cash value of a one-dollar bet on that event: that is, as the amount of money such that the agent is indifferent between getting that amount with certainty and taking the one-dollar bet. This can be represented in our framework. Getting an amount λ with certainty is represented by the 'bet' which returns λ whatever the outcome: that is, by the constant function $f(s) = \lambda$. And placing a one-dollar bet on an event E is represented by the function $f(s) = 1$ if $s \in E, f(s) = 0$ otherwise. For convenience, we write the constant function $f(s) = \lambda$ just as λ, and the function which represents a one-dollar bet on E as χ_E.

The remaining concept we need to represent is the idea that an agent might be indifferent between two bets, or might prefer one to another. Representing this simply requires a relation $\succ$ between bets: if $f \succ g$, this just represents the fact that the agent would prefer to take bet f rather than bet g. We write $x \sim y$ iff neither $x \succ y$ nor vice versa, and to represent the fact that $\succ$ is a preference relation, we require that $\succ$ is transitive and asymmetric and that $\sim$ is an equivalence relation. (We define $x \succeq y$ to mean that either $x \succ y$ or $x \sim y$; $\prec$ and $\preceq$ are defined in the obvious way.)

So in this simple framework, to say $\Pr(E) = p$ is to say that (for the agent we are considering) $\chi_E \sim p$. But to show that these probabilities satisfy the probability calculus, we need a further assumption: we need to assume that if an agent is willing to pay p dollars for one bet, and q dollars for another bet, he is willing to pay $p + q$ dollars for the combined bet. If this is so, then $\chi_E + \chi_F \sim p + q$ and so $\Pr(E \cup F) = \Pr(E) + \Pr(F)$.

Finally, if we extend this linearity assumption to also include the assumption that an agent willing to pay p dollars for a one-dollar bet on E is willing to pay λp dollars for a λ-dollar bet on E, it follows that the agent's preferences can be expressed in terms of the MEU rule. For

$$f = \sum_s f(s)\chi_{\{s\}}, \tag{B.3}$$

so that

$$f \sim \sum_s \Pr(s)f(s) \equiv EU(f). \tag{B.4}$$

Mathematically speaking, we have now proved the following theorem.[4]

Theorem 1 (Toy Representation Theorem). *Suppose that S is a finite set and $\succ$ is a total ordering on real-valued functions of S satisfying*

1. *If $f_1 \sim g_1$ and $f_2 \sim g_2$, then $\lambda f_1 + \mu f_2 \sim \lambda g_1 + \mu g_2$ for any λ and μ greater than zero.*
2. *For any f there is a constant function c such that $f \sim c$.*

Then there exists a unique function $\Pr$ from S to the nonnegative reals, such that

$$\sum_{s \in S} \Pr(s) = 1 \tag{B.5}$$

and such that for any functions f, g, $f \succ g$ iff $EU(f) > EU(g)$, where

$$EU(f) = \sum_{s \in S} \Pr(s)f(s). \tag{B.6}$$

---

[4] Mathematically inclined readers may recognize this as an elementary form of the Radon–Nikodyn theorem (cf. Rudin 1987: 121).

B.4 Representing preferences; additive decision theory

This is the first example we have so far seen of a *representation theorem*: an argument which starts off with some constraints on an agent's preferences between bets, and concludes that those preferences are uniquely (or very nearly uniquely) represented by some probability and utility, via the MEU rule. To get a bit more precise about these notions, we make the following definitions:

1. A *preference order* on a set B of bets for a (synchronic) decision problem **E** is just a total ordering of B.
2. A *probability measure on an event space* $\mathcal{O}$ is a map Pr from $\mathcal{O}$ into the non-negative real numbers, satisfying

   (a) $\Pr(E \vee F) = \Pr(E) + \Pr(F)$ whenever $E \wedge F = \emptyset$.
   (b) $\Pr(\mathcal{O}) = 1$.

   A *probability measure for a (synchronic) decision problem* is an assignment, to every chance setup $\mathcal{E}$, of a probability measure $\Pr(\cdot|\mathcal{E})$ on the event space $\mathcal{O}_\mathcal{E}$ of $\mathcal{E}$.
3. A *utility* for a decision problem is a function from the reward set of that problem to the real numbers. A utility $\mathcal{U}$ satisfying some condition is *quasi-unique* if any other utility $\mathcal{U}'$ satisfying that condition is related to $\mathcal{U}$ by a positive affine transformation: $\mathcal{U}'(r) = \alpha\mathcal{U}(r) + \beta$, with $\alpha > 0$. (The reason why *quasi*-uniqueness rather than uniqueness is the relevant condition is that positive affine transformations of $\mathcal{U}$ do not affect the ordering determined by the MEU rule: the scaling and zero points of $\mathcal{U}$ are purely conventional.)
4. The *expected utility* of a bet $\mathcal{F} = \langle \mathcal{E}, f \rangle$, with respect to a given probability measure $\Pr(\cdot|\cdot)$ and utility function $\mathcal{U}$ for the decision problem, is

$$EU(\mathcal{F}) = \sum_{r \in \mathcal{R}} \Pr(f(r)|\mathcal{E}) \times \mathcal{U}(r). \tag{B.7}$$

5. A preference order $\succ$ over a set B of bets is *represented* by a probability measure if there is a quasi-unique utility function such that with respect to that measure and that function, for any bets $\mathcal{F}$ and $\mathcal{G}$ in B

$$\mathcal{F} \succ G \text{ iff } EU(\mathcal{F}) > EU(\mathcal{G}). \tag{B.8}$$

(It is probably more normal to say that the order is represented jointly by the probability measure and the utility function; I make this choice partly for technical reasons, mostly because ultimately it is the probability measure that mainly interests us.)

So a representation theorem is a theorem which gives a set of conditions on an agent's preferences over some set of bets sufficient to require that those preferences are represented uniquely by some given probability measure.

As representation theorems go, theorem 1 is simple but not terribly illuminating. The restrictions to a single chance setup and to a finite set of atomic events are unimportant (I made them mostly for expository convenience). More seriously, the theorem effectively smuggles in a utility function in the premises by requiring, somewhat implausibly, that the set of rewards is isomorphic to the real numbers; this makes it confusing to see exactly how a utility function is *derivable* from constraints on an agent's preferences.

However, this assumption can be weakened, and the weakening is instructive. What really matters about the reward space is not its full topological structure: it is the fact that rewards can be added together. This gives us an operational notion of utility which is independent of probability: we can say that reward A is twice as valuable (to an agent) as reward B iff the agent is indifferent between receiving A and receiving two copies of B. Using an arbitrary reward as a reference point (and arbitrarily giving it utility 1) we can then define a utility function for all rewards; probabilities then follow by our earlier argument.

We might call this *additive* decision theory, since it relies essentially on this additive structure on the reward space and the interaction of that structure with the preference ordering. Formally, it looks like this.

Theorem 2 (Additive Representation Theorem). *Suppose that E is a synchronic decision problem and that $\succ$ is a preference order for some set $\mathcal{B}$ of bets for the decision problem. If*

A1 *The set $\mathcal{B}$ includes all possible bets;*

A2 *The event space of E is the set of subsets of some finite set of atomic events;*

B1 *Any two constant bets with the same payoff are equivalent: if r is a reward and $\mathcal{E}_1, \mathcal{E}_2$ are chance setups, then $C_r(\mathcal{E}_1) \sim C_r(\mathcal{E}_2)$;*

B2 *There is an associative, commutative map $\oplus$ from pairs of rewards to rewards, and some reward 0 such that $r \oplus 0 = r$ for any reward r, such that*

(i) If $\mathcal{F}_1 \succeq \mathcal{F}_2$ and $\mathcal{G}_1 \succeq \mathcal{G}_2$, then $\mathcal{F}_1 \oplus \mathcal{G}_1 \succeq \mathcal{F}_2 \oplus \mathcal{G}_2$;
(ii) If in addition $\mathcal{F}_1 \succ \mathcal{F}_2$, then $\mathcal{F}_1 \oplus \mathcal{G}_1 \succ \mathcal{F}_2 \oplus \mathcal{G}_2$;

B3 *There is a reward r such that $C_r(\mathcal{E})$ is not equivalent to $C_0(\mathcal{E})$;*

C1a *For any bets* $\mathcal{F}$, $\mathcal{G}$, *there is some positive integer* n *such that* $n\mathcal{F} \succ \mathcal{G}$ *(where* $n\mathcal{F}$ *just means* $\mathcal{F}$ *added to itself* n *times);*

C1b *For any bets* $\mathcal{F}$, $\mathcal{G}$ *such that* $\mathcal{G} \succ C_0(\mathcal{E})$, *there is some positive integer* n *such that* $n\mathcal{G} + \mathcal{F} \succ C_0(\mathcal{E})$;

C1c *For any bets* $\mathcal{F}$, $\mathcal{G}$ *such that* $\mathcal{G} \prec C_0(\mathcal{E})$, *there is some positive integer* n *such that* $n\mathcal{G} + \mathcal{F} \prec C_0(\mathcal{E})$;

C2a *For any bets* $\mathcal{F}$, $\mathcal{G}$ *such that* $\mathcal{F} \succ \mathcal{G} \succ C_0(\mathcal{E})$, *there is some positive integer* n *such that* $n\mathcal{F} \succ (n+1)\mathcal{G}$;

C2b *For any bets* $\mathcal{F}$, $\mathcal{G}$ *such that* $\mathcal{F} \prec \mathcal{G} \prec C_0(\mathcal{E})$, *there is some positive integer* n *such that* $n\mathcal{F} \prec (n+1)\mathcal{G}$;

then $\succ$ *is uniquely represented by some probability measure* Pr.

The proof of this theorem can be found in Appendix 3.

B.5 Structure axioms; almost representation

As my labelling probably indicates, the axioms of the additive representation theorem fall naturally into several categories, and in fact this is typical of decision-theoretic representation theorems. Axioms A1 and A2 are not really axioms about the agent's preferences at all: rather, they are constraints on the decision problem itself. We might call these *richness axioms*: they force the decision problem to be sufficiently rich that a representation theorem can be proved. Such axioms are invariably required: if there are (say) only three or four bets over which the preference order is defined, there will generally be nothing like enough structure to construct a unique probability measure.

Axioms B1–B3 are the main way in which this theorem formalizes what we might call the intuitively defensible constraints on an agent's preferences: in all three cases, it is reasonable to ask how we might in practice check if someone conforms to them. B1 says that if two bets each deliver some particular reward with certainty, agents should be indifferent between them: if the choice is between rolling a die and getting 100 dollars whatever the result, and flipping a coin and getting 100 dollars whatever the result, then it is irrational to prefer one to the other. This rule is really part of the definition of rewards: if the agent actually does prefer one to the other, we would have done better to represent '100 dollars, die rolled' and '100 dollars, coin flipped' as different rewards.

B3 is just a requirement of nontriviality. The Zen-like individual with no preferences between any two actions cannot really be called an agent at all, and lies outside the scope of decision theory. B2 basically expresses the fact that an agent can make more than one bet, and that his preferences between bets is unaffected by whatever bets he has already taken. Its reasonableness, or otherwise, will be discussed in section B.6.

I call axioms B1–B3 *rationality axioms*, but I should note a slight blurring of the distinction between these axioms and the richness axioms: B2 is partially a statement about the structure of the decision problem itself, but it is unavoidably stated in terms of the preference order. I should also note that the assumption that $\succ$ is an ordering relation is a further, tacit rationality axiom.

C1 and C2 are also constraints on a rational agent's preferences, but of a very different kind. Their role in the theorem is basically to ensure that the utility and probability functions have the right mathematical structure.

In particular, C1a–C1c serve to rule out the existence of rewards, or probabilities, which differ infinitesimally from one another. C2a–C2b instead rule out rewards that differ infinitely from one another. In the former case, we are ruling out the agent who prefers r to s but will accept any bet which returns s over one which returns r with lower probability, however small the probability difference. In the latter case, we are ruling out the agent who regards r as so much more valuable than s that arbitrarily small probabilities of r make it preferred to s. Axioms of this form I will call *structure axioms*. It should be fairly obvious for each of them that no realistic behaviour pattern could distinguish between agents who do and do not conform to them; fairly transparently, they are not chosen because some argument can be given for them but so as to rule out precisely the cases which they do rule out.[5] In other areas of decision theory, further structure axioms are required to handle bets with infinitely many different outcomes (I avoided this by fiat in theorem 2).

We could deal with this issue by carefully stating structure axioms for each representation theorem; we could deal with it by using a more generalized set than the real numbers to represent probabilities and utilities. But instead, I will mostly sidestep the issue of structure axioms by slightly weakening the idea of representation. First, let us define any three rewards r_1, r_2, r_3 as *comparable* iff there exist chance setups $\mathcal{E}_0, \mathcal{E}_+, \mathcal{E}_-$, and events $E_\pm \subset O_{\mathcal{E}_\pm}$, such that

$$C_r(\mathcal{E}_0) \succ C_{r,t}(E_+|\mathcal{E}_+) \succeq C_s(\mathcal{E}_0) \succeq C_{r,t}(E_-|\mathcal{E}_-) \succ C_t(\mathcal{E}_0) \qquad \text{(B.9)}$$

---

[5] I borrow the distinction between structure and richness axioms from Suppes (1974) (cf. the discussion in Joyce 1999); I use it in a slightly different way, though.

for r, s, t some permutation of r_1, r_2, r_3. (The idea of comparability is that, where rewards are comparable, we can rule out the possibility that they differ infinitely.) A comparable set of rewards is a set of rewards such that any three such rewards are comparable.

I also say that a bet $\langle \mathcal{E}, f \rangle$ is *finite* iff $f(r) = \emptyset$ for all but a finite number of rewards r. (Notice that we revert here to the general notion of a bet as an ordered pair of a chance setup and a reward-valued partition.)

I will now define a preference order $\succ$ on a set $\mathcal{B}$ of bets to be *almost represented* on a comparable set of rewards $\mathcal{T}$ by a probability measure Pr iff there is a quasi-unique utility function on $\mathcal{T}$ such that, for any finite bets $\mathcal{F}, \mathcal{G}$ which have outcomes in $\mathcal{T}$, if $EU(\mathcal{F}) > EU(\mathcal{G})$ then $\mathcal{F} \succ \mathcal{G}$. And I say that a preference order is *uniquely almost represented* iff for any comparable set of rewards it is almost represented on that set by a unique probability measure.

In other words, almost representation is like representation except that (i) we ignore nonfinite bets; (ii) we restrict our attention to sets of rewards which are not infinitely different; (iii) we are relaxed about the possibility of differences in probabilities or utilities two small to be represented by real-valued functions. The difference between representation and almost representation seems to have little physical or operational significance, so for most of my discussion I shall be content if I can establish that a given preference order is uniquely almost represented. In due course, however, we will see that the Everett interpretation does allow us to go some way to removing this inelegant aspect of decision theory.

B.6 Beyond additivity

It is time to give a little critical consideration to the additivity axiom in theorem 2. As I noted before, it implies that agents' preferences between bets are unaffected by whatever previous bets they have taken. This assumption is essential to the structure of the theorem; unfortunately, it is scarcely reasonable. We can see this directly: it is hardly irrational to suppose that my preference between, say, a vintage bottle of champagne and a hundred dollars would change if I received a billion dollars. It is even more clear by looking at its indirect consequences: additivity forces the utility of a cash reward to be proportional to its cash value, which in turn implies that it is irrational to be risk averse. But if it is irrational not to swap the mortgage for a one-in-a-million chance of winning Microsoft, I for one plead guilty. Additivity might be a plausible assumption when playing for very small stakes, but it is not reasonable in general.

This creates difficulties, though. It is only through additivity that the reward set acquires enough structure that we can define utilities prior to considerations of probability and so use those utilities to measure probabilities. Without either additivity, the reward set is just an arbitrary ordered set, and if $\mathcal{U}$ is a valid utility function, so is any increasing function of $\mathcal{U}$.

In fact, it has become generally accepted in decision theory[6] that the only meaningful way to measure utilities is via probabilities. Roughly, to say that one reward is twice as valuable as another to me is to say that I am indifferent between the second reward and a 50% chance of the first reward. More accurately, if r, s, t are three rewards with $r \succ s \succ t$, and if I am indifferent between (i) a bet which returns s with certainty and (ii) a bet which returns r with probability p and t with probability $(1-p)$, then

$$p\mathcal{U}(r) + (1-p)\mathcal{U}(t) = \mathcal{U}(s), \qquad (B.10)$$

and so

$$\mathcal{U}(r) - \mathcal{U}(t) = \frac{1}{p}(\mathcal{U}(s) - \mathcal{U}(t)), \qquad (B.11)$$

which fixes the utility up to a positive affine transformation.

But the fact that we can use probability to quantify utility is of little use if we need utility to quantify probability. Fortunately, there is another way to do the latter task, and it relies on the fact that—unlike the set of rewards—the space of events automatically has a rich algebraic structure.

Indeed, suppose that $\succ$ is an ordering on an event space which (let us say) represents likelihood. Then not only can we say of any two events which is more likely; we can say of any three events whether one is more likely than the union of the other two. And, indeed, if A and B are disjoint events each of which is as likely as the other, and if $A \cup B$ is as likely as C, it is reasonable to describe C as twice as likely as A. If C is the entire event space, then, it is reasonable to say that A has probability 0.5. For a sufficiently rich event space, arguments of this form suffice to define numerical probabilities for every event.

Where does the likelihood ordering come from, though? In fact, such orderings can be constructed from a preference ordering. For suppose r and s are any two rewards such that r is preferred to s, and E and F are events in the event spaces of chance setups $\mathcal{E}$ and $\mathcal{F}$. Then to say that the agent judges E to be more likely (given $\mathcal{E}$) than F is (given $\mathcal{F}$) is just to

---

[6] See Savage (1972: 91–104), and references therein, for some discussion of the history of this problem.

say that they prefer a bet on $\mathcal{E}$ that returns r if E obtains and s otherwise, to one on $\mathcal{F}$ that returns r if F obtains and s otherwise.

To be more formal: we define a *likelihood ordering* for a given synchronic decision problem as a weak ordering relation between ordered pairs of chance setups and events in the event spaces of those setups, and we write it thus: $E|\mathcal{E} \succ F|\mathcal{F}$. Given an preference order which generates a well-defined ordering on rewards (i.e., one satisfying $C_r(\mathcal{E}) \sim C_r(\mathcal{F})$ for all r), and any two rewards r, s with $r \succ s$, we can define a likelihood ordering by

$$E|\mathcal{E} \succ_{r,s} F|\mathcal{F} \text{ iff } C_{r,s}(E|\mathcal{E}) \succ C_{r,s}(F|\mathcal{F}). \qquad (B.12)$$

If the ordering thus generated is independent of the particular choice of r and s, we call the original preference order *probabilistic*, and write it just as $\succ$.

One further qualitative sense of probability is immediately available. If E is an event in $\mathcal{O}_\mathcal{E}$, we call E *null* (with respect to $\mathcal{E}$) iff $E|\mathcal{E} \sim \emptyset|\mathcal{E}$: that is, if E is as likely as the empty set.

Our strategy, then, is to use an agent's qualitative preferences between bets with fixed stakes to define an ordering on events, use the algebraic structure of the event space to generate a quantitative representation of that ordering, and then in turn use that representation to define a quantitative measure of utility. (This strategy basically follows Savage 1972.) As a useful intermediate step, we shall say that a probability measure Pr *represents* an likelihood ordering $\succ$ if $E|\mathcal{E} \succ F|\mathcal{F}$ if and only if $\Pr(E|\mathcal{E}) > \Pr(F|\mathcal{F})$; if $\Pr(E|\mathcal{E}) > \Pr(F|\mathcal{F})$ entails $E|\mathcal{E} \succ F|\mathcal{F}$ but not necessarily vice versa, we say that Pr *almost represents* $\succ$.

We have now used orderings on bets to define orderings on rewards and on events, in each case requiring some extra rationality assumption to do so. There is one more example of this process which will prove vital later: it will be essential to ask when an ordering can be restricted to a subalgebra of the event space. In the case of a likelihood ordering, this is trivial: the restriction of a total ordering of $\mathcal{E}$ to E is automatically a total ordering of E. Things are a bit more complicated for preference orderings, though. The restriction to E of a preference order $\succ$ will be well-defined only if, for any set of reward-valued partitions f, f' and g, g' such that

- on E, $f = f'$ and $g = g'$;
- on $\neg E$, $f = g$ and $f' = g'$,

then $f \succ g$ iff $f' = g'$. If this condition is satisfied, we can safely define a new ordering $\succ_E$ on reward-valued partitions of E: $f \succ_E g$ iff there are extensions f^*, g^* of f and g to $\mathcal{E}$ such that $f^* \succ g^*$. Orderings which restrict in a well-defined way like this to all subalgebras of $\mathcal{E}$ are said to obey the *sure thing principle*; its justification will be considered later.

Before we can (finally!) state and prove a representation theorem along these lines, we need one last ingredient. It is well known (see e.g. Fishburn 1981), and in any case fairly obvious, that qualitative assumptions alone are not enough to prove the existence of a unique representation (or almost representation). It is also necessary to assume that the setup is sufficiently rich as to allow the construction of divisions of any given event into arbitrarily many equiprobable events.

If we only have one chance setup, the only way to do this is directly: to require that for any N, any event may be partitioned into N subevents each of which is equally likely. (As a consequence, the event space must contain infinitely many elements.) This is the more common strategy, and the one adopted by Savage (1972). In our generalized setup, however, an alternative is available which is more suited to our later needs:

A likelihood ordering $\succ$ for a synchronic decision problem **E** is *partitionable* iff for any N there exists some chance setup $\mathcal{E}_N$, and some N-element partition $O_{\mathcal{E}_N} = E_N^1 \vee \cdots \vee E_N^N$, such that:
for any chance setup $\mathcal{E}$, there exists chance setup $\mathcal{E}^*$, and embeddings

$$v : O_{\mathcal{E}} \longrightarrow O_{\mathcal{E}^*}$$
$$v' : O_{\mathcal{E}_N} \longrightarrow O_{\mathcal{E}^*}$$

compatible with $\succ$, such that

$$v(E) \wedge v'(E_N^i) \sim v(E) \wedge v'(E_N^j).$$

This looks worse than it is: partitionability is intended to express the assumption that there exist actions with N outcomes such that (i) each outcome is equiprobable; (ii) the action can be performed concurrently with any other action, without affecting this equiprobability. If, for example, there exist fair dice (with arbitrarily many sides, strictly speaking!) which an agent can always choose to roll irrespective of whatever else he does, that agent's decision problem is partitionable.

B.7 Synchronic representation theorems

Our current goal is in sight: we have assembled sufficient ingredients to give, for the synchronic decision problem which we are considering, a representation theorem that does not rely on the implausible additivity assumption. In fact, we give two such theorems: one for likelihood orderings, one for preference orderings.

Theorem 3 (Synchronic Likelihood Theorem). *Suppose that $\succ$ is a likelihood ordering for some synchronic decision problem $\mathbf{E}$, satisfying:*

SL1 $O_{\mathcal{E}}|\mathcal{E} \succ \emptyset|\mathcal{E}$.
SL2 $O_{\mathcal{E}}|\mathcal{E} \succeq E|\mathcal{E} \succeq \emptyset|\mathcal{E}$ *for any $\mathcal{E}$ and any $E \in \mathcal{E}$.*
SL3 *If $E, F, G \subset O_{\mathcal{E}}$ and $E \wedge G = F \wedge G = \emptyset$, then $E|\mathcal{E} \succ F|\mathcal{E}$ iff $E \vee G|\mathcal{E} \succ F \vee G|\mathcal{E}$.*
SL4 $\succ$ *is partitionable.*

Then $\succ$ is uniquely almost represented.

Theorem 4 (Synchronic Preference Theorem). *Suppose $\succ$ is a total ordering on bets for some synchronic decision problem $\mathbf{E}$, satisfying:*

SP1 *If f, g are bets on the same chance setup $\mathcal{E}$ and $E_1, \ldots E_n$ is a partition of $O_{\mathcal{E}}$ such that $f^{-1}(E_i)$ and $g^{-1}(E_i)$ are single-valued (such a partition can always be found), then*

  1. *If $f^{-1}(E_i)|\mathcal{E} \succeq g^{-1}(E_i)|\mathcal{E}$ for all non-null E_i, then $f \succeq g$.*
  2. *If in addition $f^{-1}(E_i)|\mathcal{E} \succ g^{-1}(E_i)|\mathcal{E}$ for at least one non-null E_i, then $f \succ g$.*

  (This is sometimes called the dominance principle.)

SP2 $\succ$ *defines a well-defined ordering on the set of rewards: that is, $C_r(\mathcal{E}) \sim C_r(\mathcal{F})$.*
SP3 $\succ$ *defines a well-defined ordering $\succ_E$ on the restriction of bets to any event E: that is, it satisfies the sure thing principle.*
SP4 $\succ$ *defines a well-defined likelihood ordering: that is, it satisfies probabilism.*
SP5 *The likelihood order determined by $\succ$ is partitionable.*

Then the preference ordering is uniquely almost represented.

Most of the work required to prove these results (the proofs can as usual be found in Appendix C) goes into establishing the synchronic likelihood theorem. With this theorem in hand, a representation theorem for *preference* orderings follows quickly: stripped of the technicalities, its proof is just a formalization of the informal definition of utility in terms of probability.

What is the status of the assumptions that go into the synchronic preference theorem? As I noted before, SP2 is really just a statement of what rewards are: it is constitutive of something being a reward, in our sense, that an agent is indifferent between two sequences of events each of which returns that reward with certainty.

SP5 is largely a statement about the richness of the decision-theoretic situation. As I noted earlier, it is tantamount to the assumption that the world contains fair dice. Of course, this assumption is not quite innocent. In a foundational context where we are hoping to explicate what probability is, it is not entirely satisfactory that concepts like 'fair die' have to enter as basic premises. Nonetheless, the premise is essential. Equally essential, and equally bound up with pre-theoretic notions of what probability is, is SP4, which guarantees that if one event is judged more probable than another using one set of stakes, this judgement is not affected by a change of stakes. However, we will be able to place neither SP4 nor SP5 on a more satisfactory footing, at least until we consider the Everett interpretation, and for now we have to take them on trust.

SP1 and SP3 are more interesting. SP1, the dominance principle, says in effect that if one course of action always does at least as well as another, and does better in some circumstances, then the first course of action is to be preferred. SP3, the sure thing principle, says that if we prefer bet X to bet Y, and both bets deliver the same reward if outcome A obtains, then our preference shouldn't be changed if we alter both bets' rewards on outcome A whilst keeping them equal to each other.

SP1 probably sounds intuitively obvious. SP3 generally strikes people as considerably less obvious, and indeed intuitions can easily be made to run in the opposite direction. A famous example known as *Allais' paradox* goes like this:[7] suppose that events A, B, and C have probabilities of 0.4, 0.59 and 0.01 respectively, and that we are considering the following 4 bets.

X1 pays one million dollars whatever happens.

X2 pays five million dollars on A, one million dollars on B, and nothing on C. (So it gives a 40% chance of winning five million dollars, a 59% chance of winning one million dollars, and a 1% chance of winning nothing.)

Y1 pays one million dollars on A and C, and nothing on B. (So it gives a 41% chance of winning one million dollars, and a 59% chance of winning nothing.)

---

[7] The paradox was presented in Allais (1953); see Savage (1972: 101–3) for an English-language account and a response.

Y2 pays five million dollars on A, and nothing on B and C. (So it gives a 40% chance of winning five million dollars, and a 60% chance of winning nothing.)

Many people, on first being asked, prefer X1 to X2 and Y2 to Y1. A typical narrative might go: X1 offers a life-changing[8] million-dollar windfall. Even a small chance of losing that windfall is very serious; the chance X2 offers of a merely quantitative increase from one million to five million dollars is not adequate compensation.

On the other hand (the narrative continues), Y1 and Y2 both offer moderate-sized, and pretty similarly sized, chances of life-changing windfalls. The chance is only slightly lower in Y2 than in Y1, and it more than pays for itself via the greatly increased size of the windfall.

However, a moment's reflection shows that this pattern of preferences is inconsistent with the sure thing principle. For X1 and X2, and Y1 and Y2, are identical on C; X1 and Y1, and X2 and Y2, are identical on A and B.

Now, there are ways of redescribing the example to try to undermine the narrative and make the sure-thing-compliant choices more intuitive. But this is not a satisfactory way to proceed. We are, after all, ultimately investigating decision theory for the light it can shed on physics, and in particular on the Everett interpretation But in physics our pretheoretic intuitions are all but worthless as a guide to the truth. And in a defence of the *many-worlds interpretation* in particular, sceptics are unlikely to be impressed by appeals to intuition.[9]

So it would be nice to find a different route to defend the sure thing principle (which plays an indispensable role in the synchronic preference theorem and (through its surrogate, SL3) in the synchronic likelihood theorem), and for all that it is more straightforwardly intuitive, it would be nice to find a more principled defence of the dominance principle too. In fact, both can be (more) satisfactorily understood once we consider situations where decisions are taken over a period of time; this *diachronic* decision theory will be our concern for the remainder of the chapter.

---

[8] Readers for whom winning a million dollars would not be life-changing are encouraged to (i) increase the numbers appropriately; (ii) contact the author's academic institution for details of philanthropic opportunities.

[9] This might suggest that in mainstream philosophy, appeals to intuition *would* be a valid way of proceeding. As a sociological observation about the discipline, this is (unfortunately) true, but as a rule accounts as to why intuitions are guides to truth in philosophy are conspicuous by their absence. For criticisms of intuition-based methodology in philosophy (from very different perspectives) see Williamson (2007), Ladyman and Ross (2007), and Knobe and Nichols (2008).

B.8 The diachronic perspective

To see why a diachronic decision theory (a decision theory in which we allow for agents making multiple sequential decisions and interacting with multiple sequential random processes) can help with the justification of the sure thing principle, consider the following situation. A token between 1 and 100 is randomly drawn out of a hat, but not shown to the agent: all he is told is whether or not it is greater than 60.

If the result is between 1 and 60, a reward r is given to the agent, independent of any choices he might make. If it is between 61 and 100, he is offered a choice: he can either have one million dollars whatever is on the ticket; or he can have five million dollars if the number on the ticket is not 100, and nothing if it is.

In other words, the agent is in a situation where he has a 60% chance of winning r (whatever r turns out to be), and a 40% chance of being given a choice between (a) a million dollars and (b) a 39/40 chance of five million dollars. I claim that his choice, should he be offered the choice in the first place, is independent of r: he *has not received* r, so its value does not affect him in any way. (He could, indeed, be offered the choice without even knowing what r is.)

But in this case, the earlier choice of X1 over X2 and Y2 over Y1 is shown to be irrational. For if r is one million dollars, our choice is between X1 and X2; if r is nothing, our choice is between Y1 and Y2. In either case, the agent will prefer X1 to X2, or Y1 to Y2, if and only if he prefers getting one million dollars with certainty to getting five million dollars with probability 39/40. So the sure thing principle (at any rate in this example) can be informally derived once we allow for diachronic decision problems (and we will see shortly how to make this formal).[10]

A similar defence can be given of dominance. Suppose (for example) that a coin is tossed. If it lands heads, the agent gets nothing; if it lands tails he is offered a choice between ten dollars and twenty dollars. On the assumption that he values the latter more than the former, he will of course take it.

But this is just one way of offering an agent a choice between (a) a bet which pays ten dollars on heads, nothing on tails; and (b) a bet which pays twenty dollars on heads, nothing on tails. The agent's preference for twenty dollars with certainty over ten dollars with certainty commits him to a preference for bet (b) over bet (a). This is of course just a special case of the dominance principle.

---

[10] This defence is adapted from Savage (1972: 103).

So, diachronic decision theory casts light on *synchronic* decision theory by illuminating its premises. A very different reason for studying it comes from problems specific to the diachronic context. Consider the following, for instance: an agent knows one of A_1 through A_n is true, and his preferences between bets on them determine his probability of each. He then learns some new information: B is true. How should he update his probabilities to allow for this new information?

As so often in this subject, the answer is less controversial than its rationale. The agent should replace his probabilities by the *conditional probabilities*

$$\Pr'(A_i) \equiv \Pr(A_i|B) = \Pr(A_i \cap B)/\Pr(B). \qquad (B.13)$$

The rule is usually known as *Bayesian updating*, and is extremely familiar; indeed, I already made tacit use of it in my defence of the sure thing principle earlier in this section. But what justifies it?

Quite a number of defences have been made in the literature,[11] but from the operational perspective we are taking to probability, the most relevant proceed via variants on the Dutch book argument. Just as in synchronic decision theory the probability axioms can be derived by showing that agents who violate them are committed to accepting a set of bets which always lose money, so in diachronic decision theory we can attempt to justify the Bayesian update rule by showing that agents who violate it are committed to accepting a sequence of bets which always lose money. (See Lewis 1997 for a formal presentation of the diachronic Dutch book argument.)

Synchronic Dutch book arguments turned out to be less than satisfactory, and we were led to replace them by a more systematic synchronic decision theory, so I will construct in the remainder of this chapter a systematic *diachronic* decision theory which allows us to derive both that an agent's preferences can be represented by means of a probability measure and that this measure satisfies Bayesian updating.

In doing so, I will have to make a crucial assumption which I call *diachronic consistency*, and which already has been tacit throughout this section.[12] Informally, diachronic consistency requires that an agent's preferences at different times must not be in conflict. If an agent knows that,

---

[11] See e.g. Teller (1976), Williams (1980), van Fraassen (1989; 1999), and Greaves and Wallace (2006).

[12] I explore the principle of diachronic consistency further, largely independently of the Everett interpretation, in Wallace (2010a).

should in the future he be offered a choice between C and D he will take C, then he should not now choose D over C.

When I justified the sure thing principle by breaking the choice between bets into two steps, I assumed tacitly that the mere breaking was not decision-theoretically relevant: it was not significant for an agent's preferences at what time in the process he made a choice. This makes no sense unless we assume that the agent's preferences do not change over the course of making the bet. When I justified the dominance principle by appealing to the agent's determinate preference between rewards after a coin had been flipped, I assumed that the pre-flip and post-flip versions of the agent had their interests aligned.

That some form of diachronic consistency is essential even to make sense of decision-making follows from the fact that actual decision-making takes place over time. An agent's actions take time to carry out; his desires and goals take time to be realized. If his preferences do not remain consistent over this timescale, action is not possible at all. Such common-place decisions as agreeing to pick someone up from work, accepting a conference invitation, proposing marriage, going out to buy an ice-cream... all require that an agent can reasonably decide, at a particular moment, what he is going to do over some future period. Decision theory is hopeless if it reduced to a hedonistic calculus of which instantaneous muscle movements will lead to the most pleasurable feeling. Indeed, since almost any action I decide upon at a given instant is enacted largely through a series of actions performed at later instants, it is not really even coherent to think of agents *acting* unless their behaviour is basically diachronically consistent.

In fact (I claim), diachronic consistency is somewhat akin to the rationality principle of *ordering*. We have been assuming throughout this discussion that an agent's preference order is indeed an *order*: that it is transitive, antisymmetric, and so forth. The justification for this is not so much that we have empirical evidence, or an intuitive sense, that agents' real preferences are indeed ordered; it is rather that without this assumption we cannot really make sense of an agent having preferences at all. To be sure, in stylized cases (where an agent is simply given three options to consider and asked which one of each pair he would choose), we can extract a set of behavioural dispositions which could just about be called preferences, but in messy real-world situations where the exact set of choices available is not cleanly defined, this will break down.

This is not to say that no real people display inconsistent preferences from time to time; of course they do. Real people are not ideally rational, but perfect rationality is an idealization which we adopt in order to interpret the speech acts and physical movements of real people as the actions of agents. It would be a wild piece of psychological speculation to suppose that encoded in your head and mine are a set of real numbers which give your and my personal probabilities of each outcome, but provided that our behavioural dispositions are *approximately* in accordance with the axioms of decision theory, those dispositions will tacitly fix an *approximate* set of probabilities.[13] (This is just one more example of emergence, as discussed in Chapter 2.)

In a similar way, it seems that what makes it appropriate to treat a certain temporally extended object as a single agent is the fact that we can idealize it both as having instantaneously consistent dispositions and as having dispositions at different times which are compatible.

Arguably, there are plenty of localized violations of this idealisation. If I tell my friend not to let me order another glass of wine after my second, I acknowledge that my desires at that point will conflict with my desires now. But notice that such situations

(a) are generally not taken to be rational;
(b) are indeed analysed as situations of conflict, where my present self acts to prevent my future self having access to his preferred choice;
(c) are localized, taking place against a general assumption of diachronic consistency in myself and others (as when I assume that my friend will indeed persuade me out of that second glass, or that the morning after the night before, I'll be glad that she did).

Occasional localized violations of diachronic consistency can just be taken as localized irrationality; some more systematic violations can be analysed by dropping the assumption that an agent at two rather different times really is best analysed as a single agent; but a really widespread, generic violation of diachronic consistency, as with a really widespread, generic violation of *synchronic* consistency, corresponds not to an agent who is violating a norm of decision theory but to a physical system which is not an agent at all.

I now return to the formal development of decision theory, and in particular to the development of a formal framework for diachronic decision theory.

---

[13] Readers familiar with the philosophy of mind will notice the parallels with debates there about the normativity or otherwise of ascriptions of rationality; my position presupposes, to some extent, that ascriptions of rationality have a significant normative component, as defended famously by Davidson (1973), Lewis (1974), and Dennett (1987b), and opposed by e.g. Fodor (1985; 1987); a more recent exchange is Wedgwood (2007) and Rey (2007).

B.9 The diachronic decision problem

Informally, the setting of our *diachronic* decision theory looks as follows. Agents choose between certain actions, as before, but then depending upon the results of those actions, they make further choices. An agent's overall action, therefore, consists of a series of individual actions.

Our particular context is mildly stylized. We are to imagine an agent confronted with a series of choices between bets. The bets available at one time are potentially dependent on the outcomes of earlier bets, and the winnings of later bets replace the winnings of earlier bets (so that all but the last bet an agent makes matter only because of what other bets they make available, not what winnings they themselves provide). At the end of the whole process, the agent goes away with the winnings: it's simplest to imagine that the winnings are in the form of tokens or money that can be spent only at the conclusion of the process.

We represent the possible outcomes of bets, as before, by a Boolean algebra. Bets, which in this context we call *acts*, again have outcomes in that algebra, but now a given act is only available at some particular event. The acts available at an event E represent those actions available to an agent who knows his current situation to be E. Again, it helps to visualize the theory (though it is not strictly necessary) to imagine that events are subsets of some state space $\mathcal{S}$, and that the actual agent's state is some element of $\mathcal{S}$. Agents may have more or less accurate information about their current state: an agent whose state is $s \in \mathcal{S}$ will know, for some event E, that $s \in E$, but E might be larger or smaller according to an agent's knowledge.

To be more formal: A *diachronic decision problem* is specified by:

1. A complete Boolean algebra $\mathcal{E}$ of events.
2. A partition $\mathcal{R}$ of $\mathcal{E}$ into *rewards*.
3. For each $E \in \mathcal{E}$, a set $\mathcal{A}_E$ of *acts available at* E, such that no two acts are available at distinct events.
4. For each act A available at E, an *outcome event* O_A of A. O_A represents the smallest event such that an agent knows that it will be obtained if A is performed.
5. For any $F \subset E$ and any act A available at E, an act $A|_F$ (the *restriction of A to F*) available at F, such that:

   (a) If $G \subset F \subset E$, then $(A|_F)|_G = A|_G$.
   (b) If $F \subset E$, $G \subset E$, and $F \wedge G = \emptyset$, then $O_{A|_F} \wedge O_{A|_G} = \emptyset$.
   (c) If $\mathcal{P}$ is a partition of E, then

   $$\vee\{O_{A_F} : F \in \mathcal{P}\} = O_A. \tag{B.14}$$

(d) If A, B are both available at E, and $A|_F = B|_F$ for all F in some partition of E, then $A = B$.

6. For any two acts $A \in \mathcal{A}_E$ and $B \in \mathcal{A}_{O_A}$ (that is: any two acts A, B such that B is available at the outcome of A), a unique act $B \cdot A$ (the *composition* of A and B) available at E, with $O_{B \cdot A} = O_B$.

7. For any event E, an act $1_E \in \mathcal{A}_E$, satisfying $1_E|_F = 1_F$ and $A \cdot 1_E = A$ whenever either is defined.

with the additional requirements that

(i) composition is associative: $A \cdot (B \cdot C) = (A \cdot B) \cdot C$.

(ii) restriction commutes with composition:

$$(B \cdot A)|_F = B|_{O_{A|_F}} \cdot A|_F. \tag{B.15}$$

We say that an event E is a *possible outcome* of an act A if $E \subset O_A$.

In keeping with the mildly stylized nature of the problem, we will usually require the decision problem to satisfy

Reward availability. For each event E and any reward-valued partition f of E, there is an act $A(f)$ available at E such that $O_{A(f)|_{f}(r)} \subset r$ for each reward r.

This just corresponds to the assumption that among the acts available at any subevent of E, there is always one consisting of a particular payoff being made. This may be an appropriate point to note that to say an act is 'available' does not mean that an agent will certainly be offered it. If decision theory worked that way, a decision problem could be completely solved just be giving one particular option, the agent's favourite! Rather, the point of decision theory is that an agent considers a set of possible acts and is required to give a preference order on any subset of elements in that set, specifying how he would act if offered a choice of acts in that subset. (From that perspective, 'contemplatable' might be a better name than 'available'; it is cumbersome, though, and perhaps unhelpfully psychological.)

In diachronic decision theory, which action is preferable may well depend on what has gone before. With this in mind, we need some concept of a *history*, which we define as follows: a history is a finite sequence of alternating acts and events,

$$h = \langle E_n, A_n, E_{n-1}, \ldots E_1, A_1 \rangle \tag{B.16}$$

such that for each i, A_{i+1} is available at E_{i-1}, and E_i is a possible outcome of A_i. I call the unique act E_0 at which A_1 is available the *start* start(h) of

h; I define the *end* of h by $\text{end}(h) = E_n$. (We allow the empty sequence to count as a history.)

If h_1 and h_2 are histories with $\text{start}(h_2) = \text{end}(h_1)$, there is a natural definition of their composition $h_2 \cdot h_1$: just concatenate the two histories.

B.10 Solutions to the diachronic decision problem

We can now specify what rational action is in this framework. Whereas in the synchronic framework, a single preference order over bets was sufficient, in the diachronic framework we will need an ordering for each event E, telling us what preferences an agent has *conditional on E obtaining*. Furthermore, this preference may depend on the sequences of earlier acts chosen, and on their outcomes, since those outcomes may convey relevant information.

With this in mind, we define a *solution* to a diachronic decision problem as specified by:

- an event E, the *starting point*;
- for each history h that starts at E, a two-place relation $\succ_h$ on the set of acts available at $\text{end}(h)$.

We call the histories that start at E the *available histories*.

A concept of a zero-probability (null) event is now available, but it is defined in terms of an agent's future actions: a possible outcome of an act is null (with respect to that act) iff the agent is indifferent as to what decisions to make conditional on that outcome. So if $E \subset O_A$ is a possible outcome of the act A available at $\text{end}(h)$, if $F = O_A - E$, and if $h' = \langle F, A \rangle \cdot h$, then E is null (given A and h) iff for any B and C available at O_A, $B \cdot A \sim_h C \cdot A$ whenever $B|_F \sim_{h'} C|_F$.

We can now state our formalisation of rationality for the diachronic decision problem. We will define the following three properties of a solution:

Ordering. For each available h, $\succ_h$ is a total ordering on the acts available at $\text{end}(h)$.

Diachronic consistency. Suppose that A is an act available at h, and that B and C are acts available at O_A, and for $E \subset O_A$, write $h_E \equiv \langle E, A \rangle \cdot h$. Then:

1. If for some partition $\mathcal{P}$ of O_A, if for every element E of the partition not null with respect to A $B|_E \succeq_{h_E} C|_E$, then $B \cdot A \succeq_h C \cdot A$.
2. If in addition for some non-null $E \in \mathcal{P}$, $B|_E \succ_{h_E} C|_E$, then $B \cdot A \succ_h C \cdot A$.

Macrostate indifference. If:

- A, B are acts available at E;
- A', B' are acts available at E';
- The outcomes of A and A' are both subevents of the same reward;
- The outcomes of B and B' are both subevents of the same reward;

then for any available histories h, h' with end$(h)=E$ and end$(h')=E'$, $A \succeq_h B$ iff $A' \succeq_{h'} B'$.

It should be clear that diachronic consistency is just a formalisation of the informal notion of diachronic consistency in section B.8.

All that macrostate indifference says is that if, of two acts, we know that one gives reward r_1 whatever happens and one gives rewards r_2 whatever happens, then we already know enough to say which we prefer. At least in the classical (nonbranching) framework, this assumption (rather like our earlier axiom SP2) just guarantees that the 'rewards' are correctly so called. If, for instance, there was a partition of a reward into two events one of which was preferable to the other, that would just mean that the rewards need to be more finely grained. (In the context of the Everett interpretation there is a little more to be said: cf. the discussion of branching indifference in Chapter 5.)

Reward availability and macrostate indifference together guarantee that $\succ$ defines a total ordering on the set of rewards, which we also write as $\succ$. We will sometimes just write r as shorthand for 'some act whose outcome is a subset of r', since some such act is always available and all are decision-theoretically equivalent.

Now, suppose that A is an act available at end(h) and that f is a reward-valued partition of O_A. Assuming reward availability, macrostate indifference and diachronic consistency, it follows that

1. there exists an act B available at O_A such that $O_{B|_{f(r)}} \subset r$. (Reward availability);
2. given any two such acts B, B', $B \sim_{\langle f(r),A \rangle \cdot h} B'$ (macrostate indifference);
3. given any two such acts

$$B \cdot A \sim_h B' \cdot A. \tag{B.17}$$

(diachronic consistency).

Hence, we can without ambiguity write A_f as shorthand for $B \cdot A$ if B is such an act. In particular, if f is given by

- $f(r) = E$;
- $f(s) = O_A - E$;
- $f(t) = \emptyset$ unless $t = r$ or $t = s$

then we write A_f as $A_{r,s}(E)$. Such an act represents a bet on whether the outcome of A is E or not.

Our goal in diachronic decision theory is to show that a rational agent's preferences must be representable by a unique probability measure; this in turn requires us to generalise the notion of representation to the diachronic context. We do so as follows:

1. A probability measure Pr on a diachronic decision problem is a set of probability measures, one for each available h and each act A available at end(h), on the space of events. We write each individual measure as $\Pr(\cdot|A, h)$, and write just Pr for the set of all the measures.
2. Utilities are defined as in the synchronic case.
3. The *expected utility at h* $\mathrm{EU}_h(A)$ of some act A available at end(h), with respect to some probability Pr and some utility $\mathcal{V}$, is defined by

$$\mathrm{EU}_h(A) = \sum_{r \in R} \Pr(O_A \wedge r|A, h) \times \mathcal{V}(r). \qquad (B.18)$$

4. Three rewards r, s, t satisfying $r \succ s \succ t$ are comparable iff, for some available history h, some act A available at end(h), and some $E_+, E_- \subset O_A$,

$$r \succ_h A_{r,t}(E_+) \succeq_h s \succeq_h A_{r,t}(E_-) \succ_h t. \qquad (B.19)$$

A set of rewards is comparable if any three such rewards in the set are comparable.

We also add one concept which is specific to the diachronic problem: a probability measure is *Bayesian* if for any act A available at end(h), if B is an act available at O_A, $E \subset O_A$, and $F \subset O_{B|_E}$, then

$$\Pr(F| B|_E, \langle E, A \rangle \cdot h) = \frac{\Pr(F|B \cdot A, h)}{\Pr(E|A, h)}. \qquad (B.20)$$

Analogously to the synchronic case, then, we say that

- A Pr represents $\succ$ if there exists a quasi-unique utility such that (with respect to Pr and that utility)

$$A \succ_h B \text{ iff } EU_h(A) > EU_h(B). \tag{B.21}$$

- Pr almost represents $\succ$ for some set of comparable rewards if there exists a quasi-unique utility such that (with respect to Pr and that utility), if $EU_h(A) > EU_h(B)$ and $O_A \wedge r$ and $O_B \wedge r$ are nonempty for only finitely many r, then $A \succ_h B$.
- $\succ$ is uniquely almost represented if it is almost represented for any set of comparable rewards by a unique Pr.
- $\succ$ is Bayesian if is almost represented for any set of comparable rewards by a unique Bayesian Pr.

B.11 A diachronic representation theorem

We will establish a representation theorem for the diachronic decision problem via our existing synchronic results. Given a decision problem with event space $\mathcal{E}$, and a history h, the ordered pair $\langle \mathcal{E}, \mathcal{A}_{\text{end}(h)} \rangle$ is a synchronic decision problem: the chance setups are the acts available at end(h) and the subalgebra of $\mathcal{E}$ associated with the chance setup $A \in \mathcal{A}_{\text{end}}(h)$ is just O_A. Call this the *synchronic problem at h*.

Furthermore, we have seen that to each A and each reward-valued partition f on O_A we can construct an act A_f which represents performing A and then placing the bet represented by f on the outcomes.

This means that an agent's preferences over acts generated from A in this way determine a well-defined preference order over bets on the event space O_A, and so an agent's preference over all acts available at end(h) determines a well-defined preference order over the synchronic problem at h, which we call the *preference order at h* (and again write as $\succ_h$):

$$f|A \succ_h g|B \text{ iff } A_f \succ_h B_g. \tag{B.22}$$

Theorem 5 (Diachronic lemma). *If:*

- *$\mathcal{P}$ is a diachronic decision problem satisfying reward availability;*
- *$\succ$ is a solution to $\mathcal{P}$ satisfying ordering, macrostate indifference and diachronic consistency*

then for any available history h, the preference order at h satisfies axioms SP1 (the dominance axiom), SP2 (the well-definedness of the ordering on rewards) and SP3 (the sure thing principle) of theorem 4.

Proof. The dominance axiom and the sure thing principle are almost immediate corollaries of diachronic consistency; SP2 is an almost immediate corollary of macrostate indifference.

To establish that $\succ_h$ is uniquely almost represented, then, we need to find diachronic assumptions which entail SP4 and SP5. SP4 (probabilism) can be taken over directly as a diachronic axiom, but SP5—the assumption of partitioning—requires a little more work. Recall that the idea of partitioning was that there would always be some N-outcome process all of whose outcomes were judged equally likely, which could be performed concurrently with any other process without either affecting the other.

It is not completely straightforward to model this in our formal system. Acts are individuated by the events at which they are available, so strictly speaking the same act cannot be performed in two different contexts. We get round this as follows:

Independence. Suppose that A, B are available at end(h) and that $\mathcal{E}$ is a partition of O_A. Then $\langle A, \mathcal{E} \rangle$ is *independent of B at h* iff there exist:

1. An act A' available at O_B;
2. An embedding σ of the algebra generated by $\mathcal{E}$ into $O_{A'}$

such that, for any $E \subset O_B$ and F, G in the algebra generated by $\mathcal{E}$, and any pair of rewards (r, s), $A_{r,s}(F) \succeq_h A_{r,s}(G)$ iff

$$(A' \cdot B)_{r,s}(\sigma(F) \wedge O_{B|E}) \succ_{\langle E, A \rangle \cdot h} (A' \cdot B)_{r,s}(\sigma(G) \wedge O_{B|E}). \quad \text{(B.23)}$$

Informally, this just tells us that the probability of act A delivering a given outcome is unaffected by conditionalizing on act B. Intuitively, the existence of independent acts is obvious (again, think of tossed coins) although again it arguably builds in some strong contingent assumptions about the way our world is.

Now suppose that, for given available history h, any pair of rewards r, s, and any N, there is an act A available at end(h), and an N-element partition $E_1 \ldots E_N$ of O_A, such that

1. A is independent of any act available at end(h).
2. $A_{r,s}(E_i) \sim_h A_{r,s}(E_j)$.

It follows from our results above that the preference order at h satisfies SP3: that is, the existence of such an A is the diachronic equivalent of the partitionability requirement. This allows us to define the special acts we need: they should have N distinct outcomes each of which is equiprobable, and they should be co-performable with any other act.

We can now state and prove the central result of this appendix:

Theorem 6 (Diachronic Representation Theorem). *Suppose that*

- $\mathcal{P}$ *is a diachronic decision problem satisfying reward availability;*
- $\succ$ *is a solution to* $\mathcal{P}$ *satisfying ordering, macrostate indifference, and diachronic consistency;*
- $\succ$ *is diachronically partitionable: for any available history* h, *any pair of rewards* r, s, *and any* N, *there is an act* A^N *available at* $\mathrm{end}(h)$, *and an N-element partition* $E_1^N \ldots E_N^N$ *of* $\mathrm{end}(h)$, *such that*

  1. $\langle A^N, \{E_1^N, \ldots E_N^N\}\rangle$ *is independent of any act available at* $\mathrm{end}(h)$.
  2. $A_{r,s}^N(E_i) \sim_h A_{r,s}^N(E_j)$.

- *The preference order* $\succ_h$ *at* h *is probabilistic: that is, the likelihood order over events that it determines is independent of the bets used to define that order.*

Then the preference order $\succ_h$ *at* h *satisfies axioms SP1 — SP4 of theorem 4, and hence is uniquely almost representable; furthermore, it is Bayesian (that is, uniquely almost represented by a Bayesian probability measure on any comparable set of acts).*

Given the above comments, the proof is in large part a direct corollary of the synchronic representation theorem; its proof may be found in appendix 3.

B.12 Subproblems of a decision problem

In the diachronic decision theory that I have developed, an agent's judgements about the probability of an event change not just because the agent gains more information, but also because acts themselves cause the state of the world to change. This feature is well suited to many contexts (including the quantum-mechanical decision theory in the main part of the book), but is somewhat less appropriate for many philosophical applications. There we are usually interested in the truth values of timeless propositions, whether those propositions are general truths like 'the charge of the electron is 1.602×10^{-19} coulombs' or more parochial ones like 'at 6:31pm on Tuesday 5th July 2011 David got home'.

To partially bridge this gap—and to help with Chapter 6's formal development of quantum decision theory—let us define a *subproblem* of a diachronic decision problem $\mathcal{P}$ as an event H of $\mathcal{P}$ such that, for any event E and any act A available at E,

$$\mathcal{O}_{A|_{E \wedge H}} = \mathcal{O}_A \wedge H. \tag{B.24}$$

(Fairly clearly, if H is a subproblem, so is its complement $\neg H$.) Subproblems are events that 'stay true', so to speak: any act available at a subproblem also has range contained within that subproblem.

Given a subproblem H, an available history h with $\text{end}(h) = F$, and act A available at F, we have

$$\Pr(H \wedge O_A|A|_H, \langle H, \mathbf{1}_F \rangle \cdot h) = \frac{\Pr(H \wedge O_A|A, h)}{\Pr(H|\mathbf{1}_F, h)}. \tag{B.25}$$

But the LHS is unity, and $\Pr(X|A, h) = \Pr(X \wedge O_A|A, h)$ for any event X, so we have

$$\Pr(H \wedge O_A|A, h) = \Pr(H|\mathbf{1}_F, h), \tag{B.26}$$

and so may as well just define $\Pr(H|h) \equiv \Pr(H|A, h)$ for arbitrary A. This makes intuitive sense: the probability of H being true should not depend on an agent's choices.

Furthermore, the update rule for probability measures simplifies when we are interested in updating the probability of a subproblem. For any A available at $\text{end}(h)$ and any $E \subset \mathcal{O}_A$, we have h_E as shorthand for $\langle E, A \rangle \cdot h$:

$$\Pr(H|h_E) \equiv \Pr(H|\mathbf{1}_E, h_E) = \Pr(H \wedge E|\mathbf{1}_E, h_E) = \frac{\Pr(H \wedge E|A, h)}{\Pr(E|A, h)}, \tag{B.27}$$

which is a familiar form of Bayes' Theorem.

Since

$$\Pr(E \wedge H|A|_H, \langle H, \mathbf{1}_{\text{end}(h)} \rangle \cdot h) = \frac{\Pr(E \wedge H|A, h)}{\Pr(H|A, h)} \tag{B.28}$$

and

$$\Pr(E|A, h) = \Pr(E \wedge H|A, h) + \Pr(E \wedge \neg H|A, h), \tag{B.29}$$

we also have

$$\Pr(E|A, h) = \Pr(E|A|_H, \langle H, \mathbf{1}_{\text{end}(h)} \rangle \cdot h)\Pr(H|h)$$
$$+ \Pr(E|A|_{\neg H}, \langle \neg H, \mathbf{1}_{\text{end}(h)} \rangle \cdot h)\Pr(\neg H|h); \tag{B.30}$$

if we define $\Pr(E|A, H, h) = \Pr(E|A|_H, \langle H\widehat{\mathbf{1}}_{\text{end}(h)} \rangle \cdot h)$ this takes the familiar form

$$\Pr(E|A, h) = \Pr(E|A, H, h)\Pr(H|h) + \Pr(E|A, \neg H, h)\Pr(\neg H|h); \tag{B.31}$$

the same definition allows us to rewrite (B.27) in the equally familiar form

$$\Pr(H|h) = \frac{\Pr(E|A, H, h)\Pr(H|h)}{\Pr(E|A, H)}$$

$$= \frac{\Pr(E|A, H, h)\Pr(H|h)}{\Pr(E|A, H, h)\Pr(H|h) + \Pr(E|A, \neg H, h)\Pr(\neg H|h)}. \quad (B.32)$$

Appendix C

Formal Proofs of Decision-Theoretic Results

In this appendix, I give formal proofs (and in some cases, formal statements) of various decision-theoretic results stated in Part II and Appendix B.

C.1 Proof of the Additive Representation Theorem

Theorem 1 (Additive Representation Theorem). *Suppose that E is a synchronic decision problem and that $\succ$ is a preference order for some set $\mathcal{B}$ of bets for the decision problem. If*

A1 *The set $\mathcal{B}$ includes all possible bets;*

A2 *The event space of E is the set of subsets of some finite set of atomic events;*

B1 *Any two constant bets with the same payoff are equivalent: if r is a reward and $\mathcal{E}_1, \mathcal{E}_2$ are chance setups, then $C_r(\mathcal{E}_1) \sim C_r(\mathcal{E}_2)$;*

B2 *There is an associative, commutative map $\oplus$ from pairs of rewards to rewards, and some reward 0 such that $r \oplus 0 = r$ for any reward r, such that*

  (i) If $\mathcal{F}_1 \succeq \mathcal{F}_2$ and $\mathcal{G}_1 \succeq \mathcal{G}_2$, then $\mathcal{F}_1 \oplus \mathcal{G}_1 \succeq \mathcal{F}_2 \oplus \mathcal{G}_2$;
  (ii) If in addition $\mathcal{F}_1 \succ \mathcal{F}_2$, then $\mathcal{F}_1 \oplus \mathcal{G}_1 \succ \mathcal{F}_2 \oplus \mathcal{G}_2$;

B3 *There is a reward r such that $C_r(\mathcal{E})$ is not equivalent to $C_0(\mathcal{E})$;*

C1a *For any bets $\mathcal{F}$, $\mathcal{G}$, there is some positive integer n such that $n\mathcal{F} \succ \mathcal{G}$ (where $n\mathcal{F}$ just means $\mathcal{F}$ added to itself n times);*

C1b *For any bets $\mathcal{F}$, $\mathcal{G}$ such that $\mathcal{G} \succ C_0(\mathcal{E})$, there is some positive integer n such that $n\mathcal{G} + \mathcal{F} \succ C_0(\mathcal{E})$;*

C1c *For any bets $\mathcal{F}$, $\mathcal{G}$ such that $\mathcal{G} \prec C_0(\mathcal{E})$, there is some positive integer n such that $n\mathcal{G} + \mathcal{F} \prec C_0(\mathcal{E})$;*

C2a *For any bets $\mathcal{F}$, $\mathcal{G}$ such that $\mathcal{F} \succ \mathcal{G} \succ C_0(\mathcal{E})$, there is some positive integer n such that $n\mathcal{F} \succ (n+1)\mathcal{G}$;*

C2b *For any bets $\mathcal{F}$, $\mathcal{G}$ such that $\mathcal{F} \prec \mathcal{G} \prec C_0(\mathcal{E})$, there is some positive integer n such that $n\mathcal{F} \prec (n+1)\mathcal{G}$.*

then $\succ$ is uniquely represented by some probability measure Pr.

Proof. For simplicity, we start by constructing the utility function $\mathcal{U}$. By B1, the ordering $\succ$ determines a well-defined ordering on some subset of $\mathcal{R}$; by A1, that subset is in fact the whole of $\mathcal{R}$. We write that ordering also as $\succ$.

By B3, there is some reward r such that either $r \succ 0$ or $0 \succ r$; we will assume the former. (If actually all rewards are less desirable than 0, the proof can just be repeated with $\succ$ replaced by $\prec$.) By definition set $\mathcal{U}(0) = 0$ and $\mathcal{U}(r) = 1$.

Now for any reward $s \succ 0$, we define $\mathcal{U}(s)$ by

$$\mathcal{U}(s) = \mathrm{lub}\{m/n : ns \succ mr\}. \tag{C.1}$$

(That some such upper bound exists is a consequence of C1a and the Bolzano–Weierstrass Theorem.[1]) It is straightforward to see that this is additive: $\mathcal{U}(s+t) = \mathcal{U}(s) + \mathcal{U}(t)$. More importantly, it represents $\succ$: $\mathcal{U}(s) > \mathcal{U}(t)$ iff $s \succ t$. To see this, firstly suppose $\mathcal{U}(s) > \mathcal{U}(t)$. Then there must be some m,n such that $ns \succ mr$ but $mr \succeq nt$, so by transitivity, $ns \succ nt$. But if $t \succeq s$, by repeated application of B2 it follows that $nt \succeq ns$. It follows that $s \succ t$.

Conversely, suppose that $s \succ t$. To begin with, assume that $t \succ 0$. By C2a, there is an integer N such that $Ns \succ (N+1)t$. Now let $\{m_i\}$ and $\{n_i\}$ be sequences of positive integers such that $\{m_i/n_i\}$ is an increasing sequence and $\lim_i m_i/n_i = \mathcal{U}(t)$. Then by definition, for each i $n_i t \succ m_i r$, and by B2 again, $(N+1)n_i t \succ (N+1)m_i r$. By transitivity and B2 again, we get $Nn_i s \succ (N+1)m_i r$. So $\mathcal{U}(s) \geq (N+1)m_i/Nn_i$ for all i, and so $\mathcal{U}(s) \geq (N+1)/N\mathcal{U}(t)$.

If instead $t \prec s \prec 0$, the same argument can be run with $\prec$ and $\succ$ interchanged, using C2b instead of C2a. Finally, if $s \succ 0 \succ t$, all we need to show is that if $s \succ 0$ then $\mathcal{U}(s) > 0$ and that if $t \prec 0$ then $\mathcal{U}(t) < 0$. To show the former, we use C1a to find some N such that $Ns \succ R$; it follows that $\mathcal{U}(s) > 1/N$. To show the latter, we apply a similar argument using C1b.

---

[1] See e.g. Apostol (1974: 54).

In fact, we can extend $\mathcal{U}$ to non-constant bets very straightforwardly: the definition of $\mathcal{U}$ works as well for non-constant as for constant bets. That is, we define

$$EU(\mathcal{F}) = \mathrm{lub}\{m/n : n\mathcal{F} \succ mC_r(\mathcal{E})\} \qquad (C.2)$$

for some arbitrary $\mathcal{E}$. If we define $\Pr(E|\mathcal{E}) = EU(C_{r,0}(E|\mathcal{E}))$, it follows via A2 and B2 that $EU(F)$ is indeed the expected utility of $\mathcal{F}$ with respect to this probability measure and to $\mathcal{U}$.

This proves the existence of a representation. To prove uniqueness, let Pr be the probability measure already constructed, and suppose that $\Pr'$ is another probability measure which represents $\succ$. By definition there is some utility function $\mathcal{V}$ such that the expected utility generated by $\Pr'$ and $\mathcal{V}$ represents $\succ$ via the MEU rule. $\mathcal{V}$ must be an increasing function $f(\mathcal{U})$ of the already-constructed utility function $\mathcal{U}$.

Let $\mathcal{E}$ be a chance setup and E be an event in $\mathcal{O}_\mathcal{E}$ with $\Pr(E|\mathcal{E})$ not equal to 0, 1 or 1/2 (if there are no such events, uniqueness is trivial), and for the moment suppose that $\Pr(E)$ is rational, $\Pr(E) = m/n$ say. This means that $C_{nr,0}(E|\mathcal{E}) \sim C_{mr}(\mathcal{E})$, and so $\Pr'(E) = f(m)/f(n)$. Since $km/kn = m/n$, for consistency we must have $f(km) = f(k)f(m)$ and so (iterating) $f(m^M) = f(m)^M$. Putting $l = n - m$ and considering the complement $\neg E$ of E in $\mathcal{O}_\mathcal{E}$, we similarly get that $f(L^L) = f(l)^L$.

Now, f is an increasing function, so we require that $f(k)^K > f(l)^L$ iff $k^K > l^L$; taking logarithms of both sides, the requirement is that $K\ln f(k) > L\ln f(l)$ iff $K\ln k > L\ln l$. Since this is true for arbitrary K and L, by constructing a sequence we can show that $\ln f(k)/\ln f(l) = \ln k/\ln l$. If we write $f(k) = k^\lambda$ and $f(l) = l^\mu$ for some λ, μ, it follows immediately that $\mu = \lambda$. Hence $\Pr'(E) = \Pr(E)^\lambda$ and $\Pr'(\neg E) = \Pr(\neg E)^\lambda$. Since we require $\Pr'(E) + \Pr'(\neg E) = 1$, this forces $\lambda = 1$, i.e. $\Pr'(E) = \Pr(E)$. The extension to irrational probabilities is a straightforward if tedious argument from limiting sequences of rational numbers.

C.2 Proof of the Synchronic Likelihood and Synchronic Preference Theorems

Theorem 2 (Synchronic Likelihood Theorem). *Suppose that*

SL1 $\mathcal{O}_\mathcal{E}|\mathcal{E} \succ \emptyset|\mathcal{E}$.
SL2 $\mathcal{O}_\mathcal{E}|\mathcal{E} \succeq E|\mathcal{E} \succeq \emptyset|\mathcal{E}$ *for any $\mathcal{E}$ and any $E \in \mathcal{E}$.*

SL3 *If $E, F, G \subset O_{\mathcal{E}}$ and $E \wedge G = F \wedge G = \emptyset$, then $E|\mathcal{E} \succ F|\mathcal{E}$ iff $E \vee G|\mathcal{E} \succ F \vee G|\mathcal{E}$.*

SL4 $\succ$ *is partitionable.*

Then $\succ$ is uniquely almost represented.

Proof. Our first step is to show that, if SL3 holds, then a stronger composition rule is also possible.

Lemma 1. *(additivity lemma). Suppose that $\succ$ is an ordering for an event space E satisfying*

$$\text{If } E \wedge G = F \wedge G = \emptyset, \text{ then } E \succeq F \text{ iff } E \vee G \succeq F \vee G. \qquad \text{(C.3)}$$

and that A, B, E, F satisfy $A \wedge B = E \wedge F = \emptyset$ Then if $A \succeq E$ and $B \succeq F$, $A \vee B \succeq E \vee F$. Furthermore, the result continues to hold if $\succeq$ is replaced by $\succ$.

This extension is intuitively plausible, but rather fiddly to prove. Define C and G as the complements of $A \vee B$ and $E \vee F$ respectively; then elements like $A \wedge F$ and $C \wedge E$ form a ninefold partition of $\mathcal{E}$. For simplicity let us write elements of the partition as (e.g.) AF. The proof now proceeds by repeated use of C.3.

Start with A and E: $A \succeq E$, so $AE \vee AF \vee AG \succeq AE \vee BE \vee CE$, so $AE \vee AF \vee AG \vee BG \succeq AE \vee BE \vee CE \vee BG$.

Now do B and F: $B \succeq F$ so $BE \vee BF \vee BG \succeq AF \vee BF \vee CF$, so $BE \vee BG \succeq AF \vee CF$, so $BE \vee BG \vee AE \vee CE \succeq AF \vee CF \vee AE \vee CE$.

Combining these: $AE \vee AF \vee AG \vee BG \succeq AF \vee CF \vee AE \vee CE$, so $AE \vee AF \vee AG \vee BG \vee BE \vee BF \succeq AF \vee CF \vee AE \vee CE \vee BE \vee BF$.

Rearranging: $A(E \vee F \vee G) \vee B(E \vee F \vee G) \succeq (A \vee B \vee C)E \vee (A \vee B \vee C)F$; that is, $A \vee B \succeq E \vee F$, and the lemma is proved.

(Since this proof may appear incomprehensible at first sight, I recommend the following way of visualizing it. We represent the event space as a 3-by-3 grid, with A–C labelling the rows and E–G the columns; we represent events by shading in squares (so that $AE \vee CG$ is represented by a grid with the top left and bottom right squares shaded in). The equivalences between events are now equivalences between squares, and the application of additivity now comes down to the following rule: if two grids are identical, they remain so if we shade in a square which was unshaded in both grids, or erase a square which was shaded in in both grids. The stages of the proof can now be represented graphically as a series of shadings or unshadings of pairs of grids.)

We begin the main proof by proving that if $F_1 \vee \cdots \vee F_N = O_{\mathcal{F}}$ and $G_1 \vee \cdots \vee G_N = O_{\mathcal{G}}$ are partitions satisfying $F_i|\mathcal{F} \sim F_j|\mathcal{F}$ and $G_i|\mathcal{G} \sim G_j|\mathcal{G}$, then $F_i|\mathcal{F} \sim G_j|\mathcal{G}$. We do so via the 'preferred' N-element partition $E_1 \vee \cdots E_N = O_{\mathcal{E}_N}$ guaranteed by partitionability. The definition of partitionability tells us that there exists an event space $\mathcal{E}^*$ and an N^2-element partition E_{ij} of $O_{\mathcal{E}^*}$, such that

- $E_{ij}|\mathcal{E}^* \sim E_{i'j'}|\mathcal{E}^*$.
- $E_i^* = E_{i1} \vee \cdots \vee E_{iN}|\mathcal{E}^* \sim E_i|\mathcal{E}$.
- $F_j^* = E_{1j} \vee \cdots \vee E_{Nj}|\mathcal{E}^* \sim F_j|\mathcal{F}$.

Applying the additivity lemma repeatedly tells us that, if $E_i^*|\mathcal{E}^* \succ F_j^*|\mathcal{E}^*$, then $O_{\mathcal{E}^*}|\mathcal{E}^* \succ O_{\mathcal{E}^*}|\mathcal{E}^*$—which obviously is contradictory. So the result follows.

Given this result, it follows in particular that

$$E|\mathcal{E} \succ E_1 \vee \cdots \vee E_M|\mathcal{E}_N \text{ iff } E|\mathcal{E} \succ E_1 \vee \cdots \vee E_{kM}|\mathcal{E}_{kN}. \quad \text{(C.4)}$$

So for any rational number $\alpha = M/N$, we can consistently write $E|\mathcal{E} \succ \alpha$ to indicate that $E|\mathcal{E} \succ E_1 \vee \cdots \vee E_M|\mathcal{E}_N$.

We now define the probability function as follows:

$$\mathrm{Pr}E = \mathrm{lub}\{\alpha : E \succ \alpha\}. \quad \text{(C.5)}$$

Suppose $E \wedge F = \emptyset$. If $E \succ M/N$ and $F \succ M'/N'$, by the additivity lemma $E \vee F \succ (M/N) + (M'/N')$. So $\mathrm{Pr}(E \vee F) \geq \mathrm{Pr}(E) + \mathrm{Pr}(F)$; running the same argument for $\prec$ shows that $\mathrm{Pr}(E \vee F) \leq \mathrm{Pr}(E) + \mathrm{Pr}(F)$, so Pr is additive. Clearly, $\mathrm{Pr}(\mathcal{E}) = 1$.

Suppose $\mathrm{Pr}(E) > \mathrm{Pr}(F)$; then there is some rational number M/N such that $E \succ M/N \succ F$; hence, $E \succ F$. So Pr almost represents $\succ$.

Theorem 3 (Synchronic Preference Theorem). *Suppose $\succ$ is a total ordering on bets for some synchronic decision problem $\mathbf{E}$, satisfying:*

   SP1 *If f, g are bets on the same chance setup $\mathcal{E}$ and $E_1, \ldots E_n$ is a partition of $O_{\mathcal{E}}$ such that $f^{-1}(E_i)$ and $g^{-1}(E_i)$ are single-valued (such a partition can always be found), then*

   1. *If $f^{-1}(E_i)|\mathcal{E} \succeq g^{-1}(E_i)|\mathcal{E}$ for all non-null E_i, then $f \succeq g$.*
   2. *If in addition $f^{-1}(E_i)|\mathcal{E} \succ g^{-1}(E_i)|\mathcal{E}$ for at least one non-null E_i, then $f \succ g$.*

   (This is sometimes called the dominance principle.)

   SP2 *$\succ$ defines a well-defined ordering on the set of rewards: that is, $C_r(\mathcal{E}) \sim C_r(\mathcal{F})$.*

SP3 $\succ$ *defines a well-defined ordering* $\succ_E$ *on the restriction of bets to any event* E: *that is, it satisfies the sure thing principle.*

SP4 $\succ$ *defines a well-defined likelihood ordering: that is, it satisfies probabilism.*

SP5 *The likelihood order determined by* $\succ$ *is partitionable.*

Then the preference ordering is uniquely almost represented.

Proof. Our first step is to prove that (at least for finite bets on a fixed chance setup) a bet is completely characterized by the probabilities it assigns to each reward. This we do via SP3. For suppose E and F are two events in $\mathcal{O}_{\mathcal{E}}$ with equal probability, that f is a reward-valued partition such that $f^{-1}(E) = r$ and $f^{-1}(F) = s$, and that g is obtained from f by swapping the rewards on E and F.

Now, define f' and g' by $f'(r) = E$, $f'(s) = \mathcal{O}_{\mathcal{E}} - E$ and $g'(r) = F$, $g'(s) = \mathcal{O}_{\mathcal{E}} - F$. By SP3, if $\langle f, \mathcal{E} \rangle \succ \langle g, \mathcal{E} \rangle$, then also $\langle f', \mathcal{E} \rangle \succ \langle g', \mathcal{E} \rangle$. But this contradicts the definition of probability.

Since any two finite bets on $\mathcal{E}$ which assign the same probability to each reward can be obtained from one another by finitely many permutations, our first result follows.

The remainder of the proof will be easier to follow if we assume that for any finite set of positive real numbers $p_1, \ldots p_n$ summing to 1, $\mathcal{E}$ can be partitioned into N sets $E_1, \ldots E_n$ with $\Pr(E_i | \mathcal{E}) = p_i$. This is not in fact guaranteed to be true, but the idealisation is harmless: the partition can be simulated for $p_1 \ldots p_n$ rational by using the partitioning axiom SP5, and if the numbers are irrational we can approximate them arbitrarily well with rational numbers (this is one place where looking only for almost representations makes life simpler). I leave the technical details to the reader.

So: let r, s, t be any three rewards with $r \succ t \succ s$. For any set $E \subset \mathcal{E}$, and any $\alpha < 1$, I define αE to be some arbitrary subset of E with $\Pr(\alpha E) = \alpha \Pr(E)$; with respect to r and s, I define the E-utility of t as

$$\mathcal{U}_E(t) = \text{lub}(\alpha : C_t(E) \succ_E C_{r,s}(\alpha E)). \tag{C.6}$$

(A moment's reflection will show that in the special case where $E = O_{\mathcal{E}}$, this is a formalization of our informal definition of utility, on a scale where $\mathcal{U}(r) = 0$ and $\mathcal{U}(s) = 1$.)

By our previous result, if E and F are equiprobable then $\mathcal{U}_E(t) = \mathcal{U}_F(t)$. If E and F are *not* equiprobable, but $\Pr(E)/\Pr(F)$ is a rational number (M/N, say), we can decompose E into M sets and F into N sets such that all $M + N$ sets are equiprobable; it follows again that $\mathcal{U}_E(t) = \mathcal{U}_F(t)$. And if $\Pr(E)/\Pr(F)$ is irrational, a standard least-upper-bound argument establishes the same thing again. We conclude that for given $\mathcal{E}$, all the E-utilities are the same, and we can without ambiguity just refer to the $\mathcal{E}$-utility of t, $\mathcal{U}_{\mathcal{E}}(t)$. By definition,

$$C_t(\mathcal{E}) \succ C_{r,s}(\alpha\mathcal{O}_E|\mathcal{E}) \tag{C.7}$$

whenever $\alpha < \mathcal{U}_{\mathcal{E}}(t)$ and

$$C_t(\mathcal{E}) \prec C_{r,s}(\alpha\mathcal{O}_E|\mathcal{E}) \tag{C.8}$$

whenever $\alpha > \mathcal{U}_{\mathcal{E}}(t)$. Now let $\mathcal{E}$ and $\mathcal{F}$ be two distinct chance setups, and suppose that $\mathcal{U}_{\mathcal{E}}(t) > \mathcal{U}_{\mathcal{F}}(t)$. Then there must exist α such that

$$C_t(\mathcal{E}) \succ C_{r,s}(\alpha\mathcal{O}_E|\mathcal{E}) \tag{C.9}$$

and

$$C_{r,s}(\alpha\mathcal{O}_F|\mathcal{F}) \succ C_t(\mathcal{F}). \tag{C.10}$$

But by the definition of probability $C_{r,s}(\alpha\mathcal{O}_E|\mathcal{E}) \sim C_{r,s}(\alpha\mathcal{O}_F|\mathcal{F})$, and by SP2 $C_t(\mathcal{E}) \sim C_t(\mathcal{F})$, so this is impossible. It follows that the $\mathcal{E}-$utility is actually independent of $\mathcal{E}$ as well, and that we can talk unproblematically just of the *utility* of t (with respect to r and s).

We now consider any comparable set $\mathcal{T}$ of rewards all of which lie between r and s, and restrict our attention to finite bets with outcomes in $\mathcal{T}$. We have seen that for preference purposes any such bet can be characterized by the probabilities assigned to each reward; we can then represent a bet unambiguously as a real-valued function F on $\mathcal{T}$ satisfying

$$\sum_{t \in \mathcal{T}} F(t) = 1. \tag{C.11}$$

Our strategy is now to replace every reward except r and s themselves with a mixture of r and s. In particular, suppose that ϕ_+ and ϕ_- are functions of $\mathcal{T}$ satisfying $0 \leq \phi_{\pm}(t) \leq 1$, $\phi_+(t) > \mathcal{U}(t)$, and $\phi_-t < \mathcal{U}(t)$. If for some particular t, the bet gives reward t on outcome E, then it follows from the definition of E-utility that partitioning E into $\alpha(t)E$ and $(1 - \alpha(t))E$, and changing the bet to give reward r on $\alpha(t)E$ and s on $(1 - \alpha(t))E$, will produce a bet preferred to the original one if $\alpha(t) > \mathcal{U}(t)$ and vice versa if $\alpha(t) < \mathcal{U}(t)$.

If we make such substitutions sequentially for all outcomes of the bet, we obtain a bet whose only outcomes are r and s, and which gives r with probability

$$\sum_{t \in \mathcal{T}} \alpha(t)F(t). \tag{C.12}$$

If for all t $\alpha(t) \leq \mathcal{U}(t)$ then this bet is definitely not preferred to the original bet; if for all t $\alpha(t) > \mathcal{U}(t)$, then it definitely is preferred. It follows via transitivity that if we define

$$EU(F) = \sum_{t \in \mathcal{T}} \mathcal{U}(t) F(t) \tag{C.13}$$

then F is preferred to G whenever $EU(F) > EU(G)$. But this, of course, is the result we seek.

C.3 Proof of the Diachronic Representation Theorem

Theorem 4 (Diachronic Representation Theorem). *Suppose that*

- $\mathcal{P}$ *is a diachronic decision problem satisfying reward availability;*
- $\succ$ *is a solution to* $\mathcal{P}$ *satisfying macrostate indifference and diachronic consistency;*
- $\succ$ *is diachronically partitionable: for any available history* h, *any pair of rewards* r, s, *and any* N, *there is an act* A^N *available at* end(h), *and an* N-*element partition* $E_1^N \ldots E_N^N$ *of* end(h), *such that*

  1. *$\langle A^N, \{E_1^N, \ldots E_N^N\} \rangle$ is independent of any act available at end(h).*
  2. *$A_{r,s}^N(E_i) \sim_h A_{r,s}^N(E_j)$.*

- *The preference order* $\succ_h$ *at* h *is probabilistic: that is, the likelihood order over events that it determines is independent of the bets used to define that order.*

Then the preference order $\succ_h$ *at* h *satisfies axioms SP1 − SP4 of theorem 4, and hence is uniquely almost representable; furthermore, it is Bayesian (that is, uniquely almost represented by a Bayesian probability measure on any comparable set of acts).*

Proof. To show that $\succ_h$ is partitionable, fix N, r and s, and (in the terminology used in section B.6 to define synchronic partitioning) take $\mathcal{E}_N = A^N$. Given any other act B, we know there exist acts A', B' such that $O_{A' \cdot B} = O_{B' \cdot A^N}$ and $(A' \cdot B)_f = (B' \cdot A^N)_f$. Take $\mathcal{E}^* = O_{A' \cdot B}$.

We now need to construct embeddings of O_B and O_{A^N} into $\mathcal{E}^*$, and we do so as follows:

- If $E \subset O_B$, $\nu(E) = O_{A'|_E}$.
- If $F \subset O_{A^N}$, $\nu'(F) = O_{B'|_F}$.

$SP3$ now follows from the definition of independence. $SP1 - 2$ have been shown to follow from macrostate indifference and diachronic consistency; $SP4$ is an immediate consequence of probabilism.

It follows that $\succ$ is uniquely almost representable, but it remains to be shown that it is Bayesian. To show this, first notice that if A is an act available at $\mathrm{end}(h)$ and f is a reward partition on its outcome space taking values in some set of comparable rewards, by our earlier constructions we have

$$\mathrm{EU}_h(A_f) = \sum_r \mathrm{Pr}(f(r)|A, h) \times \mathcal{V}(r). \tag{C.14}$$

Now let B be available at O_A and $E \subset O_A$. Define $B_1 = B|_E$ and $B_2 = B|_{O_A - E}$, and let f, f' be reward partitions on O_B which agree on restriction to O_{B_2}. Then we have

$$\mathrm{EU}_h[(B \cdot A)_f] = \sum_r \mathrm{Pr}(f(r) \wedge O_{B_1}|B \cdot A, h) \times \mathcal{V}(r)$$
$$+ \sum_r \mathrm{Pr}(f(r) \wedge O_{B_2}|B \cdot A, h) \times \mathcal{V}(r) \tag{C.15}$$

and

$$\mathrm{EU}_h[(B \cdot A)_{f'}] = \sum_r \mathrm{Pr}(f'(r) \wedge O_{B_1}|B \cdot A, h) \times \mathcal{V}(r)$$
$$+ \sum_r \mathrm{Pr}(f(r) \wedge O_{B_2}|B \cdot A, h) \times \mathcal{V}(r). \tag{C.16}$$

Subtracting out the common part and dividing by a common factor, we have that $(B \cdot A)_f \succ_h (B \cdot A)_{f'}$ if

$$\sum_r \frac{\mathrm{Pr}(f(r) \wedge O_{B_1}|B \cdot A, h)}{\mathrm{Pr}(O_{B_1}|B \cdot A, h)} \times \mathcal{V}(r)$$
$$> \sum_r \frac{\mathrm{Pr}(f'(r) \wedge O_{B_1}|B \cdot A, h)}{\mathrm{Pr}(O_{B_1}|B \cdot A, h)} \times \mathcal{V}(r). \tag{C.17}$$

But by diachronic consistency, $(B \cdot A)_f \succ_h (B \cdot A)_{f'}$ iff $(B|_E)_f \succ_{\langle E, A \rangle \cdot h} (B|_E)_{f'}$. By assumption Pr uniquely almost represents $\succ$; that is, by assumption it is the *only* probability measure such that $(B \cdot A)_f \succ_h (B \cdot A)_{f'}$ if

$$\sum_r \Pr(f(r)|B|_E, \langle E, A \rangle \cdot h) \times \mathcal{V}(r)$$

$$> \sum_r \Pr(f'(r)|B|_E, \langle E, A \rangle \cdot h) \times \mathcal{V}(r). \qquad \text{(C.18)}$$

It follows (replacing B_1 with $B|_E$) that

$$\Pr(F|\, B|_E, \langle E, A \rangle \cdot h) = \frac{\Pr(F|B \cdot A, h)}{\Pr(O_{B|_E}|B \cdot A, h)}. \qquad \text{(C.19)}$$

That $\Pr(O_{B|_E}|B \cdot A, h) = \Pr(E|A, h)$ follows again from the construction of Pr (and thus, indirectly, from macrostate indifference).

C.4 Statement and proof of the Everettian Inference theorem

I take as a starting point the definition of a quantum decision problem given in Chapter 5, of a state-dependent solution to that problem as specified in the same chapter, and of a (non-state-dependent) solution to the problem as specified in chapter 6. Note also that a quantum decision problem is a plain decision problem in the sense of Appendix B, and the two definitions of solution coincide. In particular, the definitions in Appendix B of ordering, diachronic consistency, macrostate indifference and probabilism apply directly to solutions of quantum decision problems.

I now state formally some additional constraints on quantum decision problems and their solutions. Recall that for any act U available at E, and any partition $\mathcal{F}$ of $\mathcal{E}$, the set

$$\{O_r(U) : F \in \mathcal{F}\} \qquad \text{(C.20)}$$

is a positive operator valued measure (POVM) on E, which I call the $\mathcal{F}$-POVM of U, and that a POVM is available at E iff for some partition $\mathcal{F}$, some act available at E has that POVM as its $\mathcal{F}$-POVM.

I need only one additional constraint on a decision problem.

Rich structuring. Suppose that $\mathcal{F}$ is an available set of macrostates and that for each $F \in \mathcal{F}$, $\mathcal{X}_F$ is a (finite or countable) set of positive operators on F satisfying $\sum_{X \in \mathcal{X}_F} X \leq 1_F$. Then there is a compatible act function $\mathcal{U}$ for $\mathcal{F}$ such that for each $F \in \mathcal{F}$, the characteristic POVM of $\mathcal{U}(F)$ contains all the operators in $\mathcal{X}_F$.

And I need the following additional constraints on solutions to a problem.

Solution continuity. If h is an available history and U, U' are available at end(h), and $U \succ_h U'$, then in the space of unitary maps from end(h) into $\mathcal{H}$ there are neighbourhoods (in norm topology) $\mathcal{N}, \mathcal{N}'$ of U, U' respectively such that any act in $\mathcal{N}$ available at end(h) is preferred (at h) to any act in $\mathcal{N}'$ available at end(h).

Noncontextuality. For any available history h and any U, V available at end(h), if U and V have the same characteristic POVM then $U \sim_h V$.

I now state and prove, in succession, the noncontextual inference theorem and the Everettian inference theorem.

Noncontextuality Theorem. Suppose that:

- $\mathcal{P}$ is a quantum decision problem satisfying reward availability, problem continuity, branching availability, and rich structuring.
- $\succ_h$ is a solution to $\mathcal{P}$ satisfying macrostate indifference, ordering, diachronic consistency, solution continuity, and probabilism.

Then $\succ_h$ is represented by some density operator iff it is noncontextual.

Proof. If $\succ_h$ is represented by some density operator, trivially it is noncontextual and probabilistic; the trick is proving the converse. I prove it in three steps. First I will establish that $\succ_h$ is uniquely almost represented by some probability measure. Secondly I will prove that this probability function can be represented by a density operator. Finally, I prove that $\succ_h$ is represented, not just almost represented.

To prove the first step, fix an available history h and a partition $\mathcal{F} = F_1, \ldots F_n$ of $\mathcal{E}$; by branching availability, there is an $\mathcal{F}$-POVM available at end(h) whose element associated with F_i is $(1/N)\mathbf{1}_{\text{end}(h)}$. Let U be the act generating this POVM: then bets (in the technical sense of Appendix B) on $F_i \wedge \mathcal{O}_U$ and $F_j \wedge \mathcal{O}_U$ generate the same $\mathcal{R}$-POVM and so are equally likely at h (again, in the sense of Appendix B).

Now, for any act V available at end(h), let U' be an act available at $\mathcal{O}_V$ whose $\mathcal{F}$-partition at $\langle \mathcal{O}_V, V \rangle \cdot h$ associates $(1/N)\mathbf{1}_{\mathcal{O}_V}$ to each element of $\mathcal{F}$. It should now be apparent that U is independent (in the sense of Appendix B) of V, and so $\succ_h$ is diachronically partitionable. As such, it follows from the diachronic representation theorem of Appendix B that $\succ_h$ is uniquely almost represented.

The second step is essentially a variant on the POVM version of Gleason's theorem proved by Caves et al. (2004). Specialize to the trivial history $\emptyset$ and the starting-point macrostate M, define

$$\mathcal{X} = \{U^\dagger \Pi_E U : U \text{ is available at } M, E \subset \mathcal{O}_U\} \qquad \text{(C.21)}$$

and define a function Λ on $\mathcal{X}$ by

$$\Lambda(U^\dagger \Pi_E U) = \Pr(E|U, h_0). \qquad \text{(C.22)}$$

(That this is well-defined follows from noncontextuality). By the rich structuring axiom and the additivity of Pr, Λ satisfies $\Lambda(X + Y) = \Lambda(X) + \Lambda(Y)$ whenever $X + Y \le 1_M$ and hence $\Lambda(M/NX) = M/N\Lambda(X)$. Diachronic consistency entails that if $p > q$ then $\Lambda(pX) > \Lambda(qX)$, so it follows that Λ is linear on $\mathcal{X}$, and so extends uniquely to the span of $\mathcal{X}$ in the real linear space of bounded self-adjoint operators on M. If this span is not the whole space, we extend it to the whole space arbitrarily;[2] we can then extend it uniquely to an antilinear functional on the whole space of bounded self-adjoint operators on M. Since that space is a Hilbert space with inner product $\langle X, Y \rangle = \text{Tr}(X^\dagger Y)$, there is a unique operator ρ such that $\text{Tr}(\rho X) = \Lambda(X)$.

Finally, we define ρ_h recursively: if h is an available history, U is available at end(h), and h_E is shorthand for $\langle E, U \rangle \cdot h$, then

$$\rho_\emptyset = \rho \qquad \text{(C.23)}$$

and

$$\rho_{h_E} = \frac{\Pi_E U \rho_h U^\dagger \Pi_E}{\text{Tr}(\Pi_E U \rho_h U^\dagger \Pi_E)}. \qquad \text{(C.24)}$$

If $\text{Tr}(U^\dagger \Pi_E U \rho_h) = \Pr(E|U, h)$, then for any V available at $\mathcal{O}_U$ and any $F \subset \mathcal{O}_V$ we have

$$\text{Tr}(V^\dagger|_E \Pi_F V|_E \rho_{h_E}) = \frac{\text{Tr}(V^\dagger|_E \Pi_F V|_E \Pi_E U \rho_h U^\dagger \Pi_E)}{\text{Tr}(\Pi_E U \rho_h U^\dagger \Pi_E)}$$

$$= \frac{\text{Tr}(U^\dagger \Pi_E V^\dagger|_E \Pi_F V|_E \Pi_E U \rho_h)}{\text{Tr}(\Pi_E U \rho_h U^\dagger \Pi_E)}$$

---

[2] Technical detail: $\Lambda(X) \le \|X\|$, where $\|X\| = \sup\{\|X_x\| : \|x\| \le 1\}$ is the operator norm. One of the forms of the Hahn–Banach theorem (see e.g. Rudin 1991) is that any linear functional on a subspace of a normed space which is bounded by that norm may be extended to a linear functional on the whole space.

$$= \frac{\mathsf{Tr}((VU)^\dagger \Pi_F VU\rho_h)}{\mathsf{Tr}(\Pi_E U\rho_h U^\dagger \Pi_E)}$$

$$= \frac{\mathrm{Pr}(F|VU, h)}{\mathrm{Pr}(E|U, h)}$$

$$= \mathrm{Pr}(F|V|_E, h_E) \tag{C.25}$$

as required.

The third step follows from problem and solution continuity; I leave it to the reader.

Everettian Inference theorem. Suppose that:

- $\mathcal{P}$ is a quantum decision problem satisfying reward availability, problem continuity, branching availability and rich structuring;
- $\succ_h$ is a solution to $\mathcal{P}$ satisfying ordering, diachronic consistency and solution continuity;
- $\succ^\psi$ is a state-dependent solution to $\mathcal{P}$, satisfying ordering, diachronic consistency, solution continuity, branching indifference, macrostate indifference, and state supervenience, and defined on an act-closed set of states which contains all states in the starting point of $\succ_h$;
- $\succ^\psi$ and $\succ_h$ are compatible.

Then $\succ_h$ is represented by a density operator.

Proof. First, note that by the Born Rule theorem, there is some quasi-unique utility function such that $\succ^\psi$ is given by the MEU principle with respect to that utility function and the Born-rule probabilities. Noncontextuality and macrostate indifference of $\succ_h$ follow immediately.

Next, take two rewards r and s with $r \succ s$ (if there are not two such rewards, the conclusion is trivially true) and divide $\mathcal{R}$ arbitrarily into two subsets $\mathcal{R}_r$ and $\mathcal{R}_s$. If we define r^* and s^* as the direct sums of $\mathcal{R}_r$ and $\mathcal{R}_s$ respectively, there will be a new decision problem $\mathcal{P}^*$ with the same acts, events and macrostates as $\mathcal{P}$ but with rewards r^*, s^*.

For any event E there will be an act V_E available at E such that if $t \in \mathcal{R}_r$, V_E maps $t \wedge E$ into r and similarly *mutatis mutandis* for s. We can define a solution $\succ_h^*$ to the decision problem $\mathcal{P}^*$ as follows:

$$U \succ_h^* U' \text{ iff } V_{O_U} U \succ_h V_{O_{U'}} U' \tag{C.26}$$

Since there are only two rewards in $\mathcal{P}^*$, $\succ^*$ is trivially probabilistic; by the noncontextual inference theorem, it is represented by some density operator ρ. Clearly, ρ also represents the original solution so long as we restrict our attention to acts whose outcomes lie in $r \vee s$ (for any such act U, $V_{\mathcal{O}_U} = 1_{\mathcal{O}_U}$).

Now let $\mathcal{U}$ be a utility function for the state-dependent solution satisfying $\mathcal{U}(r) = 1$ and $\mathcal{U}(s) = 0$ (there is exactly one such function, of course). For any available h and any reward t satisfying $r \succeq t \succeq s$, suppose that U_t and $U_{r,s}$ are acts available at end(h) whose $\mathcal{R}$-POVMs are defined as follows: the $\mathcal{R}$-POVM of U_t assigns $1_{\text{end}(h)}$ to t and 0 to all other rewards; the $\mathcal{R}$-POVM of $U_{r,s}$ assigns $\mathcal{U}(t)1_{\text{end}(h)}$ to r, $(1 - \mathcal{U}(t))1_{\text{end}(h)}$ to s, and 0 to all other rewards. For any $\psi \in$ end(h) $U_t \sim^\psi U_{r,s}$, so by compatibility, $U_t \sim_h U_{r,s}$.

Now for any available h, and any act U available at end(h) whose possible rewards all lie between r and s, I write the $\mathcal{R}$-POVM of U as a function from rewards between r and s to positive operators: $O_t(U)$ is the positive operator assigned to t. By the above result, if U' is another such act available at end(h) with

$$O_r(U') = \sum_{t \in \mathcal{R}: r \succeq t \succeq s} O_t(U')\mathcal{U}(t)$$

$$O_s(U') = \sum_{t \in \mathcal{R}: r \succeq t \succeq s} O_t(U')(1 - \mathcal{U}(t))$$

$$O_t(U') = 0 \text{ otherwise} \tag{C.27}$$

then by the above result and diachronic consistency, $U \sim_h U'$; by branching availability, such an act is available. Since $\text{EU}_{\rho_h}(U) = \text{EU}_{\rho_h}(U')$, it follows that ρ represents $\succ_h$ for any acts whose outcomes lie between r and s. Since r and s were arbitrary, the result follows.

C.5 Statement and proof of the classical inference theorems

I define a *classical decision problem* as specified by:

- A topological space $\mathcal{H}$. Given a countable set $\mathcal{S}$ of subsets of $\mathcal{H}$, I write $\vee \mathcal{S} \equiv \cup \mathcal{S}$ and $\wedge \mathcal{S} \equiv \cap \mathcal{S}$; Given subspaces E and F, I define $E \vee F = \vee\{E, F\}$ and likewise for $\wedge$.
- A complete Boolean algebra $\mathcal{E}$ of Borel subsets of $\mathcal{H}$, the *event space*. (So $\mathcal{E}$ contains $\mathcal{H}$ and is closed under $\vee$, $\wedge$, and taking the complement.) I define a *partition* of an event E to be a countable set of mutually disjoint events whose conjunction is E.

- For each $E \in \mathcal{E}$, a set $\mathcal{U}_E$ of continuous maps from E into $\mathcal{H}$, which we call the set of *acts available at* E. We write $\mathcal{O}_U$ for the smallest event containing the range of the act U and require that the choice of available acts satisfies:

  1. *Restriction*: If $E, F \in \mathcal{E}$ and $F \subset E$, then if U is available at E then the map $U|_F$, defined by $U(x) = U|_F(x)$ whenever $x \in F$, is available at F.
  2. *Composition*: If U is available at E, and V is available at $\mathcal{O}_U$, then VU is available at E.
  3. *Indolence*: For any event E, if there are any acts available at E then the identity 1_E is available at E. (More precisely, the embedding map of E into $\mathcal{H}$ is available at E.)
  4. *Continuation*: If U is available at some E, then there is some act available at $\mathcal{O}_U$.
  5. *Irreversibility*: If U is available at $E \vee F$ and $E \wedge F = \emptyset$, $\mathcal{O}_{U|_E} \wedge \mathcal{O}_{U|_F} = \emptyset$.
  6. *Extension*: If $F \subset E$ and U is available at F, there is some act V available at E such that $U = V|_F$.

- A partition $\mathcal{R}$ of $\mathcal{E}$ (that is, a set of mutually disjoint elements of $\mathcal{E}$ whose disjunction is $\mathcal{H}$), the set of *rewards*.

A classical decision problem is a diachronic decision problem: as such, many of the concepts of general diachronic decision theory (in particular, the concepts of solution and history) carry over to it.

If E is the starting point of a solution $\succ_h$, and ρ is a [finite] Borel probability measure on E, then we define ρ_h as a measure on end(h) recursively by

$$\rho_\emptyset = \rho \tag{C.28}$$

and

$$\rho_{\langle E, U \rangle \cdot h}(X) = \frac{\rho_h(U^{-1}(X \wedge \mathcal{O}_U))}{\rho_h(U^{-1}(E \wedge \mathcal{O}_U))} \tag{C.29}$$

whenever the right-hand side is defined (here $\mathcal{O}_U$ is the range of U, as in the quantum case). And we say that ρ [almost] represents $\succ_h$ provided that the probability measure

$$\Pr(E|U, h) = \rho_h(U^{-1}(E \wedge \mathcal{O}_U)) \tag{C.30}$$

[almost] represents $\succ_h$.

For any act U available at E, and any partition $\mathcal{F}$ of $\mathcal{H}$ by events, I define the $\mathcal{F}$-partition generated by U to be a map f from elements of $\mathcal{F}$ to subsets of E, defined by

$$f(F) = U^{-1}(F \wedge \mathcal{O}_U). \qquad (C.31)$$

(I will conflate f and the range of f where no confusion would be caused by this.) A subset of E is *available at* E iff it is a member of the characteristic partition of some act available at E.

I say that a classical decision problem is *richly structured* if, given any set $\mathcal{X}$ of disjoint subsets of an event all of which are available at the event, there is some act available at that event and some partition $\mathcal{F}$ of $\mathcal{H}$ such that all elements of $\mathcal{X}$ are elements of the $\mathcal{F}$-partition generated by the act. And I say that a solution to the decision problem is *classically noncontextual* if for any two acts U and V available at end(h), if U and V have the same $\mathcal{R}$-partition then $U \sim_h V$.

Finally, a classical decision problem is *extension-friendly* if, given any event E and any real function f of the available subsets of E which is additive on disjoint sets and satisfies $f(\emptyset) = 0$, f is extendible, though not necessarily uniquely, to a Borel measure on E. (This will be satisfied if, for instance, the set of available acts is large enough that all Borel subsets of E are available at E, although of course this is a very strong idealization.)

We can now prove the

Classical noncontextual inference theorem. Suppose that:

- $\mathcal{P}$ is a classical decision problem satisfying reward availability, rich structuring and extension-friendliness;
- $\succ_h$ is a solution to $\mathcal{P}$ satisfying ordering, diachronic consistency, and macrostate indifference;
- $\succ_h$ is diachronically partitionable;
- $\succ_h$ is probabilistic;
- $\succ_h$ is classically noncontextual.

Then $\succ_h$ is almost represented by some finite Borel measure.

Proof. By the diachronic representation theorem, there is a probability measure Pr on $\mathcal{P}$ which almost represents $\succ_h$. $\Pr(E|U, h) = \Pr(F|V, h)$ whenever $U^{-1}(E \wedge \mathcal{O}_U) = V^{-1}(F \wedge \mathcal{O}_V)$.

Specialize to the starting-point event M of $\succ_h$, and the trivial history h_0. If $E \subset M$ is available at M then there is some event F and some act U available at M such that $U^{-1}(F \wedge \mathcal{O}_U) = E$; without contradiction we can define $\mu(E) = \Pr(F|U, h_0)$. By rich structuring, if $E_1, \dots E_n$ are available at M, there is some single act U and some events $F_1 \dots F_n$ such that $E_i = U^{-1}(F_i \wedge \mathcal{O}_U)$; hence,

$$\sum_i \mu(E_i) = \sum_i \Pr(F_i|U, h_0) = \Pr(\vee_i F_i|U, h_0) = \mu(\vee_i E_i); \qquad (C.32)$$

so μ is additive on disjoint sets. By extension-friendliness, there is some Borel measure ρ on M which agrees with μ on available subsets of M.

We now prove recursively that $\Pr(E|U,h) = \rho_h(U^{-1}(E \wedge \mathcal{O}_h))$. Suppose it true for given h and suppose U is available at $\text{end}(h)$, that E is a subevent of $\mathcal{O}_U$ and that W is available at E. Then there is some act V available at $\mathcal{O}_U$ such that $V|_E = W$, and we have (writing h_E as shorthand for $\langle E, U \rangle \cdot h$)

$$\rho_{h_E}(V|_E^{-1}(F \wedge \mathcal{O}_{V|_E})) = \frac{\rho_h(U^{-1}(V|_E^{-1}(F \wedge \mathcal{O}_{V|_E}) \wedge \mathcal{O}_U))}{\rho_h(U^{-1}(E \wedge \mathcal{O}_U))}$$

$$= \frac{\text{Tr}((VU)^\dagger \Pi_F VU \rho_h)}{\text{Tr}(\Pi_E U \rho_h U^\dagger \Pi_E)}$$

$$= \frac{\Pr(F|VU,h)}{\Pr(E|U,h)}$$

$$= \Pr(F|V|_E, h_E). \tag{C.33}$$

We now define a *state-dependent solution* to a classical decision problem as an association, to each event E and each $x \subset E$, of a preference order $\succ^x$ over the acts available at E (E will usually be taken as tacit). A state-dependent solution satisfies *ordering* if $\succ^x$ is a total ordering for each x, *diachronic consistency* if $V \succeq^{U(x)} W$ iff $VU \succeq^x WU$, *macrostate indifference* if $U \sim^x V$ whenever $U(x)$ and $V(x)$ are in the same reward, and *state supervenience* if $U \sim^x V$ whenever $U(x) = V(x)$. A state-dependent solution is compatible with a solution iff, whenever $U \succeq^x V$ for all x, $U \succeq V$.

A classical decision problem satisfies *erasure* iff for any reward r, any subevents $E_1, E_2 \subset r$, and any $x_1, x_2 \in E_1, E_2$, there are acts U_1, U_2 such that U_i is available at E_i and $U_1(x_1) = U_2(x_2) \in r$.

Now we can prove the

Classical realist inference theorem. Suppose that:

- $\mathcal{P}$ is a classical decision problem satisfying reward availability, rich structuring, extension-friendliness, and erasure;
- $\succ^x$ is a state-dependent solution to $\mathcal{P}$ satisfying ordering, diachronic consistency, macrostate indifference, and state supervenience;
- $\succ_h$ is a solution to $\mathcal{P}$ satisfying ordering and diachronic consistency;
- $\succ_h$ is diachronically partionable;
- $\succ_h$ is probabilistic;
- $\succ_h$ is compatible with $\succ^x$.

Then $\succ_h$ is almost represented by some finite Borel measure.

Proof. It suffices to prove that $\succ^h$ is macrostate-indifferent and noncontextual; macrostate indifference is immediate from compatibility and the macrostate-indifference of $\succ^x$. For noncontextuality, suppose that U and V are available at end(h) and have the same $\mathcal{R}$-partition. For any $x \in \text{end}(h)$, $U(x)$ and $V(x)$ are in the same reward (r, say); hence by erasure there are acts U' and V' available at $\mathcal{O}_U$ and $\mathcal{O}_V$ respectively such that $U'U(x) = V'V(x) \in r$. By macrostate indifference, $U' \sim^{U(x)} 1_{\mathcal{O}_U}$ and so by diachronic consistency $U'U \sim^x U$; the same is true *mutatis mutandis* for V and V'. State supervenience means that $U'U \sim^x V'V$, hence $U \sim^x V$. Since x was arbitrary, noncontextuality now follows from the compatibility of $\succ^x$ and $\succ_h$.

C.6   Statement and proof of the Everettian Epistemic theorem

A *quantum structure* for a diachronic decision problem is specified by:

- a Hilbert space $\mathcal{H}$;
- a Boolean algebra isomorphism $E \to E_Q$ of the event space $\mathcal{E}$ onto a Boolean algebra $\mathcal{E}_Q$ of subspaces of $\mathcal{H}$ including $\mathcal{H}$ itself;
- For each event E, a map $A \to U_A$ from acts available at E to unitary maps from E_Q into $\mathcal{H}$, satisfying

   (i) $(\mathcal{O}_A)_Q$ is the smallest element of $\mathcal{E}_Q$ containing the range of U_A.
   (ii) $U_{AB} = U_A U_B$.
   (iii) $U_{A|E} = (U_A)|_{E_Q}$.

Fairly clearly, this makes the diachronic decision problem into a quantum decision problem whose events and macrostates coincide.

A *partially quantum decision problem* is a triple $\langle \mathcal{P}, Q, \mathcal{X} \rangle$, where $\mathcal{P}$ is a diachronic decision problem, Q is a subproblem of $\mathcal{P}$, and $\mathcal{X}$ is a quantum structure for Q.

Everettian Epistemic theorem.   Suppose that:

- $\langle \mathcal{P}, Q, \mathcal{X} \rangle$ is a partially quantum decision problem satisfying reward availability;
- $\succ_h$ is a solution to $\langle \mathcal{P}, Q, \mathcal{X} \rangle$ satisfying ordering, diachronic consistency, macrostate indifference, probabilism, diachronic partitioning and noninfinitesimality;

- The quantum decision problem $\langle Q, \mathcal{X} \rangle$ satisfies branching availability and rich structuring;
- $\succ^{\psi}$ is a solution to $\langle Q, \mathcal{X} \rangle$ satisfying ordering, diachronic consistency, macrostate indifference, and state supervenience;
- The restriction of $\succ_h$ to $\langle Q, \mathcal{X} \rangle$ is compatible with $\succ^{\psi}$.

Then:

(i) $\succ_h$ is uniquely represented by a probability measure Pr;

(ii) If $\text{end}(h) \wedge Q \neq \emptyset$, there is a density operator ρ_h on $\text{end}(h) \wedge Q$ such that if we write $E_Q \equiv E \wedge Q$ and $E_C \equiv E \wedge \neg Q$, we have

$$\Pr(E|A, h) = \Pr(E_C|A|_{E_C}, \langle C, \mathbf{1}_{\text{end}(h)} \rangle \cdot h)\Pr(\neg Q|\widehat{\mathbf{1}}_{\text{end}(h)}, h)$$

$$+ \,\text{Tr}(\rho_h U^{\dagger}_{A|_{E_Q}} \Pi_{E_Q} U_{A|_{E_Q}})\Pr(Q|\mathbf{1}_{\text{end}(h)}, h). \quad \text{(C.34)}$$

(iii) ρ_h obeys the update rule

$$\rho_{h_E} = \frac{\Pi_{E_Q} U_{A|_{\text{end}(h)_Q}} \rho_h U^{\dagger}_{A|_{\text{end}(h)_Q}} \Pi_{E_Q}}{\text{Tr}(\Pi_{E_Q} U_{A|_{\text{end}(h)_Q}} \rho_h U^{\dagger}_{A|_{\text{end}(h)_Q}})} \quad \text{(C.35)}$$

Proof. (i) is just a special case of the Diachronic Representation Theorem. For the rest, it follows from the Everettian Inference theorem that the restriction of $\succ_h$ is represented by some density operator ρ_h, and (C.35) is just a special case of (B.31).

Appendix D

Proof of the Utility Equivalence Lemma

Recall our starting point: we assume that an agent (i) has a preference order between quantum games (conditional on the state being ψ) via maximizing utility with respect to a utility function $\mathcal{U}$ and to the quantum-mechanical weights; (ii) has a preference order between bets whose outcomes depend on unknown information by maximizing utility with respect to a utility function $\mathcal{V}$ and to some probability measure quantifying the agent's degrees of belief in the objectively unknown information. It is not immediately obvious that $\mathcal{U} = \mathcal{V}$.

Indeed, a sceptic might argue:

> From the point of view of ordinary, nonbranching decision theory, bets on known states are deterministic processes, and the entire post-bet branching structure is a single outcome. So the *expected* utility of that outcome is no more or less than the utility of that outcome.
>
> Now, granted: you've argued that U is preferable to U' (given that the state is ψ) iff $\mathrm{EV}(U|\psi) > \mathrm{EV}(U'|\psi)$. So it had better be the case that $\mathcal{U}(U|\psi) > \mathcal{U}(U'|\psi)$ iff $\mathrm{EV}(U|\psi) > \mathrm{EV}(U'|\psi)$; or put another way, $\mathcal{U}$ must be an increasing function of $\mathcal{V}$. But for the Everettian to make use of Bayesian inference via your arguments, you don't just need $\mathcal{U}$ to be an increasing function of $\mathcal{V}$, you need it to equal $\mathcal{V}$ (or at any rate, I suppose, to be a positive affine function of $\mathcal{V}$).

Now, I think it is actually a mistake to see (6.10) and (6.11) as fundamentally different in this way, as I show in section 6.6 (and argue from a much more philosophical standpoint in Chapter 7). Here, though, I wish to respond to the sceptic on their own terms by showing that in fact $\mathcal{U}$ must be a positive affine function of $\mathcal{V}$. All that is required is a simple thought experiment (and a certain amount of messy mathematics).

To do this, first note that, as the sceptic argues, there must be an increasing real function f such that $\mathcal{U}(U|\psi) = f[(\mathcal{V}(U|\psi)]$. Now suppose that s and t are rewards with t preferred to s: by exploiting the affine ambiguities in both $\mathcal{U}$ and $\mathcal{V}$, I can require that $\mathcal{V}(t) = 1$, that $\mathcal{V}(s) = 0$, that $f(0) = 0$, and that $f(1) = 1$.

Now, a little terminology: I will use 'Q-probability' (which I write as Pr_Q) to refer to the relative weight of some branch in the agent's future), and 'C-probability' (Pr_C) to refer to the agent's personal probability function over states of the whole Universe (or at least, of all branches in his future). I call the utility defined by $\mathcal{U}$ and the C-probability the C-utility, and the utility defined by $\mathcal{V}$ and the Q-probability the Q-utility.

Finally, for any x between 0 and 1, let $K(x)$ be some quantum process with Q-utility x. (Such processes can always be found assuming that we have the freedom to prepare and measure arbitrary superpositions.)

I now consider the following scenarios:

Scenario A. Some random process[1] is arranged such that the agent assigns a certain outcome E of that process probability $\mathrm{Pr}_C(E) = \lambda$. If E does occur, then a quantum measurement is made such that a particular outcome F has weight $\mathrm{Pr}_Q(F) = \mu$, and the agent's future selves gets to perform process $K(x)$ if they are in an F-branch, $K(y)$ otherwise. If E does not occur, the agent just gets to perform $K(y)$.

Scenario B. A quantum measurement is made, such that a particular outcome E' has weight $\mathrm{Pr}_Q(E') = \mu$. In the E'-branches, some random process is arranged such that the agent's future selves in those branches assign a certain outcome F' of that event probability $\mathrm{Pr}_C(F') = \lambda$; if it does occur, the agent's future selves get to perform $K(x)$ and if it doesn't, they get to perform $K(y)$. In the non-E' branches, the agent's future selves just get to perform $K(y)$.

Now, the C-utility EU_A of scenario A is $(1 - \lambda)$ times the C-utility of getting to perform $K(y)$ (which is $f(y)$), plus λ times the C-utility of getting to perform a process which leads to $K(x)$ on branches of weight μ and to $K(y)$ on the other branches. The Q-utility of this latter process is $\mu x + (1 - \mu)y$, so its C-utility is $f(\mu x + (1 - \mu)y)$, and so

$$\mathrm{EU}_A = \lambda f(\mu x + (1 - \mu)y) + (1 - \lambda)f(y). \qquad (\mathrm{D.1})$$

---

[1] Candidates include pseudo-random number generators and observations of the outcomes of previous quantum measurements whose results are determinate in the agent's branch but currently unknown.

And the Q-utility EV_B of scenario B is $(1 - \mu)$ times the Q-utility of getting to perform $K(y)$ (which is just y), plus μ times the Q-utility of getting to perform $K(x)$ with personal probability λ and $K(y)$ otherwise. The C-utility of this latter process is $\lambda f(x) + (1 - \lambda)f(y)$, so its Q-utility is $f^{-1}(\lambda f(x) + (1 - \lambda)f(y)))$, and so

$$EV_B = \mu f^{-1}(\lambda f(x) + (1 - \lambda)f(y)) + (1 - \mu)y. \tag{D.2}$$

As the observant reader my have noted, though, the net effects of scenarios A and B are the same: with C-probability λ the agent gets to perform a measurement which leads to process $K(x)$ with relative weight μ and $K(y)$ with relative weight $1 - \mu$; with C-probability $1 - \lambda$ the agent gets to perform process $K(y)$ with certainty. So the agent ought really to be indifferent between the two processes, and so we must have $f(EV_B) = EU_A$, and hence

$$f^{-1}(\lambda f(\mu x + (1 - \mu)y) + (1 - \lambda)f(y))$$
$$= \mu f^{-1}(\lambda f(x) + (1 - \lambda)f(y)) + (1 - \mu)y. \tag{D.3}$$

Given that this must hold for arbitrary values of x, y, μ and λ, it places a very severe constraint on f: severe enough, in fact, that it forces $f(x) = x$.

Lemma. Suppose that $f : [0, 1] \to \text{R}$ is twice differentiable and increasing, that $f(0) = 0$ and $f(1) = 1$, and that

$$\mu f^{-1}\left[\lambda f(v) + (1 - \lambda)f(w)\right] + (1 - \mu)w$$
$$= f^{-1}\left[\lambda f(\mu v + (1 - \mu)w) + (1 - \lambda)f(w)\right] \tag{D.4}$$

holds for all $v, w, \lambda, \mu \in \mathcal{R}$. Then $f(x) = x$.

Proof. First take $w = 0$. The equation then simplifies to

$$\mu f^{-1}(\lambda f(v)) = f^{-1}(\lambda f(\mu v)). \tag{D.5}$$

Setting $g = f^{-1}$ and differentiating both sides with respect to v gives

$$\lambda\mu\frac{g'[\lambda g^{-1}(v)]}{g'(g^{-1}(v))} = \lambda\mu\frac{g'[\lambda g^{-1}(\mu v)]}{g'(g^{-1}(\mu v))}. \tag{D.6}$$

It follows that $g'[\lambda g^{-1}(v)] = k(\lambda)g'(g^{-1}(v))$, and hence that $g'(\lambda x) = k(\lambda)g'(x)$ for $x \in [0, 1]$. In particular, taking $x = 1$ yields $g'(\lambda) = k(\lambda)g'(1)$, so we have $g'(xy) = g'(x)g'(y)/g'(1)$.

Defining $h(x) = g'(x)/g'(1)$ yields $h(xy) = h(x)h(y)$; by requiring this formula to hold by definition, we can uniquely extend h to all positive real numbers. Iterating it, we have $h(x^n) = [h(x)]^n$, and in particular

$$h[(1 + y/n)^n] = [h(1 + y/n)]^n. \tag{D.7}$$

For sufficiently large n, $h(1 + y/n) \simeq h(1) + h'(1)y/n$, with the approximation becoming arbitrarily accurate for sufficiently large n. So

$$h[(1 + y/n)^n] \to_{x \to \infty} [h(1) + h'(1)y/n]^n = (1 + h'(1)y/n)^n \tag{D.8}$$

since $h(1) = 1$. Since $(1 + y/n)^n$ tends to e^y as $n \to \infty$, the limiting case of this formula is

$$h[e^y] = e^{h'(1)y}, \tag{D.9}$$

so $h(x) = x^\beta$ for some β. Backtracking through our definitions, it follows that $f(x) = Ax^\alpha$ for some α, and in fact since we required $f(1) = 1$, actually $f(x) = x^\alpha$.[2]

Now returning to the original equation D.3, we put $v = kw$. The constraint now becomes

$$\mu \left[\lambda k^\alpha + (1 - \lambda) \right]^{1/\alpha} + (1 - \mu) = \left[\lambda(k\mu + (1 - \mu))^\alpha + (1 - \lambda) \right]^{1/\alpha}. \tag{D.10}$$

Now, suppose that $\alpha \neq 0$ and $\alpha \neq 1$. Differentiating both sides by k, cancelling terms, and raising both sides to the power of $\alpha/(1 - \alpha)$, we get

$$(\lambda k^\alpha + (1 - \lambda))k^{-\alpha} = (\lambda \kappa^\alpha + (1 - \lambda))\kappa^{-\alpha} \tag{D.11}$$

where $\kappa = \mu k + (1 - \mu)$. It follows that the function $F(x, \lambda) = (\lambda x^\alpha + (1 - \lambda))x^{-\alpha}$ is actually independent of x, and so we have

$$\lambda k^\alpha + (1 - \lambda) = \Lambda(\lambda)k^\alpha \tag{D.12}$$

which obviously requires $\alpha = 0$. So we conclude that $\alpha = 0$ or $\alpha = 1$, and the boundary conditions rule out $\alpha = 0$.

This suffices to prove that $\mathcal{U}$ and $\mathcal{V}$ agree, up to a positive affine transformation.

---

[2] Prima facie $f(x) \propto \ln(x)$ is also possible, but the condition that $f(0) = 0$ rules it out.

References

Aaronson, S. (2005). NP-complete problems and physical reality. Available online at: arxiv.org/abs/quant-ph/0502072.

—— and J. Watrous (2008). Closed timelike curves make quantum and classical computing equivalent. Available online at: arxiv.org/abs/0808.2669.

Abrikosov, A. A., L. P. Gorkov, and I. E. Dzyalohinski (1963). *Methods of Quantum Field Theory in Statistical Physics*. New York: Dover. Revised English edn; translated and edited by R. A. Silverman.

Adams, D. (1992). *Mostly Harmless*. London: Heinemann. Book 5 of *The Hitchhiker's Guide to the Galaxy*.

Aharonov, Y., and D. Albert (1980). States and observables in relativistic quantum field theories. *Physical Review* D 21, 3316–24.

—— and D. Bohm (1959). Significance of electromagnetic potentials in the quantum theory. *Physical Review* 115, 485–91.

Albert, D. Z. (1996). Elementary quantum metaphysics. In J. T. Cushing, A. Fine, and S. Goldstein (eds), *Bohmian Mechanics and Quantum Theory: An Appraisal*. Dordrecht: Kluwer Academic, 277–84.

—— (2000). *Time and Chance*. Cambridge, Mass: Harvard University Press.

—— (2007). Narrativity. Lecture delivered at 'Descrying the World in Metaphysics' Summer school, Budapest, July 2006.

—— (2010). Probability in the Everett picture. In S. Saunders, J. Barrett, A. Kent, and D. Wallace (eds), *Many Worlds? Everett, Quantum Theory and Reality*. Oxford: Oxford University Press.

—— and B. Loewer (1988). Interpreting the many worlds interpretation. *Synthese* 77, 195–213.

Allais, M. (1953). Le comportement de l'homme rationnel devant le risque: critique des postulats et axiomes de l'école américaine. *Econometrica* 21, 503–46.

Allori, V., S. Goldstein, R. Tumulka, and N. Zanghi (2008). On the common structure of Bohmian mechanics and the Ghirardi-Rimini-Weber theory. *British Journal for the Philosophy of Science* 59, 353–89.

—— —— —— —— (2011). Many worlds and Schrödinger's first quantum theory. *British Journal for the Philosophy of Science* 62(1), 1–27.

Anderson, P. W. (1972). More is different: broken symmetry and the nature of the hierarchical structure of science. *Science* 177, 393.

Apostol, T. M. (1974). *Mathematical Analysis*, 2nd edn. Reading, Mass: Addison-Wesley.

Arntzenius, F. (2006). Time travel: double your fun. *Philosophy Compass* 1, 599–616.

—— (n.d.). Quantum mechanics, narratability and relativity. Unpublished MS.

—— and T. Maudlin (2010). Time travel and modern physics. In E. N. Zalta (ed.), *The Stanford Encyclopedia of Philosophy* (Spring 2010 edn). Available online at: http://plato.stanford.edu/archives/spr2010/entries/time-travel-phys/.

Aspect, A., P. Grangier, and G. Roger (1982). Experimental test of Bell's inequalities using time-varying analysers. *Physical Review Letters* 49, 1804–07.

Bacciagaluppi, G. (2002). Remarks on space-time and locality in Everett's interpretation. In J. Butterfield and T. Placek (eds), *Non-locality and Modality*. Dordrecht: Kluwer.

—— (2007). Is logic empirical? In K. Engesser, D. Gabbay, and D. Lehmann (eds), *The Handbook of Quantum Logic and Quantum Structures*. Amsterdam: Elsevier. Available online at: http://philsci-archive.pitt.edu/archive/00003380/.

—— and A. Valentini (2010). *Quantum Theory at the Crossroads*. Cambridge: Cambridge University Press.

Bacon, D. (2004). Quantum computational complexity in the presence of closed timelike curves. *Physical Review A* A70, 032309.

Baierlein, R. (1971). *Atoms and Information Theory: An Introduction to Statistical Mechanics*. San Francisco Calif.: W. H. Freeman.

Baker, D. (2007). Measurement outcomes and probability in Everettian quantum mechanics. *Studies in the History and Philosophy of Modern Physics* 38, 153–69.

Ballentine, L. E. (1970). The statistical interpretation of quantum mechanics. *Reviews of Modern Physics* 42, 358–81.

—— (1990). *Quantum Mechanics*. Englewood Cliffs, NJ: Prentice Hall.

Barbour, J. B. (1994). The timelessness of quantum gravity II: The appearance of dynamics in static configurations. *Classical and Quantum Gravity* 11, 2875–97.

—— (1999). *The End of Time*. London: Weidenfeld & Nicolson.

Barrett, J. A. (1999). *The Quantum Mechanics of Minds and Worlds*. Oxford: Oxford University Press.

Barrow, J. D., and F. J. Tipler (1986). *The Anthropic Cosmological Principle*. Oxford: Oxford University Press.

Bassi, A., and G. Ghirardi (2003). Dynamical reduction models. *Physics Reports* 379, 257.

Baxter, S. (2004). *Exultant*. London: Gollancz.

Bell, J. S. (1966). On the problem of hidden variables in quantum mechanics. *Reviews of Modern Physics* 38, 447–52. Reprinted in Bell (1987: 1–13).

—— (1981a). Bertlmann's socks and the nature of reality. *Journal de physique* 42, C2 41–61. Reprinted in Bell (1987: 139–58).

—— (1981b). Quantum mechanics for cosmologists. In C. J. Isham, R. Penrose, and D. Sciama (eds), *Quantum Gravity 2: A Second Oxford Symposium*. Oxford:

Clarendon Press. Reprinted in Bell (1987: 117–38); page references are to that version.

—— (1987). *Speakable and Unspeakable in Quantum Mechanics*. Cambridge: Cambridge University Press.

Belnap, N. (1992). Branching space time. *Synthese* 92, 385–434.

—— (2002). Branching histories approach to indeterminism and free will. Available online at: http://philsci-archive.pitt.edu.

Belot, G. (1998). Understanding electromagnetism. *British Journal for the Philosophy of Science* 49, 531–55.

Benenti, G., G. Casati, and G. Strini (2004). *Principles of Quantum Computation and Information*. Singapore: World Scientific.

Bennett, C. H., G. Brassard, C. Crepeau, R. Jozsa, A. Peres, and W. Wootters (1993). Teleporting an unknown state via dual classical and EPR channels. *Physical Review Letters* 70, 1895–9.

Berry, M. V., and N. L. Balzas (1979). Evolution of semiclassical quantum states in phase space. *Journal of Physics* A 12, 625–42.

Bevers, B. (2011). Everett's 'many-worlds' proposal. *Studies in History and Philosophy of Modern Physics* 43, 3–12.

Binney, J., and S. Tremaine (2008). *Galactic Dynamics*, 2nd edn. Princeton, NJ: Princeton University Press.

Blume-Kohout, R., H. K. Ng, D. Poulin, and L. Viola (2008). Characterizing the structure of preserved information in quantum processes. *Physical Review Letters* 100, 030501.

Bohm, D. (1952). A suggested interpretation of quantum theory in terms of 'hidden' variables. *Physical Review* 85, 166–93.

Bostrom, N. (2002). *Anthropic Bias: Observation Selection Effects in Science and Philosophy*. New York: Routledge.

Brown, H. (1996a). Bovine metaphysics: remarks on the significance of the gravitational phase effect in quantum mechanics. In *Perspectives on Quantum Reality: Non-Relativistic, Relativistic, and Field-Theoretic*. Dordrecht: Kluwer, 183–93.

—— (1996b). Mindful of quantum possibilities. *British Journal for the Philosophy of Science* 47, 189–200.

—— (2005). *Physical Relativity*. Oxford: Oxford University Press.

—— (2010). Reply to Valentini: 'de Broglie–Bohm theory: many worlds in denial?' In S. Saunders, J. Barrett, A. Kent, and D. Wallace (eds), *Many Worlds? Everett, Quantum Theory, and Reality*. Oxford: Oxford University Press.

—— C. Dewdney, and G. Horton (1995). Bohm particles and their detection in the light of neutron interferometry. *Foundations of Physics* 25, 329–47.

—— and D. Wallace (2005). Solving the measurement problem: de Broglie–Bohm loses out to Everett. *Foundations of Physics* 35, 517–40.

Bruzda, W., V. Cappellini, H.-J. Sommers, and K. Zyckowski (2009). Random quantum operations. *Physics Letters* A 3, 320–24.

Bub, J. (1997). *Interpreting the Quantum World*. Cambridge: Cambridge University Press.

—— and I. Pitowsky (2010). Two dogmas about quantum mechanics. In S. Saunders, J. Barrett, A. Kent, and D. Wallace (eds), *Many Worlds? Everett, Quantum Theory, and Reality*. Oxford: Oxford University Press.

Busch, P., P. J. Lahti, and P. Mittelstaedt (1996). *The Quantum Theory of Measurement*, 2nd rev. edn. Berlin: Springer.

Butterfield, J. N. (1992). Bell's theorem: what it takes. *British Journal for the Philosophy of Science* 43, 41–83.

—— (2006a). Against pointillisme in mechanics. *British Journal for the Philosophy of Science* 57, 709–54.

—— (2006b). Against pointillisme in geometry. In F. Stadler and F. Stoeltzner (eds), *Time and History: Proceedings of 28th International Wittgenstein Conference*. Frankfurt: Ontos, 181–222.

Byrne, P. (2010). *The Many Worlds of Hugh Everett III: Multiple Universes, Mutual Assured Destruction, and the Meltdown of a Nuclear Family*. Oxford: Oxford University Press.

Caldeira, A. O., and A. J. Leggett (1983). Path integral approach to quantum Brownian motion. *Physica* A 121, 587–616.

Callender, C. (2009). The past hypothesis meets gravity. In G. Ernst and A. Hütteman (eds), *Time, Chance and Reduction: Philosophical Aspects of Statistical Mechanics*. Cambridge. Cambridge University Press. Available online at: http://philsci-archive.pitt.edu/archive/00004261.

Carr, B. (ed.) (2007). *Universe or Multiverse?* Cambridge: Cambridge University Press.

Cartwright, N. (1983). *How the Laws of Physics Lie*. Oxford: Oxford University Press.

Caves, C., C. Fuchs, K. Manne, and J. Renes (2004). Gleason-type derivations of the quantum probability rule for generalized measurements. *Foundations of Physics* 34, 193.

—— —— and R. Schack (2002). Quantum probabilities as Bayesian probabilities. *Physical Review* A 65, 022305.

—— and R. Schack (2005). Properties of the frequency operator do not imply the quantum probability postulate. *Annals of Physics* 315, 123–46.

Cheng, T.-P., and L.-F. Li (1984). *Gauge Theory of Elementary Particle Physics*. Oxford: Oxford University Press.

Cirkovic, M. M. (2006). Is quantum suicide painless? On an apparent violation of the Principal Principle. *Foundations of Science* 11, 287–96.

Clifton, R., J. Bub, and H. Halvorson (2003). Characterizing quantum theory in terms of information theoretic constraints. *Foundations of Physics* 33, 1561.

Cohen-Tannoudji, C., B. Diu, and F. Laloë (1977). *Quantum Mechanics*, vol. 1. New York: Wiley-Interscience.

Colin, S. (2003). Beables for quantum electrodynamics. Available online at: http://arxiv.org/abs/quant-ph/0310056.

—— and W. Struyve (2007). A Dirac sea pilot-wave model for quantum field theory. *Journal of Physics* A 40, 7309–42.

Cushing, J. T., A. Fine, and S. Goldstein (eds) (1996). *Bohmian Mechanics and Quantum Theory: An Appraisal*. Dordrecht: Kluwer Academic.

Cvitanovi, P., R. Artuso, R. Mainieri, G. Tanner, and G. Vattay (2009). *Chaos—Classical and Quantum*. Copenhagen: Neils Bohr Institute. Available at: ChaosBook.org.

Davidson, D. (1973). Radical interpretation. *Dialectica* 27, 313–28.

—— (2001). *Inquiries into Truth and Interpretation*, 2nd edn. Oxford: Oxford University Press.

Davies, P., and J. Brown (eds) (1986). *The Ghost in the Atom*. Cambridge. Cambridge University Press.

Dawkins, R. (1989). *The Selfish Gene*, 2nd rev. edn. Oxford: Oxford University Press.

de Finetti, B. (1931). Sul significato soggestivo della probabilità. *Fundamenta Mathematicae* 17, 298–329.

—— (1937). La prévision: ses lois logiques, ses sources subjectives. *Annales d'Institut Henri Poincaré* 7, 1–68. English translation in H. E. Kyburg, and H. E. Smokler (eds), *Studies in Subjective Probability*. New York: Wiley, 1964.

Dennett, D. C. (1981). Three kinds of intentional psychology. In R. Healey (ed.), *Reduction, Time, and Reality*. Cambridge University Press. Reprinted and expanded in Dennett (1987b: 43–81).

—— (1984). *Elbow Room: The Varieties of Free Will Worth Wanting*. Oxford: Oxford University Press.

—— (1987a). Evolution, error, and intentionality. In Dennett (1987b: 287–322).

—— (1987b). *The Intentional Stance*. Cambridge, Mass.: MIT Press.

—— (1991a). *Consciousness Explained*. London: Penguin.

—— (1991b). Real patterns. *Journal of Philosophy* 87, 27–51. Reprinted in *Brainchildren* (London: Penguin, 1998), 95–120.

—— (1995). *Darwin's Dangerous Idea: Evolution and the Meanings of Life*. New York: Simon & Schuster.

—— (2000). With a little help from my friends. In D. Ross, A. Brook, and D. Thompson (eds), *Dennett's Philosophy: A Comprehensive Assessment*. Cambridge, Mass.: MIT Press, 327–88.

—— (2003). *Freedom Evolves*. London: Allen Lane.

—— (2005). *Sweet Dreams: Philosophical Objections to a Science of Consciousness*. Cambridge, Mass.: MIT Press.

d'Espagnat, B. (1976). *Conceptual Foundations of Quantum Mechanics*, 2nd edn. New York: Addison-Wesley.

Deutsch, D. (1985a). Quantum theory as a universal physical theory. *International Journal of Theoretical Physics* 24(1), 1–41.

—— (1985b). Quantum theory, the Church–Turing principle and the universal quantum computer. *Proceedings of the Royal Society of London* A 400, 97–117.

—— (1991). Quantum mechanics near closed timelike lines. *Physical Review* D 44(10), 3197–217.

—— (1996). Comment on Lockwood. *British Journal for the Philosophy of Science* 47, 222–8.

—— (1997). *The Fabric of Reality*. London: Penguin.

—— (1999). Quantum theory of probability and decisions. *Proceedings of the Royal Society of London* A455, 3129–37.

—— (2002). The structure of the multiverse. *Proceedings of the Royal Society of London* A458, 2911–23.

—— (2010). Apart from universes. In S. Saunders, J. Barrett, A. Kent, and D. Wallace (eds), *Many Worlds? Everett, Quantum Theory, and Reality*. Oxford: Oxford University Press, 542–52.

—— and P. Hayden (2000). Information flow in entangled quantum systems. *Proceedings of the Royal Society of London* A456, 1759–74.

DeWitt, B. (1970). Quantum mechanics and reality. *Physics Today* 23(9), 30–35. Reprinted in DeWitt and Graham (1973).

—— and N. Graham (eds) (1973). *The Many-Worlds Interpretation of Quantum Mechanics*. Princeton, NJ: Princeton University Press.

Diacu, F., and P. Holmes (1999). *Celestial Encounters: The Origins of Chaos and Stability*. Princeton, NJ: Princeton University Press.

Dickson, M. (2001). Quantum logic is alive ∧ (it is true ∨ it is false). *Philosophy of Science* 68, S274–7.

Diósi, L. (2004). Anomalies of weakened decoherence criteria for quantum histories. *Physical Review Letters* 92, 170401.

Donald, M. J. (1990). Quantum theory and the brain. *Proceedings of the Royal Society of London* A 427, 43–93.

—— (1992). A priori probability and localized observers. *Foundations of Physics* 22, 1111–72.

—— (2002). Neural unpredictability, the interpretation of quantum theory, and the mind-body problem. Available online at: http://arxiv.org/abs/quant-ph/0208033.

Dowker, F., and A. Kent (1996). On the consistent histories approach to quantum mechanics. *Journal of Statistical Physics* 82, 1575–646.

Drake, S. (1978). *Galileo at Work: His Scientific Biography.* Chicago: University of Chicago Press.

Dudman, V. (1985). Thinking about the future. *Analysis* 45, 183–6.

——(1992). A popular presumption refuted. *Journal of Philosophy* 89, 431–2.

Dummett, M. (1976). Is logic empirical? In H. D. Lewis (ed.), *Contemporary British Philosophy* (4th series), 45–68. London: Allen & Unwin. Reprinted in M. Dummett, *Truth and Other Enigmas.* London: Duckworth, 1978, 269–89.

——(1993). *The Seas of Language.* Oxford: Oxford University Press.

Dürr, D., S. Goldstein, R. Tumulka, and N. Zanghi (2004). Bohmian mechanics and quantum field theory. *Physical Review Letters* 93, 090402.

————————(2005). Bell-type quantum field theories. *Journal of Physics* A38, R1.

————and N. Zanghi (1992). Quantum equilibrium and the origin of absolute uncertainty. *Journal of Statistical Physics* 67, 843–907.

Earman, J. (2006). The 'past hypothesis': not even false. *Studies in the History and Philosophy of Modern Physics* 37, 399–430.

——and C. Wuthrich (2008). Time machines. In E. N. Zalta (ed.), *The Stanford Encyclopedia of Philosophy* (Fall 2008 edn). Available online at: http://plato.stanford.edu/archives/fall2008/entries/time-machine/.

Egan, G. (1994). *Permutation City.* London: Millennium.

——(1995). The hundred-light-year diary. In *Axiomatic.* London: Millenium.

——(2002). Singleton. *Interzone* 176. Reprinted in G. Egan, *Oceanic.* London: Gollancz, 2009, 145–205.

Ekert, A., and R. Jozsa (1997). Quantum computation and Shor's factoring algorithm. *Reviews of Modern Physics* 68(3), 733–53.

Elitzur, A. C., and L. Vaidman (1993). Quantum mechanical interaction-free measurements. *Foundations of Physics* 23, 987–97.

Everett, H. I. (1957). Relative state formulation of quantum mechanics. *Review of Modern Physics* 29, 454–62. Reprinted in DeWitt and Graham (1973).

Farhi, E., J. Goldstone, and S. Gutmann (1989). How probability arises in quantum-mechanics. *Annals of Physics* 192, 368–82.

Feynman, R. P. (1967). *The Character of Physical Law.* Cambridge, Mass.: MIT Press.

Fishburn, P. C. (1981). Subjective expected utility: a review of normative theories. *Theory and Decision* 13, 139–99.

Fisher, R. A. (1925). *Statistical Methods for Research Workers.* Edinburgh: Oliver & Boyd. (14th edn, revised and enlarged, 1970).

Fodor, J. A. (1975). *The Language of Thought.* Hassocks: Harvester Press.

——(1985). Fodor's guide to mental representation: the intelligent auntie's vade-mecum. *Mind* 94, 76–100. Reprinted in *A Theory of Content and Other Essays*. Cambridge, Mass.: MIT Press, 1992.

——(1987). *Psychosemantics: The Problem of Meaning in the Philosophy of Mind*. Cambridge, Mass.: MIT Press.

——(1990). A theory of content I: the problem. In *A Theory of Content and Other Essays*. Cambridge, Mass.: MIT Press, 51–88.

Ford, L. H., and A. Vilenkin (1981). A gravitational analogue of the Aharonov–Bohm effect. *Journal of Physics* A 14, 2353–7.

Frege, G. (1892). Über sinn und bedeutung. *Zeitschrift für Philosophie und philosophiche Kritik* 100, 25–50. Reprinted in English translation in P. Geach and M. Black (eds), *Translations from the Philosophical Writings of Gottlob Frege*. Oxford: Blackwell, 1952.

Frigg, R. (2008). Typicality and the approach to equilibrium in Boltzmannian statistical mechanics. Available online at: http://philsci-archive.pitt.edu.

Fuchs, C. (2002). Quantum mechanics as quantum information (and only a little more). Available online at: http://arXiv.org/abs/quant-ph/0205039.

——and A. Peres (2000a). Quantum theory needs no 'interpretation'. *Physics Today* 53(3), 70–71.

————(2000b). Fuchs and Peres reply. *Physics Today* 53, 14.

Gell-Mann, M., and J. B. Hartle (1990). Quantum mechanics in the light of quantum cosmology. In W. H. Zurek (ed.), *Complexity, Entropy and the Physics of Information*. Redwood City, Calif.: Addison-Wesley, 425–9.

————(1993). Classical equations for quantum systems. *Physical Review* D 47, 3345–82.

————(2007). Quasiclassical coarse graining and thermodynamic entropy. *Physical Review* A 76, 022104.

Geroch, R. (1984). The Everett interpretation. *Nous* 18, 617–33.

Ghirardi, G., A. Rimini, and T. Weber (1986). Unified dynamics for micro and macro systems. *Physical Review* D 34, 470–91.

Gleason, A. (1957). Measures on the closed subspaces of a Hilbert space. *Journal of Mathematics and Mechanics* 6, 885–93.

Glymour, C. (1981). Why I am not a Bayesian. In *Theory and Evidence*. Chicago: University of Chicago Press.

Goldstein, S. (2001). Boltzmann's approach to statistical mechanics. In J. Bricmont, D. Dürr, M. Galavotti, F. Petruccione, and N. Zanghi (eds), *Chance in Physics: Foundations and Perspectives*. Berlin: Springer, 39. Available online at: http://arxiv.org/abs/cond-mat/0105242.

Graham, N. (1973). The measurement of relative frequency. In DeWitt and Graham (1973).

Grandy, W. T. (1987). *Foundations of Statistical Mechanics*. Dordrecht: Reidel.

Grandy, W. T., and L. H. Schick (eds) (1990). *Maximum Entropy and Bayesian Methods*. Dordrecht: Kluwer.

Greaves, H. (2004). Understanding Deutsch's probability in a deterministic multiverse. *Studies in the History and Philosophy of Modern Physics* 35, 423–56.

—— (2007). On the Everettian epistemic problem. *Studies in the History and Philosophy of Modern Physics* 38, 120–52.

—— and W. Myrvold (2010). Everett and evidence. In S. Saunders, J. Barrett, A. Kent, and D. Wallace (eds), *Many Worlds? Everett, Quantum Theory and Reality*. Oxford: Oxford University Press.

—— and D. Wallace (2006). Justifying conditionalization: conditionalization maximizes expected epistemic utility. *Mind* 115(459), 607–32.

Griffiths, R. B. (1984). Consistent histories and the interpretation of quantum mechanics. *Journal of Statistical Physics* 36, 219–72.

—— (1993). Consistent interpretation of quantum mechanics using quantum trajectories. *Physical Review Letters* 70, 2201–4.

—— (1996). Consistent histories and quantum reasoning. *Physical Review* A 54, 2759–73.

—— (2002). *Consistent Quantum Theory*. Cambridge: Cambridge University Press.

Haag, R. (1996). *Local Quantum Theory: Fields, Particles, Algebras*. Berlin: Springer.

Hacking, I. (1983). *Representing and Intervening: Introductory Topics in the Philosophy of Science*. Cambridge: Cambridge University Press.

Haldane, J. B. S. (1928). *Possible Worlds, and Other Papers*. New York: Harper & Brothers.

Halliwell, J. J. (1998). Decoherent histories and hydrodynamic equations. *Physical Review* D 35, 105015.

—— (2010). Macroscopic superpositions, decoherent histories and the emergence of hydrodynamic behaviour. In S. Saunders, J. Barrett, A. Kent, and D. Wallace (eds), *Many Worlds? Everett, Quantum Theory, and Reality*. Oxford: Oxford University Press.

Hamilton, P. F. (2002). *Fallen Dragon*. New York: Warner Books.

Hanfling, O. (1996). Logical positivism. In S. Shanker (ed.), *Philosophy of Science, Logic and Mathematics in the Twentieth Century*. London: Routledge, 193–213.

Hanson, R. (2003). When worlds collide: quantum probability from observer selection? *Foundations of Physics* 33, 1129–50.

—— (2006). Drift-diffusion in mangled worlds quantum mechanics. *Proceedings of the Royal Society of London* 462, 1619–27.

Hartle, J. (2010). Quasiclassical realms. In S. Saunders, J. Barrett, A. Kent, and D. Wallace (eds), *Many Worlds? Everett, Quantum Theory, and Reality*. Oxford: Oxford University Press.

Hawking, S. W. (1976). Black holes and thermodynamics. *Physical Review* D 13, 191–7.

——and G. F. R. Ellis (1973). *The Large Scale Structure of Space-Time*. Cambridge: Cambridge University Press.

Healey, R. (1991). Holism and nonseparability. *Journal of Philosophy* 88, 393–421.

——(1994). Nonseparable processes and causal explanation. *Studies in History and Philosophy of Science* 25, 337–74.

——(2007). *Gauging What's Real*. Oxford: Oxford University Press.

Hemmo, M., and I. Pitowsky (2007). Quantum probability and many worlds. *Studies in History and Philosophy of Modern Physics* 38, 333–50.

Hillery, M., R. F. O'Connell, M. O. Scully, and E. P. Wigner (1984). Distribution functions in physics: fundamentals. *Physics Reports* 106, 121–67.

Hofstadter, D. R., and D. C. Dennett (eds) (1981). *The Mind's I: Fantasies and Reflections on Self and Soul*. London: Penguin.

Hogan, J. P. (1986). *The Proteus Operation*. London: Century.

Home, D., and M. A. B. Whitaker (1997). A conceptual analysis of quantum Zeno: paradox, measurement and experiment. *Annals of Physics* 258, 237–85.

Howson, C., and P. Urbach (1993). *Scientific Reasoning: The Bayesian Approach*. Chicago: Open Court.

Hoyle, F., and J. V. Narlikar (1995). Cosmology and action-at-a-distance electro-dynamics. *Review of Modern Physics* 67, 113–55.

Husimi, K. (1940). Some formal properties of the density matrix. *Proc. Phys. Math. Soc. Japan* 22, 264.

Ismael, J. (2003). How to combine chance and determinism: thinking about the future in an Everett universe. *Philosophy of Science* 70, 776–90.

Jaynes, E. T. (1957a). Information theory and statistical mechanics. *Physical Review* 106, 620.

——(1957b). Information theory and statistical mechanics ii. *Physical Review* 108, 171.

——(1968). Prior probabilities. *IEEE Transactions on Systems Science and Cybernetics* SSC-4, 227.

——(1973). The well-posed problem. *Foundations of Physics* 3, 477–93.

Jeffrey, R. C. (1983). *The Logic of Decision*, 2nd edn. Chicago: University of Chicago Press.

Joos, E., H. D. Zeh, C. Kiefer, D. Giulini, J. Kupsch, and I. O. Stamatescu (2003). *Decoherence and the Appearence of a Classical World in Quantum Theory*, 2nd edn. Berlin: Springer.

Joyce, J. N. (1999). *The Foundations of Causal Decision Theory*. Cambridge: Cambridge University Press.

Kaloyerou, P. N., and H. R. Brown (1992). On neutron interferometer partial absorption experiments. *Physica* B 176, 78–92.

Kane, R. (ed.) (2002). *Free Will*. Oxford: Blackwell.

Kaplan, D. (1989). Demonstratives. In J. Almog, J. Perry, and H. Wettstein (eds), *Themes from Kaplan*. Oxford: Oxford University Press, 481–563.

Kaplan, M. (1996). *Decision Theory as Philosophy*. Cambridge: Cambridge University Press.

Kent, A. (1990). Against many-worlds interpretations. *International Journal of Theoretical Physics* A5, 1764. Expanded and updated version available at: http://www.arxiv.org/abs/gr-qc/9703089.

——— (1996). Quasiclassical dynamics in a closed quantum system. *Physical Review* A 54, 4670–75.

——— (2010). One world versus many: the inadequacy of Everettian accounts of evolution, probability, and scientific confirmation. In S. Saunders, J. Barrett, A. Kent, and D. Wallace (eds), *Many Worlds? Everett, Quantum Theory, and Reality*, Oxford: Oxford University Press.

Keynes, J. M. (1921). *A Treatise on Probability*. London: Macmillan.

Kim, J. (1998). *Mind in a Physical World*. Cambridge, Mass.: MIT Press.

Kincaid, H. (2004). There are laws in the social sciences. In C. Hitchcock (ed.), *Contemporary Debates in the Philosophy of Science*. Oxford: Blackwell.

Knobe, J. M., and S. Nichols (eds) (2008). *Experimental Philosophy*. Oxford: Oxford University Press.

Knox, E. (2009a). Geometrizing gravity and vice versa: the force of a formulation. Available online at: philsci-archive.pitt.edu.

——— (2009b). Geometry, inertia and spacetime structure. Ph.D. thesis, University of Oxford.

Kochen, S., and E. Specker (1967). The problem of hidden variables in quantum mechanics. *Journal of Mathematics and Mechanics* 17, 59–87.

Kripke, S. A. (1981). *Naming and Necessity*, revised and enlarged edn. Oxford: Blackwell.

Kwiat, P., H. Weinfurter, T. Herzog, and A. Zeilinger (1995). Interaction-free measurement. *Physical Review Letters* 74, 4763–6.

Ladyman, J. (1998). What is structural realism? *Studies in History and Philosophy of Science* 29, 409–24.

——— and D. Ross (2007). *Every Thing Must Go: Metaphysics Naturalized*. Oxford: Oxford University Press.

Landsman, N. P. (2009). Algebraic quantum mechanics. In D. Greenberger, K. Hentschel, and F. Weinert (eds), *Compendium of Quantum Physics*. Berlin: Springer, 6–10.

Lebowitz, J. (2007). From time-symmetric microscopic dynamics to time-asymmetric macroscopic behavior: an overview. Available online at: http://arxiv.org/abs/0709.0724.

Leggett, A. J. (2002). Testing the limits of quantum mechanics: motivation, state of play, prospects. *Journal of Physics: Condensed Matter* 14, R415–51.

Leslie, J. (1989). *Universes*. London: Routledge.

Lewis, C. S. (1987). *The Lion, the Witch, and the Wardrobe*. London: Scholastic.

Lewis, D. (1973). *Counterfactuals*. Oxford: Blackwell.

—— (1974). Radical interpretation. *Synthese* 23, 331–44. Reprinted in *Philosophical Papers*, vol. 1 Oxford: Oxford University Press, 1983.

—— (1976). The paradoxes of time travel. *American Philosophical Quarterly* 13, 145–52. Reprinted in Lewis (1986c).

—— (1979). Attitudes *de dicto* and *de se*. *Philosophical Review* 88, 513–43. Reprinted in Lewis (1983d).

—— (1980). A subjectivist's guide to objective chance. In R. C. Jeffrey (ed.), *Studies in Inductive Logic and Probability*, vol. 2. Berkeley: University of California Press. Reprinted, with postscripts, in Lewis (1986c). Page numbers refer to this version.

—— (1983a). Languages and language. In Lewis (1983d).

—— (1983b). New work for a theory of universals. *Australasian Journal of Philosophy* 61, 343–77. Reprinted in *Papers in Metaphysics and Epistemology*. Cambridge: Cambridge University Press, 1999, 8–55.

—— (1983c). Survival and identity. In Lewis (1983d).

—— (1986a). Causal explanation. In Lewis (1986c).

—— (1986b). *On the Plurality of Worlds*. Oxford: Blackwell.

—— (1986c). *Philosophical Papers*, vol. 2. Oxford: Oxford University Press.

—— (1994). Chance and credence: Humean supervenience debugged. *Mind* 103, 473–90. Reprinted in *Papers in Metaphysics and Epistemology*. Cambridge: Cambridge University Press, 1999, 224–47.

—— (1997). Why conditionalize? In *Papers in Metaphysics and Epistemology*. Cambridge: Cambridge University Press, 1999.

—— (2004). How many lives has Schrödinger's cat? *Australasian Journal of Philosophy* 81, 3–22.

Lewis, P. J. (2000). What is it like to be Schrödinger's cat? *Analysis* 60, 22–9.

—— (2007a). How Bohm's theory solves the measurement problem. *Philosophy of Science* 74, 749–60.

—— (2007b). Uncertainty and probability for branching selves. *Studies in the History and Philosophy of Modern Physics* 38, 1–14.

Lloyd, S., L. Maccone, R. Garcia-Patron, V. Giovannetti, and Y. Shikano (2010). The quantum mechanics of time travel through post-selected teleportation. Available online at: http://arxiv.org/abs/1007.2615.

Lockwood, M. (1989). *Mind, Brain and the Quantum: The Compound 'I'*. Oxford: Blackwell.

—— (1996a). 'Many minds' interpretations of quantum mechanics. *British Journal for the Philosophy of Science* 47, 159–88.

—— (1996b). 'Many minds' interpretations of quantum mechanics: replies to replies. *British Journal for the Philosophy of Science* 47, 445–61.

Lucretius (1947). *De Rerum Natura*. Oxford: Oxford University Press. Translated by C. Bailey.

Luper, S. (2009). Death. In E. N. Zalta (ed.), *Stanford Encyclopedia of Philosophy* (Summer 2009 edn). Available online at: http://plato.stanford.edu/archives/sum2009/entries/death.

Maudlin, T. (1990). Time travel and topology. *Philosophy of Science Association* 1, 303–15.

—— (2002). *Quantum Non-locality and Relativity: Metaphysical Intimations of Modern Physics*, 2nd edn. Oxford: Blackwell.

—— (2006). Review of *Quantum Entanglements: Selected Papers*, by Rob Clifton. *Mind* 115, 1111–20.

—— (2007). *The Metaphysics Within Physics*. Oxford: Oxford University Press.

—— (2010). Can the world be only wavefunction? In S. Saunders, J. Barrett, A. Kent, and D. Wallace (eds), *Many Worlds? Everett, Quantum Theory, and Reality*. Oxford: Oxford University Press.

Maynard Smith, J. (1982). *Evolution and the Theory of Games*. Cambridge: Cambridge University Press.

McCaffrey, A. (1968). *Dragonflight*. New York: Ballantine.

McCall, S. (1984). Counterfactuals based on real possible worlds. *Nous* 18, 463–77.

—— (1994). *A Model of the Universe: Space-Time, Probability, and Decision*. Oxford: Clarendon Press.

Mellor, D. H. (1971). *The Matter of Chance*. Cambridge: Cambridge University Press.

Mermin, N. D. (1998). What is quantum mechanics trying to tell us? *American Journal of Physics* 66, 753–67.

Misner, C. W., K. S. Thorne, and J. A. Wheeler (1973). *Gravitation*. New York: W. H. Freeman.

Misra, B., and E. C. G. Sudarshan (1977). The Zeno's paradox in quantum theory. *Journal of Mathematical Physics* 18, 756.

Monton, B. (2002). Wave function ontology. *Synthese* 130, 265–77.

—— (2004). Quantum mechanics and 3N-dimensional space. Available online at: http://philsci-archive.pitt.edu/archive/00002090/; revised version in *Philosophy of Science* 73 (2006), 778–89.

—— (2006). Quantum mechanics and 3N-dimensional space. *Philosophy of Science* 73, 778–89.

Moyal, J. E. (1949). Quantum mechanics as a statistical theory. *Mathematical Proceedings of the Cambridge Philosophical Society* 45, 99–124.

Myrvold, W. (2002). On peaceful coexistence: is the collapse postulate incompatible with relativity? *Studies in History and Philosophy of Modern Physics* 33, 435–66.

Newton-Smith, W. S. (1981). *The Rationality of Science*. London: Routledge.

Nielsen, M. A., and I. L. Chuang (2000). *Quantum Computation and Quantum Information*. Cambridge: Cambridge University Press.

Norton, J. D. (2010). Time really passes. *Humana.Mente: Journal of Philosophical Studies* 13, 23–34.

—— (2011). Challenges to Bayesian confirmation theory. In Prasanta S. Bandyopadhyay and Malcolm Forster (eds), *Philosophy of Statistics*. Amsterdam: Elsevier.

Nozieres, P., and D. Pines (1999). *The Theory of Quantum Liquids*. Cambridge, Mass: Perseus.

Omnés, R. (1988). Logical reformulation of quantum mechanics, I: Foundations. *Journal of Statistical Physics* 53, 893–932.

—— (1992). Consistent interpretations of quantum mechanics. *Reviews of Modern Physics* 64, 339–82.

—— (1994). *The Interpretation of Quantum Mechanics*. Princeton, NJ: Princeton University Press.

Page, D. N., and C. D. Geilker (1981). Indirect evidence for quantum gravity. *Physical Review Letters* 47, 979–82.

Papineau, D. (1996). Many minds are no worse than one. *British Journal for the Philosophy of Science* 47, 233–41.

—— (2003). Why you don't want to get into the box with Schrödinger's cat. *Analysis* 63, 51–8.

Parfit, D. (1984). *Reasons and Persons*. Oxford: Oxford University Press.

Penrose, R. (1989). *The Emperor's New Mind: Concerning Computers, Brains and the Laws of Physics*. Oxford: Oxford University Press.

—— (1994a). On the second law of thermodynamics. *Journal of Statistical Physics* 77, 217–21.

—— (1994b). *Shadows of the Mind*. Oxford: Oxford University Press.

—— (2004). *The Road to Reality: A Complete Guide to the Laws of the Universe*. London: Cape.

Peres, A. (1993). *Quantum Theory: Concepts and Methods*. Dordrecht: Kluwer Academic.

Perry, J. (1979). The problem of the essential indexical. *Nous* 13, 3–21.

Politzer, H. D. (1992). Simple quantum systems in spacetimes with closed timelike curves. *Physical Review* D 46, 4470–76.

Polkinghorne, J. (1986). *One World: The Interaction of Science and Theology*. London: SPCK.

Popper, K. (1959). The propensity interpretation of probability. *British Journal for the Philosophy of Science* 10, 25–42.

—— (1963). *Conjectures and Refutations*. London: Routledge & Kegan Paul.

Preskill, J. (1999). Quantum information and computation. Unpublished lecture notes; available online at: http://www.theory.caltech.edu/people/preskill/ph229/#lecture.

Price, H. (1996). *Time's Arrow and Archimedes' Point*. Oxford: Oxford University Press.

Price, H. (2010). Probability in the Everett picture. In S. Saunders, J. Barrett, A. Kent, and D. Wallace (eds), *Many Worlds? Everett, Quantum Theory and Reality*. Oxford: Oxford University Press.

Prigogine, I. (1984). *Order Out of Chaos*. New York: Bantam.

Prior, A. N. (1957). *Time and Modality*. Oxford: Clarendon Press.

——(1967). *Past, Present and Future*. Oxford: Clarendon Press.

Putnam, H. (1968). Is logic empirical? In R. Cohen and M. Wartofsky (eds), *Boston Studies in the Philosophy of Science*, vol. 5. Dordrecht: Reidel, 216–41. Reprinted as 'The logic of quantum mechanics' in *Mathematics, Matter and Method: Philosophical Papers*, vol. 1. Cambridge: Cambridge University Press, 1975, 174–97.

——(1973). Meaning and reference. *Journal of Philosophy* 70, 699–711.

——(1990). *Realism with a Human Face*. Cambridge, Mass.: Harvard University Press.

Quine, W. V. O. (1953). Two dogmas of empiricism. In *From a Logical Point of View*. Cambridge, Mass.: Harvard University Press.

——(1960). *Word and Object*. Cambridge, Mass.: MIT Press.

——(1968). Propositional objects. *Critica* 2, 3–22. Reprinted in *Ontological Relativity and other essays*. New York: Columbia University Press, 1969.

——(1969). Epistemology naturalized. In *Ontological Relativity and other essays*. New York: Columbia University Press.

Ramsey, F. P. (1926). Truth and probability. In R. B. Braithwaite (ed.), *The Foundations of Mathematics and other logical essays*. New York: Harcourt, Brace, 1931, 156–98. Reprinted in H. E. Kyburg and H. E. Smokler (eds), *Studies in Subjective Probability*. New York: Wiley, 1964.

Ray, C. (2000). Logical positivism. In W. H. Newton-Smith (ed.), *A Companion to the Philosophy of Science*. Oxford: Blackwell, 243–51.

Redhead, M. (1975). Symmetry in inter-theory relations. *Synthese* 32, 77–112.

——(1980). Models in physics. *British Journal for the Philosophy of Science* 31, 145–163.

——(1987). *Incompleteness, Nonlocality and Realism: A Prolegomenon to the Philosophy of Quantum Mechanics*. Oxford: Oxford University Press.

——(1995). More ado about nothing. *Foundations of Physics* 25(1), 123–39.

Rey, G. (2007). Resisting normativism in psychology. In B. P. McLauglin and J. Cohen (eds), *Contemporary Debates in Philosophy of Mind*. Oxford: Blackwell, 69–84.

Roberts, J. T. (2004). There are no laws of the social sciences. In C. Hitchcock (ed.), *Contemporary Debates in the Philosophy of Science*. Oxford: Blackwell.

Ross, D. (2000). Rainforest realism: a Dennettian theory of existence. In D. Ross, A. Brook, and D. Thompson (eds), *Dennett's Philosophy: A Comprehensive Assessment*. Cambridge, Mass.: MIT Press, 147–68.

Rovelli, C. (2004). *Quantum Gravity*. Cambridge: Cambridge University Press.

Rowling, J. K. (1999). *Harry Potter and the Prisoner of Azkaban*. London: Bloomsbury.

Rudin, W. (1987). *Real and Complex Analysis*, 3rd edn. New York: McGraw-Hill.

—— (1991). *Functional Analysis*, 2nd edn. New York: McGraw-Hill.

Sakurai, J. J. (1994). *Modern Quantum Mechanics*. Reading, Mass.: Addison-Wesley.

Salmon, W. (2000). Logical empiricism. In W. S. Newton-Smith (ed.), *A Companion to the Philosophy of Science*. Oxford: Blackwell, 233–42.

Saunders, S. (1995). Time, decoherence and quantum mechanics. *Synthese* 102, 235–66.

—— (1996). Time, quantum mechanics and tense. *Synthese* 107, 19–53.

—— (1997). Naturalizing metaphysics. *The Monist* 80(1), 44–69.

—— (1998). Time, quantum mechanics, and probability. *Synthese* 114, 373–404.

—— (2003). Physics and Leibniz's principles. In K. Brading and E. Castellani (eds), *Symmetries in Physics: Philosophical Reflections*. Cambridge: Cambridge University Press, 289–308.

—— (2005). Complementarity and scientific rationality. *Foundations of Physics* 35, 347–72.

—— (2010). Chance in the Everett interpretation. In S. Saunders, J. Barrett, A. Kent, and D. Wallace (eds), *Many Worlds? Everett, Quantum Theory, and Reality*. Oxford. Oxford University Press.

—— and D. Wallace (2008). Branching and uncertainty. *British Journal for the Philosophy of Science* 59, 293–305.

Savage, L. J. (1972). *The Foundations of Statistics*, 2nd edn. New York: Dover.

Schack, R. (2010). The principal principle and probability in the many worlds interpretation. In S. Saunders, J. Barrett, A. Kent, and D. Wallace (eds), *Many Worlds? Everett, Quantum Theory, and Reality*. Oxford: Oxford University Press.

Schaffer, J. (2009). Spacetime the one substance. *Philosophical Studies* 145, 131–48.

Schauder, J. (1930). Der fixpunktsatz in funktionalräumen. *Studia Mathematica* 2, 171–80.

Schlosshauer, M. (2006). Experimental motivation and empirical consistency in minimal no-collapse quantum mechanics. *Annals of Physics* 321, 112–49.

—— (2007). *Decoherence and the Quantum-to-Classical Transition*. Berlin: Springer.

Shor, P. W. (1994). Algorithms for quantum computation: discrete logarithms and factoring. In *Proceedings of the 35th Annual Symposium on the Foundations of Computer Science*. Los Alamitos, Calif.: IEE Computer Society Press, 124–34.

Sider, T. (2001). *Four-Dimensionalism: An Ontology of Persistence and Time*. Oxford: Clarendon Press.

—— (2002). Time travel, coincidences and counterfactuals. *Philosophical Studies* 110, 115–38.

Sklar, L. (1993). *Physics and Chance: Philosophical Issues in the Foundations of Statistical Mechanics*. Cambridge: Cambridge University Press.

Spekkens, R. W. (2007). In defense of the epistemic view of quantum states: a toy theory. *Physical Review* A 75, 032110.

Squires, E. (1986). *The Mystery of the Quantum World*. Bristol: Hilger.

Stalnaker, R. C. (2003). *Ways a World Might Be: Metaphysical and Anti-Metaphysical Essays*. Oxford: Oxford University Press.

Stannard, R. (1993). *Doing Away with God? Creation and the Big Bang*. London: Pickering.

Steane, A. (2003). A quantum computer only needs one universe. *Studies in History and Philosophy of Modern Physics* 34, 469–78.

Stich, S. P. (1983). *From Folk Psychology to Cognitive Science: The Case Against Belief.* Cambridge, Mass.: MIT Press.

Struyve, W. (2007). Field beables for quantum field theory. Available online at: http://arxiv.org/abs/0707.3685.

—— and H. Westman (2007). A minimalist pilot-wave model for quantum electrodynamics. *Proceedings of the Royal Society of London* A 463, 3115–29.

Styer, D., S. Sobottka, W. Holladay, T. A. Brun, R. B. Griffiths, and P. Harris (2000). Quantum theory: interpretation, formulation, inspiration. *Physics Today* 53, 11.

Suppes, P. (1974). The measurement of belief. *Journal of the Royal Statistical Society*, Series B (Methodological) 36, 160–91.

Susskind, L. (2008). *The Black Hole War: My Battle With Steven Hawking to Make the World Safe for Quantum Mechanics*. New York: Little, Brown.

Talbott, W. (2008). Bayesian epistemology. In E. N. Zalta (ed.), *The Stanford Encyclopedia of Philosophy* (Fall 2008 edn). Stanford University. Available online at: http://plato.stanford.edu/archives/fall2008/entries/epistemology-bayesian.

Tappenden, P. (2000). Identity and probability in Everett's multiverse. *British Journal for the Philosophy of Science* 51, 99–114.

—— (2004). The ins and outs of Schrödinger's cat box: a response to Papineau. *Analysis* 64, 157–64.

—— (2008). Saunders and Wallace on Everett and Lewis. *British Journal for the Philosophy of Science* 59, 307–14.

—— (2011). Evidence and uncertainty in Everett's multiverse. *British Journal for the Philosophy of Science*, 62(1), 99–123.

Taylor, C., and D. Dennett (2001). Who's afraid of determinism? Rethinking causes and possibilities. In R. Kane (ed.), *Oxford Handbook of Free Will*. Oxford: Oxford University Press.

Tegmark, M. (1998). The interpretation of quantum mechanics: many worlds or many words? *Fortschrift für Physik* 46, 855–62.

Teller, P. (1976). Conditionalization, observation, and change of preference. In W. Harper and C. A. Hooker (eds), *Foundations of Probability Theory, Statistical Inference, and Statistical Theories of Science*. Dordrecht: Reidel.

Thomason, R. H. (1970). Indeterminist time and truth-value gaps. *Theoria* 3, 264–81.

Thorne, K. S. (1994). *Black Holes and Time Warps: Einstein's Outrageous Legacy*. New York: Norton.

Timpson, C. (2006). The grammar of teleportation. *British Journal for the Philosophy of Science* 57, 587–621.

—— (2008). Philosophical aspects of quantum information theory. In D. Rickles (ed.), *The Ashgate Companion to Contemporary Philosophy of Physics*. Burlington, Vt.: Ashgate, 16–98.

—— (2010). *Quantum Information Theory and the Foundations of Quantum Mechanics*. Oxford: Oxford University Press.

—— and H. Brown (2005). Proper and improper separability. *International Journal of Quantum Information* 3, 679–90. Available online at: http://arxiv.org/abs/quant-ph/0402094; page numbers refer to the online version.

Townsend, J. S. (1992). *A Modern Approach to Quantum Mechanics*. New York: McGraw-Hill.

Tsvelik, A. M. (2003). *Quantum Field Theory in Condensed Matter Physics*, 2nd edn. Cambridge: Cambridge University Press.

Tumulka, R. (2006). Collapse and relativity. In A. Bassi, T. Weber, and N. Zanghi (eds), *Quantum Mechanics: Are There Quantum Jumps? and on the Present Status of Quantum Mechanics*. American Institute of Physics Conference Proceedings, 340. Available online at: http://arxiv.org/abs/quant-ph/0602208.

Twain, M. (1996). *Life on the Mississippi*. Oxford University Press.

Vaidman, L. (1998). On schizophrenic experiences of the neutron or why we should believe in the many-worlds interpretation of quantum theory. *International Studies in the Philosophy of Science* 12, 245–61.

—— (2002). The many-worlds interpretation of quantum mechanics. In E. N. Zalta (ed.), *The Stanford Encyclopedia of Philosophy* (Summer 2002 edn). Available online at: http://plato.stanford.edu/archives/sum2002/entries/qm-manyworlds.

Valentini, A. (1996). Pilot-wave theory of fields, gravitation and cosmology. In Cushing et al. (1996: 45–67).

—— (2001). Hidden variables, statistical mechanics and the early universe. In J. Bricmont, D. Dürr, M. C. Galavotti, G. Ghirardi, and F. Petruccione (eds), *Chance in Physics: Foundations and Perspectives*, London: Springer, 165–81. Available online at: http://arxiv.org/abs/quant-ph/0104067.

—— (2004). Extreme test of quantum theory with black holes. Available online at: http://arxiv.org/abs/astro-ph/0412503.

Valentini, A. (2010). De Broglie-Bohm pilot wave theory: many worlds in denial? In S. Saunders, J. Barrett, A. Kent, and D. Wallace (eds), *Many Worlds? Everett, Quantum Theory and Reality*. Oxford: Oxford University Press.

——and H. Westman (2004). Dynamical origin of quantum probabilities. Available online at: http://arxiv.org/abs/quant-ph/0403034.

Van Fraassen, B. C. (1980). *The Scientific Image*. Oxford: Oxford University Press.

——(1989). *Laws and Symmetry*. New York: Oxford University Press.

——(1999). A new argument for conditionalization. *Topoi* 18, 93–6.

Vedral, V. (2006). *Introduction to Quantum Information Science*. Oxford: Oxford University Press.

Vihvelin, K. (1996). What time travelers cannot do. *Philosophical Studies* 81, 315–30.

von Mises, L. (1957). *Probability, Statistics, and Truth*, revised English edn. New York: Macmillan.

von Neumann, J. (1955). *Mathematical Foundations of Quantum Mechanics*. Princeton, NJ: Princeton University Press.

Vulovic, V. Z., and R. E. Prange (1986). Randomness of a true coin toss. *Physical Review* A 33, 576–82.

Wald, R. M. (1994). *Quantum Field Theory in Curved Spacetime and Black Hole Thermodynamics*. Chicago: University of Chicago Press.

Wallace, D. (2001a). Emergence of particles from bosonic quantum field theory. Available online at: http://arxiv.org/abs/quant-ph/0112149.

——(2001b). Implications of quantum theory in the foundations of statistical mechanics. Available online at: http://philsci-archive.pitt.edu.

——(2002a). Quantum probability and decision theory, revisited. Available online at: http://arxiv.org/abs/quant-ph/0211104. (This is a long (70-page) and not-formally-published paper, now entirely superseded by published work and included only for historical purposes.)

——(2002b). Worlds in the Everett Interpretation. *Studies in the History and Philosophy of Modern Physics* 33, 637–61.

——(2003a). Everett and structure. *Studies in the History and Philosophy of Modern Physics* 34, 87–105.

——(2003b). Everettian rationality: defending Deutsch's approach to probability in the Everett interpretation. *Studies in History and Philosophy of Modern Physics* 34, 415–39.

——(2003c). Quantum probability from subjective likelihood: improving on Deutsch's proof of the probability rule (v1). Earlier version of Wallace (2007); available online at: http://arxiv.org/abs/quant-ph/0312157v1.

——(2005). Language use in a branching universe. Available online at: http://philsci-archive.pitt.edu.

—— (2006a). Epistemology quantized: circumstances in which we should come to believe in the Everett interpretation. *British Journal for the Philosophy of Science* 57, 655–89.

—— (2006b). In defence of naiveté: the conceptual status of Lagrangian quantum field theory. *Synthese* 151, 33–80.

—— (2007). Quantum probability from subjective likelihood: improving on Deutsch's proof of the probability rule. *Studies in History and Philosophy of Modern Physics* 38, 311–32.

—— (2008). The interpretation of quantum mechanics. In D. Rickles (ed.), *The Ashgate Companion to Contemporary Philosophy of Physics*. Burlington, Vt.: Ashgate, 197–261.

—— (2009). QFT, antimatter, and symmetry. *Studies in the History and Philosophy of Modern Physics* 40, 209–22.

—— (2010a). Diachronic rationality and prediction-based games. *Proceedings of the Aristotelian Society*, 110, 243–66.

—— (2010b). How to prove the Born rule. In S. Saunders, J. Barrett, A. Kent, and D. Wallace (eds), *Many Worlds? Everett, Quantum Theory and Reality*. Oxford: Oxford University Press. Available online at: http://arxiv.org/abs/0906.2718.

—— (2010c). Gravity, entropy, and cosmology: in search of clarity. *British Journal for the Philosophy of Science* 61(3), 513–40.

—— (2011). Taking particle physics seriously: a critique of the algebraic approach to quantum field theory. *Studies in History and Philosophy of Modern Physics* 42, 116–25.

—— and C. Timpson (2007). Non-locality and gauge freedom in Deutsch and Hayden's formulation of quantum mechanics. *Foundations of Physics* 37(6), 951–5.

—— —— (2010). Quantum mechanics on spacetime I: Spacetime state realism. *British Journal for the Philosophy of Science* 61, 697–727.

Wedgwood, R. (2007). Normativism defended. In B. P. McLauglin and J. Cohen (eds), *Contemporary Debates in Philosophy of Mind*. Oxford: Blackwell, 85–102.

Weinberg, S. (2002). *Facing Up: Science and its Cultural Adversaries*. Cambridge, Mass.: Harvard University Press.

—— (2008). *Cosmology*. Oxford: Oxford University Press.

Wheeler, G. (2007). A review of the lottery paradox. In W. Harper and G. Wheeler (eds), *Probability and Inference: Essays in honor of Henry E. Kyburg, Jr.* London: King's College Publications, 1–31.

Wheeler, J. A., and R. P. Feynman (1945). Interaction with the absorber as the mechanism of radiation. *Review of Modern Physics* 17, 157–81.

Whitman, W. (1855). *Leaves of Grass*. New York: Rome Brothers.

Wigner, E. (1932). On the quantum correction for thermodynamic equilibrium. *Physical Review* 40, 749–59.

Williams, P. (1980). Bayesian conditionalisation and the principle of minimum information. *British Journal for the Philosophy of Science* 31, 131–44.

Williamson, T. (1994). *Vagueness*. London: Routledge.

——(2007). *The Philosophy of Philosophy*. Oxford: Blackwell.

Wilson, A. (2010a). Macroscopic ontology in Everettian quantum mechanics. *Philosphical Quarterly* 60, 1–20.

——(2010b). Modality naturalized: the metaphysics of the Everett interpretation. Ph.D. thesis, University of Oxford.

Wilson, M. (1998). Mechanics, classical. In E. Craig (ed.), *Routledge Encyclopedia of Philosophy*. London: Routledge.

Wittgenstein, L. (1953). *Philosophical Investigations,* 3rd edn. Oxford: Blackwell. Translated by Elizabeth Anscombe.

Woodhouse, N. M. J. (1997). *Geometric Quantization*. Oxford: Oxford University Press.

Worrall, J. (1989). Structural realism: the best of both worlds? *Dialectica* 43, 99–124.

Zeilinger, A. (2005). The message of the quantum. *Nature* 438, 743.

Zurek, W. H. (1991). Decoherence and the transition from quantum to classical. *Physics Today* 43, 36–44. Revised version available online at: http://arxiv.org/abs/quant-ph/0306072.

——(1998). Decoherence, einselection, and the quantum origins of the classical: the rough guide. *Philosophical Transactions of the Royal Society of London* A356, 1793–820. Available online at: http://arxiv.org./abs/quant-ph/98050.

——(2003). Decoherence, einselection, and the quantum origins of the classical. *Reviews of Modern Physics* 75, 715.

——and J. P. Paz (1994). Decoherence, chaos and the second law. *Physical Review Letters* 72(16), 2508–11.

————(1995a). Quantum chaos: a decoherent definition. *Physica D* 83, 300–308.

——(1995b). Zurek and Paz reply. *Physical Review Letters* 75, 351.

Index

Made in the USA
Monee, IL
28 September 2021

78977902R00319